Genetics
in
Medicine

Third Edition

JAMES S. THOMPSON, M.D.

Department of Anatomy,
University of Toronto

MARGARET W. THOMPSON, Ph.D.

Departments of Medical Genetics and Paediatrics,
University of Toronto and
The Hospital for Sick Children, Toronto

1980
W. B. SAUNDERS COMPANY
Philadelphia London Toronto

W. B. Saunders Company: West Washington Square
Philadelphia, PA   19105

1 St. Anne's Road
Eastbourne, East Sussex NB21 3UN, England

1 Goldthorne Avenue
Toronto, Ontario M8Z 5T9, Canada

Library of Congress Cataloging in Publication Data

Thompson, James S

Genetics in medicine.

Bibliography: p. 361.

Includes index.

1. Medical genetics.   I. Thompson, Margaret W.,
   joint author.   II. Title. [DNLM: 1. Genetics, Human.
   QH431 T473g]

RB155.T5 1979   616'.042   78–64726

ISBN 0–7216–8857–8

Listed here are the latest translated editions of this book together with
the language of the translation and the publisher.

Italian (*2nd Edition*) — UTET (Unione Tipografica Editrice Torinese)
                          Turin, Italy

Spanish (*2nd Edition*) — Salvat Editores, S. A., Barcelona, Spain

Portuguese (*2nd Edition*) — Livraria Atheneu, Rio de Janeiro, Brazil

French (*2nd Edition*) — Doin Editeurs, Paris, France

Genetics in Medicine   ISBN   0-7216-8857-8

© 1980 by W. B. Saunders Company.   Copyright 1966 and 1973 by W. B. Saunders Company.
Copyright under the International Copyright Union.   All rights reserved.   This book is protected by a copyright.   No part of it may be reproduced, stored in a retrieval system, or transmitted in any form or by any means, electronic, mechanical, photocopying, recording, or otherwise, without permission from the publisher.   Made in the United States of America.   Press
of W. B. Saunders Company.   Library of Congress catalog card number 78-64726.

Last digit is the print number: 9   8   7   6   5   4   3   2   1

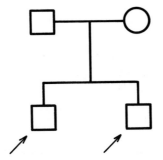

To our own F1

PREFACE

When the first edition of this book appeared in 1966, the place of genetics in medicine was not at all assured. In the intervening years, virtually all medical schools, at least in North America, have included genetics in the undergraduate medical curriculum and steps are underway in both the United States and Canada to establish genetics as a clinical specialty.

These changes indicate widespread recognition by the medical profession and by other health scientists of the crucial role of genetics in the scientific basis of medicine and its increasing significance in clinical practice. Advances in such fields as human cytogenetics, and the mapping of the human genome and the manipulation of human cells in tissue culture are yielding much information that can be promptly applied to patient management. Simultaneously there has been much progress in the delineation of genetic syndromes and in prenatal diagnosis of genetic disorders, and clinical genticists are achieving increasing recognition of their particular skills. This book has been written primarily to introduce medical students to the principles and language of human genetics and to indicate some of its fruitful clinical applications.

A major problem in teaching genetics to medical students is that in any one class there is a wide range in background knowledge of the scientific basis of the field, though there is rarely much information about its medical application. We have assumed that the student using this book will have little or no knowledge of genetics and its vocabulary, though we realize that some students will not require such an elementary approach. However, we believe that the latter will find much to assist them in understanding the genetic basis of disease.

Many advances in both the basic science and clinical applications of genetics are beyond the scope of an introductory text such as this, but we have attempted to provide a background that will make the literature in medical genetics accessible to readers who require more specialized information. Because the field is in such a rapid phase of growth, we have had to be highly selective in choosing material to elucidate principles without swamping the reader in minutiae or expanding the text far beyond the limits of the time available for its use. We regret that this has forced us to omit discussion of many notable contributions to genetics.

We have had help from numerous sources in the preparation of this edition. We are grateful to the many colleagues in Toronto and elsewhere who have assisted us with information and illustrations, and to students who have helped us to recognize which parts of the previous editions were erroneous or unclear. In particular we wish to thank medical artist Margo Simonovitch for many of the new illustrations used in this edition; Dr. Ron Worton and his assistant Chin Chin Ho, who prepared a number of cytogenetic illustrations especially for our use; and other members of the staff of The Hospital for Sick Children, especially Eva Struthers of the Department of Visual Education, medical librarian Irene Jeryn, and secretaries Pauline Kowal and Baba Torres. We have relied heavily on their helpfulness, knowledge and special skills. The preparation of this edition began while one of us (MWT) was on sabbatical leave at the Galton Laboratory, University College, London, and our debt to the Galton staff for intellectual stimulation and practical assistance is gratefully acknowledged. Finally, we wish to thank the W. B. Saunders Company and editor Roberta Kangilaski for their continuing support.

CONTENTS

1

INTRODUCTION

The place of genetics in medicine was not always as obvious as it is today. Though the significance of genetics both for the conceptual basis of medicine and for clinical practice is now generally appreciated, not many years ago the subject was thought to be concerned only with the inheritance of trivial, superficial and rare characteristics, and the fundamental role of the gene in basic life processes was not understood.

The discovery of the principles of heredity by the Austrian monk Gregor Mendel in 1865 received no recognition at all from medical scientists and virtually none from other biologists. Instead, his work lay unnoticed in the scientific literature for 35 years. Charles Darwin, whose great work *The Origin of Species* (published in 1859) emphasized the hereditary nature of variability among members of a species as an important factor in evolution, had no idea how inheritance worked. At that time inheritance was regarded as blending of the traits of the two parents, and Lamarck's idea of the inheritance of acquired characteristics was still accepted. Mendel's work could have clarified Darwin's concept of the mechanism of inheritance of variability, but Darwin seems never to have been aware of its significance or even of its existence. Darwin's cousin, Francis Galton, one of the great figures of early medical genetics, also remained ignorant of Mendel's work despite its relevance to his own studies of "nature and nurture." Mendel himself, perhaps discouraged by the results of later, less favorably designed experiments, eventually took the course followed by many successful scientists — he abandoned research and became an administrator.

Mendel's laws, which form the cornerstone of the science of genetics, were derived from his experiments with garden peas, in which he crossed pure lines differing in one or more clear-cut characteristics and followed the progeny of the crosses for at least two generations. The three laws he derived from the results of his experiments may be stated as follows:

1. **Unit inheritance.** Prior to Mendel's time, the characteristics of the parents were believed to blend in the offspring. Mendel clearly stated that blending did not occur, and the characteristics of the parents, though they might not be expressed in the first-generation offspring, could reappear quite unchanged in a later generation. Modern teaching in genetics places little stress upon this law, but in Mendel's time it was an entirely new concept.

2. **Segregation.** The two members of a *single* pair of genes are never found in the same gamete but instead always segregate and pass to different

1

gametes. In exceptional circumstances, when the members of a chromosome pair fail to segregate normally, this rule is broken, but the typical consequence of such a failure is severe abnormality.

3. **Independent assortment.** Members of *different* gene pairs assort to the gametes independently of one another. In other words, there is random recombination of the paternal and maternal chromosomes in the gametes.

With the dawn of the new century, the rest of the scientific community was ready to catch up with Mendel. By a curious coincidence, three workers (de Vries in Holland, Correns in Germany and Tschermak in Austria) independently and simultaneously rediscovered Mendel's laws. The development of genetics as a science dates not from Mendel's own paper but from the papers that reported the rediscovery of his laws.

The universal nature of Mendelian inheritance was soon recognized, and as early as 1902 Garrod, who ranks with Galton as a founder of medical genetics, could report in alcaptonuria the first example of mendelian inheritance in man. In his paper Garrod generously admitted his debt to the biologist Bateson, who had seen the genetic significance of consanguineous marriage in the parents of recessively affected persons. This is the first clear evidence of the interaction in research between medical and nonmedical geneticists, which has continued to the present day.

A growing understanding of the universal nature of the biochemical structure and functioning of living organisms has brought about an awareness of the crucial role of genes in living organisms. The work of Garrod foreshadowed this knowledge, though in the early years of genetics its fundamental significance was not apparent. The concept was formulated clearly by Beadle and Tatum in 1941 as the "one gene–one enzyme" hypothesis.

The growth in genetic knowledge and in its applications during recent years has had fruitful consequences for clinical medicine. It is estimated that today one-third of the children in pediatric hospitals are there because of genetic disorders. This is a great change from the early years of this century, and even from the preantibiotic era. Before the days of immunization, improved nutrition and antibiotics, many children were in the hospital because of infectious diseases or nutritional disorders such as rickets. Today some of those with infections have genetic defects that impair their resistance, and at least in developed countries, most cases of rickets arise not from faulty nutrition but from deleterious genes. Life-saving advances in clinical techniques (transfusion, tube feeding, maintenance of body fluids by intravenous drip) for the management of medical emergencies also play a part in increasing the prevalence of genetic defects. Improvements in surgical procedures have also contributed to the profound alteration that has been effected during the twentieth century.

Though medical genetics grew up in close association with pediatrics, it is also relevant to many other branches of medicine. One of the most recent applications of medical genetics has been in obstetrics, in which prenatal diagnosis of certain genetic defects has become an important aspect of adequate prenatal care. In adult medicine, it is increasingly obvious that many common conditions, such as coronary heart disease, hypertension, and diabetes mellitus, have important genetic components and that preventive medicine could be much more efficient if it could be directed toward special high-risk groups rather than toward the general population.

CLASSIFICATION OF GENETIC DISORDERS

In medical practice, the chief significance of genetics is its role in the etiology of a large number of disorders. Virtually any trait is the result of the combined action of genetic and environmental factors, but it is convenient to distinguish between those disorders in which defects in the **genetic information** are of prime importance, those in which **environmental hazards** (including hazards of the intrauterine environment) are chiefly to blame and those in which a **combination** of genetic constitution and environment is responsible.

Broadly speaking, genetic disorders are of three main types:

1. Single-gene disorders
2. Chromosome disorders
3. Multifactorial disorders

The initial step in analysis of the genetics of a given disorder is to determine to which of these three categories it belongs.

Single-gene defects are caused by mutant genes. The mutation may be present on only one chromosome of a pair (matched with a normal gene on the partner chromosome) or on both chromosomes of a pair. In either case, the cause of the defect is a single major error in the genetic information. Single-gene disorders usually exhibit obvious and characteristic pedigree patterns. Most such disorders are rare, the upper limit of frequency being about 1 in 2000.

In **chromosome disorders,** the defect is not due to a single mistake in the genetic blueprint but to developmental confusion arising from an excess or deficiency of whole chromosomes or chromosome segments, which upsets the normal balance of the genome. For example, the presence of a specific extra chromosome, chromosome 21, produces a characteristic disorder, Down syndrome, even though all the genes on the extra chromosome may be quite normal. Usually chromosome disorders do not run in families, though there are exceptions. On the whole these disorders are very common, affecting about seven individuals per thousand births and accounting for about half of all spontaneous first-trimester abortions.

Multifactorial inheritance is seen in a number of common disorders, especially developmental disorders resulting in congenital malformations. Here again there is no one major error in the genetic information but rather a combination of small variations that together can produce a serious defect. Multifactorial disorders tend to cluster in families but do not show the clear-cut pedigree patterns of single-gene traits.

Each of these three types of inheritance is discussed in some detail later in this book.

Not all disorders that affect more than one member of a family are genetic. On the contrary, occasionally a clearly definable environmental cause (for example, an infection or teratogen) may affect more than one member of a family at a time. Since it is not always obvious whether a particular problem is genetically determined, Neel and Schull (1954) have provided a useful list of indications of a genetic etiology:

1. The occurrence of the disease *in definite proportions* among persons related by descent, when environmental causes can be ruled out.

2. The failure of the disease to appear in unrelated lines (e.g., in spouses or in-laws).

3. A characteristic onset age and course, in the absence of known precipitating factors.

4. Greater concordance in monozygotic than in dizygotic twins.

The foregoing list was prepared some years before the role of chromosomal disorders was known. Now it is possible to add the following criterion:

5. The presence in the propositus of a characteristic phenotype (usually including mental retardation) and a demonstrable chromosomal abnormality, with or without a family history of the same or related disorders.

THE FAMILY HISTORY

Taking an adequate family history is an essential part of the assessment of a patient with a genetic disorder; it is also helpful in eliminating the possibility that a condition has a genetic basis (Fraser, 1963). Because few of us know our pedigress in any detail, family history details are often difficult and time-consuming to obtain and verify.

A genetic history should always deal with data *pertinent to the patient's condition.* (Many hospital histories record the family history only with respect to a few defects such as diabetes, asthma and "mental retardation," regardless of their relevance to the patient's problem.) Negative information, i.e., the absence of a disorder in any relative, may be as important as a positive finding. To record "family history negative" after one or two brief questions to the patient or his nearest available relative may give a completely erroneous impression. It is necessary to ask specifically about age, sex and health (present and past) of parents, sibs and other near relatives and to ask about each person separately. Miscarriages and stillbirths should be listed. If any relative of the patient has or has had a similar condition, every effort should be made to confirm the diagnosis. The age at death of deceased relatives and the causes of their deaths (if known) should be recorded. If the cause was established by autopsy, this fact should also be noted. Usually the pedigree need not extend over many generations; the more remote the relative, the less accurate the information. It is especially important to check whether there is any consanguinity in the pedigree, especially in the parents of the propositus, and to determine whether both parents are from the same geographic area or ethnic isolate.

The exact relationship of the relatives to the propositus and to one another should be established, and for this purpose it is useful to construct a pedigree chart (see Chapter 4), which will show at a glance the relationship of affected relatives to the propositus and to one another.

MAN AS AN OBJECT OF GENETIC RESEARCH

A mouse can complete a generation within two months, a fruit fly within two weeks and a microorganism within 20 minutes; but man has a generation time of at least 20 years. In lower forms it is possible to make test matings to acquire desired information or to test hypotheses, but in man Nature makes the experiment and the investigator can only record the outcome. A mouse can produce scores of offspring in its lifetime, a fruit fly hundreds and a microorganism millions; human families average about three children each. Faced with these formidable obstacles, we must ask ourselves what compensations man has to offset his disadvantages as a suitable animal for genetic research.

Probably most medical geneticists would subscribe to Pope's dictum that "the proper study of mankind is man." Man's fascination with himself has facilitated research into his genetics. Since we consider man so important (in Pope's phrase, "the glory, jest, and riddle of the world"), we have expended more effort upon him than upon some of the more suitable research organisms. Man seems more variable genetically than many other species, or at least his variants are better explored and documented, and since many of them are deleterious they tend to come to medical attention. Though individual families are small and becoming smaller, the total population is very large and becoming even larger; and though we cannot ethically perform experimental matings, we often find that somewhere in the population Nature has performed the experiments for us, or that new tricks such as those of somatic cell genetics will allow us to approach problems in human genetics from a different angle. Exploration of the genetics of human variation has already been so successful that some variants, notably the hemoglobinopathies and the blood groups, have even provided models upon which part of the conceptual structure of genetics has been erected. Some of the many areas in which genetics and medicine mutually illuminate and enrich one another are described in the following chapters.

GENERAL REFERENCES

Bodmer, W. F., and Cavalli-Sforza, L. L. 1976. *Genetics, Evolution, and Man.* W. H. Freeman and Company, Publishers, San Francisco.

Carter, C. O. 1977. *Human Heredity.* 2nd ed. Penguin Books, New York.

Cavalli-Sforza, L. L., and Bodmer, W. F. 1971. *The Genetics of Human Populations.* W. H. Freeman and Company, Publishers, San Francisco.

Claiborne, R., and McKusick, V. A., eds. 1973. *Medical Genetics.* Hospital Practice Publishing Company, New York.

Emery, A. E. H. 1975. *Elements of Medical Genetics.* 4th ed. Churchill Livingstone, Edinburgh.

Fraser, G. R., and Mayo, O. 1975. *Textbook of Human Genetics.* Blackwell Scientific Publications, Oxford.

Harris, H. 1975. *The Principles of Human Biochemical Genetics.* 2nd ed. North-Holland Publishing Company, Amsterdam and London; Elsevier North-Holland, Inc., New York.

Harris, H., and Hirschhorn, K., eds. *Advances in Human Genetics.* Vols. 1–8, 1970–1977. Plenum Publishing Co., New York.

Levitan, M., and Montagu, A. 1977. *Textbook of Human Genetics.* 2nd ed. Oxford University Press, Inc., New York and London.

McKusick, V. A. 1969. *Human Genetics.* 2nd ed. Prentice-Hall, Englewood Cliffs, New Jersey. (A *Study Guide* to this book was published in 1972.)

McKusick, V. A. 1978. *Mendelian Inheritance in Man: Catalogs of Autosomal Dominant, Autosomal Recessive and X-linked Phenotypes.* 5th ed. The Johns Hopkins Press, Baltimore.

Moody, P. A. 1975. *Genetics of Man.* 2nd ed. W. W. Norton, New York.

Nora, J. J., and Fraser, F. C. 1974. *Medical Genetics: Principles and Practice.* Lea & Febiger, Philadelphia.

Novitski, E. 1977. *Human Genetics.* Macmillan Publishing Company, Inc., New York.

Roberts, J. A. F., and Pembrey, M. E. 1978. *An Introduction to Medical Genetics.* 7th ed. Oxford University Press, Inc., New York and London.

Steinberg, A. G., ed. 1974. *Progress in Medical Genetics.* Vol. 10. Grune and Stratton, Inc. New York.

Steinberg, A. G., and Bearn, A. G., eds. *Progress in Medical Genetics.* Vols. 1–9. 1961–1973, Grune and Stratton, Inc. New York.

Steinberg, A. G., et al., eds. *Progress in Medical Genetics, New Series.* Vols. 1 and 2. 1976–1977. W. B. Saunders Company, Philadelphia.

Stern, C. 1973. *Principles of Human Genetics.* 3rd ed. W. H. Freeman and Company, Publishers, San Francisco.

Sutton, H. E., 1975. *An Introduction to Human Genetics.* 2nd ed. Holt, Rinehart & Winston, New York.

Watson, J. D. 1976. *Molecular Biology of the Gene.* 3rd ed. W. A. Benjamin, Inc., Menlo Park, Calif.

2

THE CHROMOSOMAL BASIS OF HEREDITY

When a cell divides, the nuclear material (chromatin) loses the relatively homogeneous appearance characteristic of nondividing cells, and condenses to form a number of rod-shaped organelles which are called **chromosomes** (*chromos;* color; *soma,* body) because they stain deeply with certain biological stains. Units of genetic information (**genes**) are encoded in the deoxyribonucleic acid (**DNA**) of the chromosomes.

Each species has a characteristic chromosome constitution (**karyotype**), not only with respect to chromosome number and morphology but also with respect to the genes on each chromosome and their locations (the **gene map**). The genes are arranged along the chromosomes in linear order, each gene having a precise position or **locus**. Genes which have their loci on the same chromosome are said to be **linked** or, more precisely, **syntenic** (in synteny). Alternative forms of a gene which can occupy the same locus are called **alleles.** Any one chromosome bears only a single allele at a given locus, though in the population as a whole there may be multiple alleles, any one of which can occupy that locus.

The **genotype** of an individual is his genetic constitution, usually with reference to a single locus. The **phenotype** is the expression of the genotype as a morphological, biochemical or physiological trait. The term **genome** refers to the full DNA content of the chromosome set.

Little was known about human cytogenetics until 1956 when Tjio and Levan developed effective techniques for chromosome study and found the normal human chromosome number to be 46, not 48 as had been previously believed. Since that time much has been learned about the human chromosomes, their molecular composition and their numerous and varied abnormalities. An appreciable proportion of the human genes have already been assigned a place on the chromosome map. Chromosome abnormalities are clinically important because they are major causes of birth defects, mental

7

retardation and spontaneous abortion. The development of methods for determining the karyotype during fetal life has given rise to the important new field of prenatal diagnosis.

THE HUMAN CHROMOSOMES

The 46 chromosomes of normal human somatic cells constitute 23 homologous pairs. The members of a homologous pair match with respect to the genetic information each carries; i.e., they have the same gene loci in the same sequence, though at any one locus they may have either the same or different alleles. One member of each chromosome pair is inherited from the father, the other from the mother, and one of each pair is transmitted to each child. Twenty-two pairs are alike in males and females and hence are called **autosomes.** The two **sex chromosomes,** the remaining pair, differ in males and females and are of major importance in sex determination. Normally the members of a pair of autosomes are microscopically indistinguishable, and the same is true of the female sex chromosomes, the X chromosomes. In the male the members of the pair of sex chromosomes are different from one another; one is an X, identical to the X's of the female, and the other, which is known as the Y chromosome, is smaller than the X and appears not to be homologous to it except with respect to a few genes (see Chapter 7).

There are two kinds of cell division — mitosis and meiosis. **Mitosis** is ordinary cell division by which the body grows and replaces dead or injured cells. It results in two daughter cells that are precisely identical to the parent cell in chromosome complement and genetic information. **Meiosis** occurs only once in a life cycle and results in the production of reproductive cells (gametes), each of which has a complement of 23 chromosomes. Somatic cells are said to have the **diploid** or $2n$ chromosome number (*diploos*, double), whereas gametes have the **haploid** or n chromosome number (*haploos*, single). Though a few specialized cell types are polyploid and abnormal chromosome numbers can arise both in somatic cells and in gametes by accidents of cell division, the general rule is that somatic cells are diploid and gametes are haploid.

Because human females are XX, all ova carry a single X chromosome; in contrast, males are XY and produce two kinds of sperm, X-bearing and Y-bearing. Hence, we speak of the human female as the **homogametic** sex and the male as the **heterogametic** sex. This arrangement is characteristic of many living forms but not of all; in birds the female is the heterogametic sex.

MITOSIS

In mitotic division, the cytoplasm of the cell simply cleaves into two approximately equal halves, but the nucleus undergoes a complicated sequence of activities. Four stages of mitosis can be distinguished — prophase, metaphase, anaphase and telophase. These stages are shown diagrammatically in Figure 2–1. In this figure, each homologous chromosome pair has one member outlined and one in solid black to signify that one homologue is derived from the father, the other from the mother.

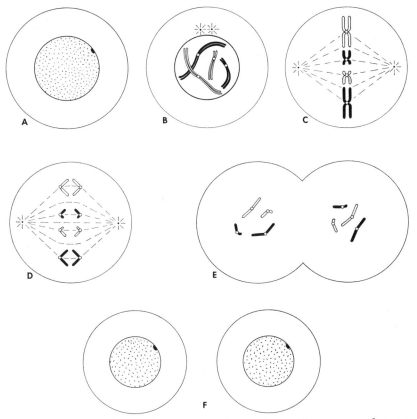

Figure 2–1　Mitosis. Only two chromosome pairs are shown. Chromosomes from one parent are shown in outline, chromosomes from the other parent, in black. A, interphase; B, prophase; C, metaphase; D, anaphase; E, telophase; F, interphase. For further details, see text.

Interphase (Fig. 2–1A).　A cell which is not actively dividing is said to be in interphase. Most of the metabolic activities of the cell, including DNA replication, occur during this stage of the cell cycle. The length and diameter of the chromosomes at this stage have been estimated as 22 cm. for the total length of DNA and 200 Å for the diameter of the chromatin fiber, of which the DNA is a component (Bahr, 1977). In female cells the Barr body or sex chromatin (an inactive X chromosome) appears at this stage as a compacted mass of chromatin, though the other chromosomes are metabolically active and not individually distinguishable. As the cell prepares to divide, the chromosomes begin to condense by a complicated process of folding and coiling and thus become visible as deeply staining bodies. As soon as the appearance of the nucleus changes and the chromosomes begin to become visible, the cell has entered the first stage of cell division, prophase.

Prophase (Fig. 2–1B).　When the chromosomes are discernible, but before there is any obvious pattern in their arrangement, the cell is in prophase. The DNA content has doubled during interphase, and each chromosome can be seen to consist of a pair of long, thin parallel strands, or **chromatids** (sister chromatids), which are held together at one spot, the **centromere.** The position of the centromere is constant for any one chromosome and is in fact one of the morphological features by which chromosomes are classified (see later discussion in this chapter). The nuclear membrane disappears, and the nucleus begins to lose its identity. Meanwhile the **centriole,** an organelle just outside the nuclear membrane, duplicates itself, and its two products migrate toward opposite poles of the cell.

Metaphase (Fig. 2–1C). When the chromosomes have reached their maximal contraction and maximal staining density, they move to the equatorial plane of the cell, which is now in metaphase. This is the stage at which chromosomes are most easily studied, because they are highly contracted, densely stained and arranged in a more or less two-dimensional **metaphase plate** along the equatorial plane of the cell. Meanwhile, the **spindle** has formed; this is a mechanism consisting of microtubules of protein (spindle fibers) that radiate from the centrioles at either pole of the cell up to the equatorial plane and from the **kinetochores,** attachment sites which are associated with the centromeres of the chromosomes.

Anaphase (Fig. 2–1D). The cell enters anaphase when the centromeres divide and the paired chromatids of each original chromosome disjoin, becoming new **daughter chromosomes.** The spindle fibers contract and draw the daughter chromosomes, centromere first, to the poles of the cell. The molecular mechanism by which the spindle fibers draw the chromosomes apart is not fully understood. It is known that spindles contain actin, so one possibility is that actin-myosin interaction between the fibers is involved.

Telophase (Fig. 2–1E). The arrival of the daughter chromosomes at the poles of the cell signifies the beginning of telophase, the final stage of cell division. Concurrently with the onset of telophase, the division of the cytoplasm (**cytokinesis**) begins with the formation of a furrow in the area of the equatorial plane. Eventually a complete membrane is formed across the cell, which is thereby divided into two new cells with identical chromosome complements. Meanwhile, the chromosomes are unwinding and consequently stain less densely. Eventually they no longer stain as individual entities and again become enclosed by a nuclear membrane. Each daughter cell now appears as a typical interphase cell (Fig. 2–1F).

THE MITOTIC CYCLE

Mitotic division takes up only a small part of the life cycle of a cell. Three other stages of the mitotic cycle are recognized (Fig. 2–2). After division the new cell enters a postmitotic period during which there is no DNA synthesis. This stage is called G_1 (Gap 1). The next stage is the S stage, the period of DNA synthesis, in which the DNA content of the cell doubles, each DNA molecule serving as a template to make a complementary copy of itself. There is then a premitotic nonsynthetic period, G_2 or Gap 2, which is ended by the onset of mitosis. Typical studies of cultured human cells have shown that the complete cycle may last 12 to 24 hours, about one hour of which involves mitosis.

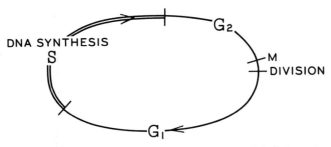

Figure 2–2 The mitotic cycle, described in the text. Modified from Stanners and Till, Biochim. Biophys. Acta 37:406–419, 1960.

Information about the timing of DNA synthesis in cultured cells can be obtained by autoradiography or by the bromodeoxyuridine technique described later in this chapter. Autoradiography involves adding to the culture thymidine that has been labeled with the radioactive isotope of hydrogen, tritium (^{3}H). Tritiated thymidine is taken up only by cells that are actively synthesizing DNA. Cells are cultured in the presence of ^{3}H-thymidine, harvested and prepared for chromosome analysis. The chromosome slides are then covered with a photographic emulsion and left in the dark for a time.

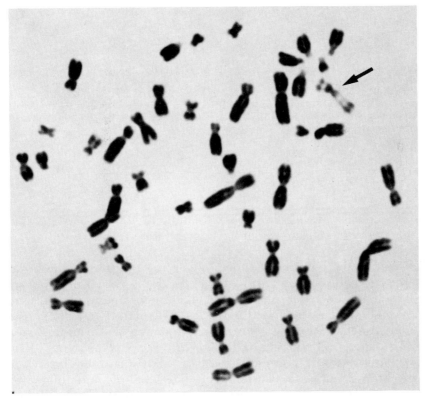

Figure 2–3 Metaphase plate from lymphocyte culture of human female after autoradiography. The arrow indicates the late-replicating X chromosome. From Hamerton, *Human Cytogenetics*. Vol. I., *General Cytogenetics*. Academic Press, New York, 1971, by permission.

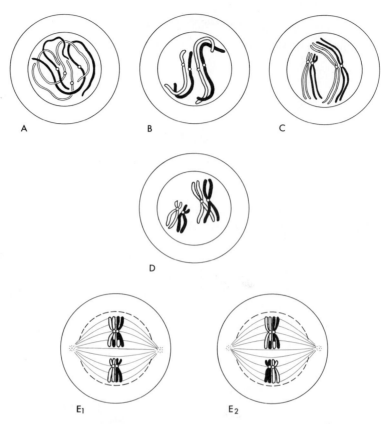

Figure 2–4   The first meiotic division. Only two of the 23 chromosome pairs are shown; chromosomes from one parent are shown in outline, chromosomes from the other parent, in black. A, leptotene; B, zygotene; C, pachytene; D, diplotene; E_1 and E_2, metaphase; F_1 and F_2, early anaphase; G_1 and G_2, late anaphase; H_{1a}, H_{1b}, H_{2a}, H_{2b}, telophase. One possible distribution of the two parental chromosome pairs is shown in illustrations E_1 to H_1, the alternative combination, in illustrations E_2 to H_2.

Illustration Continued on Opposite Page

When the slides are developed to make "autoradiographs," silver granules are seen only over chromosomes that have incorporated ^{3}H-thymidine, i.e., chromosomes that were actually synthesizing DNA while the radioactive material was present in the culture medium. By varying the timing and duration of exposure of cell cultures of ^{3}H-thymidine, the duration of the various stages of the cell cycle has been determined. Not all chromosomes replicate simultaneously. In particular, as shown in Figure 2–3, one of the two X chromosomes of the female is "late-labeling"; i.e., it incorporates ^{3}H-thymidine later in S phase than does its homologue. This late-labeling X is the same X that forms the Barr body in interphase cells (see later). Several other chromosomes or parts of chromosomes are also relatively late-labeling, and autoradiography was therefore commonly used to assist in identification of individual chromosomes before the development of banding techniques.

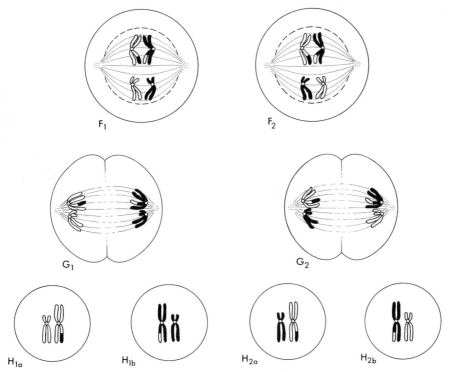

Figure 2–4 *Continued.*

MEIOSIS

Meiosis is the special type of cell division by which gametes are formed. Each daughter cell formed by meiosis has the haploid chromosome number, with one representative of each chromosome pair. This is in contrast to mitosis, in which each daughter cell is identical to the parent cell in its chromosome complement. Some of the stages distinguished in meiosis are illustrated diagrammatically in Figures 2–4 and 2–5.

There are two successive meiotic divisions. In meiosis I, the reduction division, homologous chromosomes pair during prophase and disjoin from one another during anaphase, each chromosome's centromere remaining intact. Meiosis II follows meiosis I without DNA replication, but as in ordinary mitosis, the centromere of each chromosome divides and the chromatids segregate from one another.

THE FIRST MEIOTIC DIVISION (MEIOSIS I)

First Meiotic Prophase

The prophase of meiosis I is a complicated process, with a number of important differences from the prophase of a mitotic division. Several stages can be distinguished.

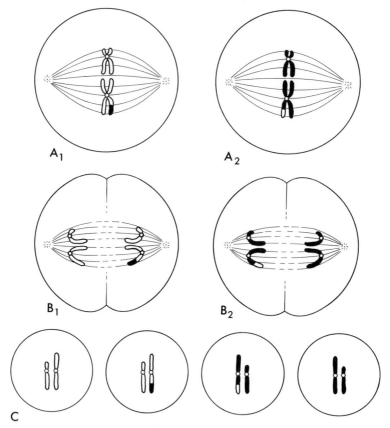

Figure 2–5 The second meiotic division. A, metaphase; B, anaphase; C, telophase. A_1 and A_2 represent H_{1a} and H_{1b} of Figure 2–4.

Leptotene (Fig. 2–4A). Leptotene is characterized by the first appearance of the chromosomes, seen as thin threads that are beginning to condense. Although the DNA has duplicated prior to this stage, the threads still appear to be single on microscopic examination. Unlike mitotic chromosomes, they are not smooth in outline but consist of alternating thicker and thinner regions; the thicker regions, which are known as **chromomeres** and may be areas where coiling and looping of the chromosome is especially pronounced, have a characteristic pattern for each meiotic chromosome.

Zygotene (Fig. 2–4B). Zygotene, as the name implies, is the stage of pairing (synapsis) of homologous chromosomes. The two members of each homologous pair lie parallel to one another in intimate, point-for-point association to form **bivalents**. *Pairing of homologous chromosomes does not occur in mitosis*. Unlike the homologous pairs, the X and Y chromosomes are associated only at the tips of their short arms (Fig. 2–6).

Pachytene (Fig. 2–4C). Pachytene is the main stage of chromosomal thickening. The chromosomes coil more tightly and stain more deeply, and the chromomeres become more pronounced. The bivalents (paired chromosomes) are in close association, and each chromosome is now seen to consist of two chromatids, so that each bivalent is a tetrad of four strands.

Diplotene (Fig. 2–4D).   Diplotene is recognizable by the longitudinal separation that begins to appear between the two components of each bivalent. Although the two chromosomes of each bivalent separate, the centromere of each remains intact, so the two chromatids of each chromosome remain together. During the longitudinal separation the two members of each bivalent are seen to be in contact in several places, called **chiasmata** (singular, chiasma; *chiasma*, cross), only one of which is shown in Figure 2–4D. Chiasmata may mark the locations of crossovers, where chromatids of homologous chromosomes have exchanged material (see later discussion in this chapter). Eventually the chromosomes draw apart and the chiasmata begin to terminalize (draw to the ends of the chromosome arms).

Diakinesis.   Diakinesis, the final stage of prophase, is marked by even tighter coiling and deeper staining of the chromosomes and by terminalization of some chiasmata.

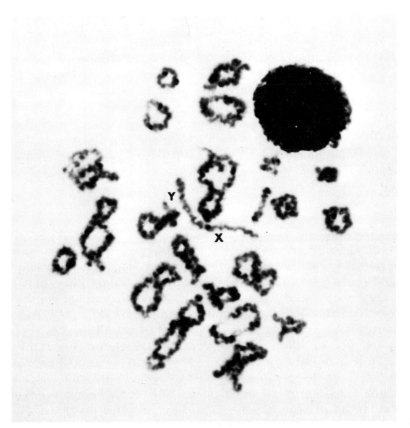

Figure 2–6   Meiosis in the human male. Note 23 chromosome pairs (bivalents), chiasmata in the bivalents, centromeres of individual chromosomes and terminal association of the X and Y. Photomicrograph courtesy of A. Chen.

First Meiotic Metaphase, Anaphase and Telophase

Metaphase I begins, as in mitosis, when the nuclear membrane disappears and the chromosomes move to the equatorial plane (Fig. 2–4E). At anaphase I (Fig. 2–4 F and G) the two members of each bivalent disjoin, one member going to each pole. The bivalents assort themselves *independently of one another* so that the chromosomes received originally as a paternal and a maternal set are now sorted into random combinations of paternal and maternal chromosomes, *with one representative of each pair going to each pole.* The disjunction of paired homologous chromosomes is the physical basis of *segregation,* and the random recombination of paternal and maternal chromosomes in the gametes is the basis of *independent assortment;* thus the behavior of the chromosomes at the first meiotic division provides the physical basis for Mendelian inheritance. The parallel between the behavior of chromosomes and the transmission of inherited traits was first noted independently by Sutton (1903) and Boveri (1904) soon after the rediscovery of Mendel's laws.

By the end of meiosis I (Fig. 2–4H) each product has the haploid chromosome number; hence meiosis I is often referred to as the **reduction division.**

THE SECOND MEIOTIC DIVISION (MEIOSIS II)

The second meiotic division follows upon the first without DNA replication and without a normal interphase. It resembles ordinary mitosis in that in each cell formed by meiosis I, the centromeres now divide and the sister chromatids disjoin, passing to opposite poles to produce two daughter cells (Figs. 2–5A through C). With the exception of areas in which crossovers occurred during meiosis I, the daughter cells have identical chromosomes.

Thus the end result of the two successive meiotic divisions is the production of four haploid daughter cells, formed by only one doubling of the chromosomal material.

CROSSING OVER

A constant feature of meiosis I is the presence of chiasmata, which hold paired chromosomes together from the diplotene stage through metaphase. As mentioned above, chiasmata are believed to mark the positions of crossovers, sites where prior to metaphase chromatids of homologous chromosomes have exchanged segments by breakage and recombination (Fig. 2–6). Though only two chromatids take part in any one crossover event, all four may be simultaneously involved in different crossovers. Chiasmata may play an important biological role in holding bivalents together and in preventing premature disjunction.

Because crossing over causes genes to become reorganized into new combinations, it increases genetic variability. As noted previously, the paternal and maternal chromosomes assort independently at meiosis, allowing for 2^{23} (about 8 million) different chromosome combinations in the gametes of a single individual. Crossing over markedly increases the number of different combinations possible, thus increasing the likelihood of favorable new combinations; it also allows favorable genes to become separated from deleterious new mutations that may arise in the same chromosome.

Linkage and crossing over are discussed further in Chapter 11.

Somatic Crossing Over

Somatic crossing over (crossing over between homologous chromosomes in mitosis) is much less common than meiotic crossing over, because homologous chromosomes are not usually closely associated except at meiosis. However, it is known from work in experimental organisms such as corn or the fruit fly that somatic crossing over does occasionally take place. The consequences are shown in Figure 2–7. In a heterozygous individual of genotype Aa, if nonsister but homologous chromatids that carry A and a undergo a crossover event between the centromere and the A locus, at anaphase the two chromatids with A alleles may pass to the same daughter cell and the two with a alleles to the other daughter cell. (This also applies to any other heterozygous

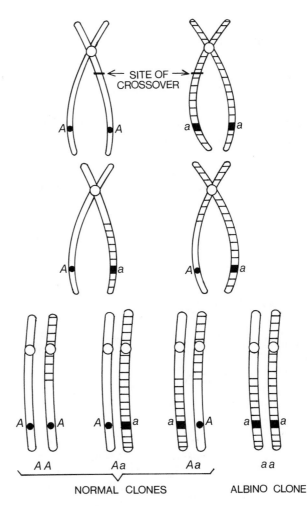

Figure 2–7 A possible consequence of somatic crossing over, described in the text.

locus between the crossover and the end of the chromosome; each such locus would become homozygous as a result of the crossover.) The clone of cells descended from each daugher cell would all have the same genotype as the original recombinant. In experimental organisms, this mechanism produces "twin spots," each homozygous for one of the two alleles of the parent. The same also happens, though rarely, in man. For example, in oculocutaneous albinism (an autosomal recessive trait), occasionally a heterozygote has a small albino patch. Somatic crossing over could explain this observation.

Sister Chromatid Exchange

Crossing over between the sister chromatids of a single chromosome is a phenomenon that was not recognized until recently, after a special technique was developed (Latt, 1973). Cultured cells are allowed to replicate twice in bromodeoxyuridine (BUdR), allowing incorporation of BUdR into new synthesized DNA in the place of thymine. BUdR modifies the staining properties of the chromatids; for example, the fluorescent stain Hoechst 33258 stains the chromatid in which both DNA strands contain BUdR less brightly than its sister chromatid, in which only one strand is BUdR-substituted. If a sister chromatid exchange (SCE) has occurred, this is readily recognized by the bright and dim fluorescence patterns along the chromatids (Fig. 2–8). The frequency of SCE is greatly increased in a particular genetic disorder, Bloom syndrome. The Bloom syndrome phenotype is very distinctive (Fig. 2–9). Affected children have a low birth weight; dwarfism; hypersensitivity to sunlight, causing skin rash; a characteristic facial appearance; a high frequency of chromosome breakage and rearrangement in cell culture; and a predisposition to leukemia. At present, however, the relationship of the curious increase in chromosome breakage and SCE frequency to the other features of the syndrome is unknown.

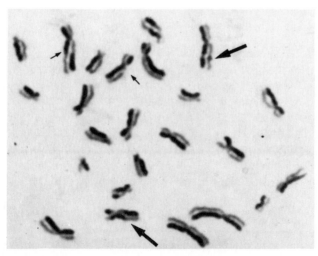

Figure 2–8 Sister chromatid exchange (SCE). These chromosomes are from a cultured lymphocyte prepared by growth in BUdR, as described in the text. Small arrows indicate two of the five chromosomes in which single exchanges have occurred, and larger arrows indicate chromosomes with two exchanges. Photomicrograph courtesy of R. G. Worton.

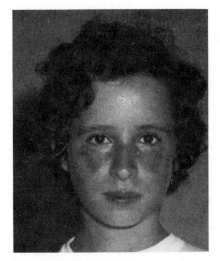

Figure 2–9   A child with Bloom syndrome. Photograph courtesy of J. L. German.

HUMAN GAMETOGENESIS

According to the old theory of "continuity of the germ plasm," germ cells are potentially immortal and are set aside for their special role at the very beginning of development. This is probably not strictly true; in the mouse, mutations at the W (dominant spotting) locus lead not only to reduced numbers of germ cells but also to reduced numbers of hematopoietic stem cells and defective functioning of pigment cells, indicating that these three cell types share a common ancestry. The earliest embryonic history of the human primordial germ cells is unknown, but by the fourth week of development they are known to lie outside the embryo proper, in the endoderm of the yolk sac. From there they migrate to the genital ridges and associate with somatic cells to form the primitive gonads, which later (at about 46 days after fertilization) differentiate into testes or ovaries in accordance with their genetic constitution.

Spermatogenesis

Spermatogenesis occurs in the seminiferous tubules of the testis of the male from the time of sexual maturity onward. The process is shown diagrammatically in Figure 2–10. At the periphery of the tubules are **spermatogonia** of several types, ranging from the self-renewing stem cells of the series to more specialized forms committed to the pathway of sperm formation; the latter are derived from the stem cells by a series of mitoses. The last stage in this developmental sequence is the **primary spermatocyte,** the cell that undergoes meiosis I. It divides to form two **secondary spermatocytes,** each with 23 chromosomes. These cells rapidly undergo meiosis II, each forming two **spermatids.** The spermatids mature without further division into **sperm** (spermatozoa) that are released into the lumen of the tubule. The total time involved in all of the stages from the beginning of meiosis to the formation of mature sperm is about 64 days (Clermont, 1972).

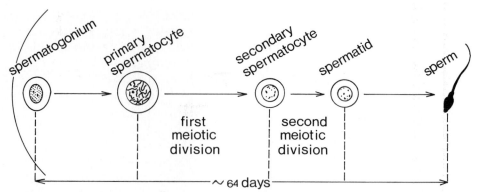

Figure 2–10   Diagram to illustrate human spermatogenesis. For discussion, see text.

Sperm are produced in enormous numbers, as many as 200 million per ejaculate. To provide such numbers over a long period of time, several hundred successive cell divisions are necessary. The older a man is when he becomes a father, the greater the number of DNA replications in the history of the germ cell he gives to the child.

Male meiotic chromosomes can be studied in material obtained by testicular biopsy. Figure 2–6 shows a human male cell in metaphase of the first meiotic division.

OOGENESIS

In contrast to spermatogenesis, the process of oogenesis is largely complete at birth. The ova develop from **oogonia,** cells in the cortical tissue of the ovary that have originated from the primordial germ cells by a series of mitoses. Each oogonium is the central cell in a developing follicle. By about the third month of prenatal development the oogonia of the embryo have begun to develop into **primary oocytes,** and some primary oocytes have already entered first meiotic prophase. The process is not synchronized, with early and later stages coexisting in the fetal ovary. The primary oocytes remain in "suspended prophase" (**dictyotene**) for at least several years, until sexual maturity is reached. Then, as each individual follicle begins to mature, the meiotic division of its oocyte resumes. Meiosis I is completed at about the time of ovulation, which may be more than 40 years after the beginning of the division.

As shown diagrammatically in Figure 2–11, the primary oocyte completes meiosis I in such a way that, while each daughter cell receives 23 chromosomes, one receives most of the cytoplasm and becomes the **secondary oocyte** and the other becomes a **polar body.** The second meiotic division commences almost immediately and proceeds as the ovum passes into and down the Fallopian tube. It is not completed until after fertilization, which usually takes place (if it takes place at all) before the ovum reaches the uterus. The second meiotic division produces the mature **ovum** with virtually all the cytoplasm and a second polar body. The first polar body may also divide. The polar bodies are ordinarily incapable of forming embryos, though rare exceptions occur (see later discussion of dispermic chimeras).

The maximum number of germ cells present in the human female is 6.8 million, found in the five-month fetus (Baker, 1963). By birth the number has dropped to 2 million, and by puberty less than 200,000, only 3 percent of the original number, still remain. The total possible number of germ cells present is the number that could be formed by no more than 23 mitoses (Vogel and Rathenberg, 1975).

The differences between spermatogenesis and oogenesis probably have genetic significance. The long duration of meiotic prophase in females may be causally related to the increasing risk of meiotic nondisjunction (failure of paired chromosomes to disjoin) with increasing maternal age, whereas the opportunity for error in the numerous replications of genetic information that take place during spermatogenesis in mature males may account for the fact that late paternal age is often observed in the fathers of new mutants.

FERTILIZATION

The process of fertilization normally takes place within the Fallopian tube within a day or so of ovulation. Ordinarily only a single ovum is released in any one menstrual cycle, whereas very large numbers of sperm may be present. Nevertheless, when one sperm penetrates the ovum a series of biochemical events is set into motion that ordinarily prevents the entry of any other sperm.

After entering the ovum, the sperm head rounds up to form the **male pronucleus.** Meanwhile the second meiotic division of the ovum is being completed, producing the **female pronucleus** and the second polar body. The male and female pronuclei approach one another, lose their membranes and combine to form the **zygote,** with the restored diploid chromosome number. Though DNA synthesis has been entirely shut off in the ovum, soon after the entry of the sperm it recommences, each haploid chromosome set replicates and the combined total of 46 chromosomes divide as at any mitosis to form two 46-chromosome daughter cells. This is the first of the series of cleavage divisions that initiate the process of embryonic development.

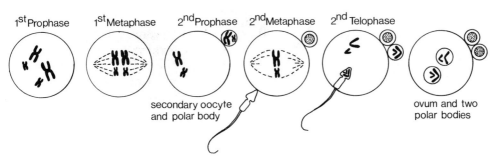

Figure 2-11   Diagram to illustrate human oogenesis and fertilization. For discussion, see text.

There is considerable speculation but little hard information about whether "aging" of the ovum or sperm in the Fallopian tube is deleterious in any way to the development of the resulting child.

Though human embryos have been fertilized in the laboratory and followed through a few cell divisions to the **morula** stage, most of our knowledge of the molecular events surrounding fertilization comes from studies of the sea urchin, a common echinoderm. In this organism it appears that there are two separate barriers to penetration of the ovum by more than one sperm (polyspermy) (Epel, 1977). When one sperm binds to a receptor site on the vitelline membrane of the ovum, there is an immediate change in the membrane potential associated with influx of sodium ions. A slower, enzymatically controlled block begins 25 to 30 seconds later, with specific alteration of the sperm-binding sites and formation of a fertilization membrane derived from the vitelline layer. A number of metabolic changes are then initiated, oxygen consumption and the rate of protein synthesis rise, and by 20 to 25 minutes after insemination DNA synthesis begins. In the sea urchin the first cleavage of the zygote occurs about two hours after the sperm first binds to the ovum.

CHROMOSOME CLASSIFICATION

When prepared for analysis, the chromosomes of a human metaphase cell appear under the microscope as a **chromosome spread** (Fig. 2–12). To analyze such a spread, the chromosomes are cut out from a photomicrograph and arranged in pairs in a standard classification. This process is called **karyotyping** and the completed picture is a karyotype.

The original classification was devised in 1960 at a meeting of cytogeneticists in Denver, Colorado. The "Denver classification" distinguished

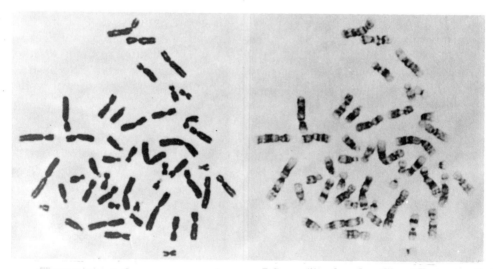

Figure 2–12 A chromosome spread prepared from a lymphocyte culture. The same cell is shown with solid staining (left) and Giemsa banding (right). Photomicrograph courtesy of R. G. Worton.

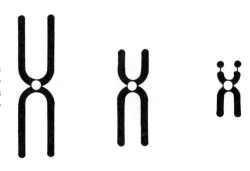

Figure 2–13 The three types of human chromosomes. Left to right: metacentric, submetacentric and acrocentric. Note the satellites on the short arms of the acrocentric chromosome.

seven chromosome groups, identified by the letters A through G on the basis of their overall length and centromere position.

The Centromere. The location of the centromere, or primary constriction, is a constant feature of each chromosome. Human chromosomes can be classified by centromere position into three types. If the centromere is central, the chromosome is **metacentric** (or mediocentric); if it is somewhat off-center, the chromosome is **submetacentric** (submediocentric); and if it is near one end, the chromosome is **acrocentric** (Fig. 2–13). In man, chromosomes 1, 3, 16, 19 and 20 are metacentric or nearly so, the D (13–15) and G (21, 22 and Y) chromosomes are acrocentric, and the remainder are submetacentric. A fourth type, telocentric, which has a terminal centromere, does not occur in man.

Satellites. The human acrocentric chromosomes (except the Y) have small masses of chromatin known as satellites attached to their short arms by narrow stalks (secondary constrictions). In metaphase spreads satellites are often seen in satellite association; this association apparently reflects the participation of the stalks in the organization of the **nucleolus,** an intracellular organelle involved in the synthesis of ribosomal ribonucleic acid (rRNA).

The seven chromosome groups in the Denver classification, in order of decreasing size, are illustrated in Figure 2–14 and are listed as follows:

Group A	Chromosomes 1, 2 and 3
Group B	Chromosomes 4 and 5
Group C	Chromosomes 6 to 12 and the X chromosome
Group D	Chromosomes 13 to 15 (the "large acrocentrics," with satellited short arms)
Group E	Chromosomes 16 to 18
Group F	Chromosomes 19 and 20
Group G	Chromosomes 21 and 22 (the "small acrocentrics," with satellited short arms) and the Y chromosome

Since the Denver conference there have been several other conferences that have attempted to standardize chromosome nomenclature. At the most recent, held in Paris in 1971, the nomenclature was extended by the

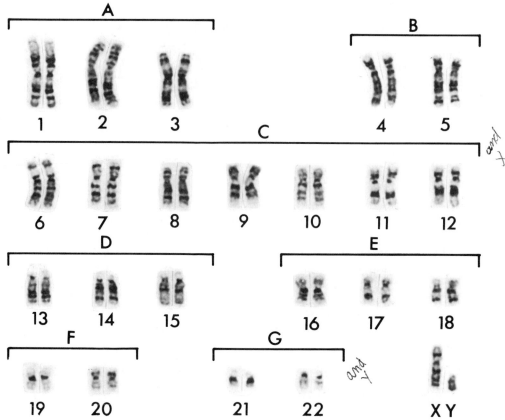

Figure 2–14 Normal male karyotype with Giemsa banding ("G banding"). The chromosomes are individually labeled, and the seven groups A to G are indicated. Photomicrograph courtesy of R. G. Worton.

use of banding techniques to identify each chromosome unequivocally by its banding pattern and to number the bands of each chromosome. The standard classification agreed to at the Paris Conference is shown in Figure 2–15.

SYMBOLS FOR CHROMOSOME NOMENCLATURE

With the accumulation of knowledge concerning numerical and structural aberrations of the chromosomes, it has become necessary to devise a set of symbols to designate certain features of the karyotype. Table 2–1 lists some of the more commonly used symbols, most of which will be used in later chapters to describe chromosome abnormalities of various types.

CHROMOSOME TECHNIQUES

Cells for chromosome analysis must be capable of growth and rapid division in culture. The most readily accessible cells that meet this require-

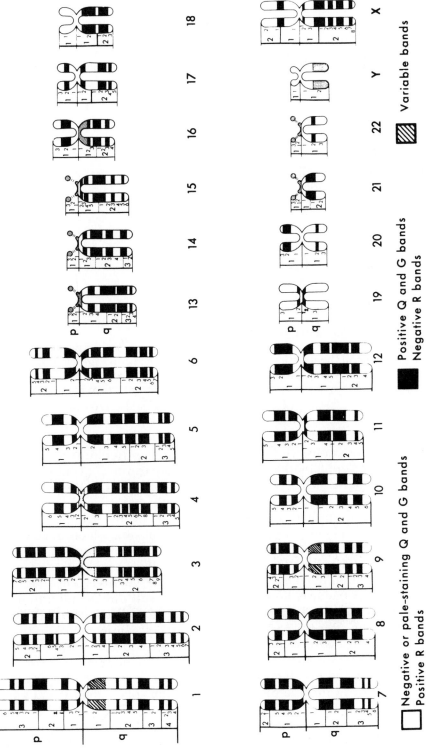

Figure 2-15 A diagrammatic representation of the human chromosomes, showing the banding patterns and numbering scheme adopted at the Paris Conference. Reprinted from Paris Conference (1971): Standardization in human cytogenetics. Birth Defects: Orig. Art. Ser. 8(7), 1972.

TABLE 2–1 SYMBOLS FOR CHROMOSOME NOMENCLATURE (PARTIAL LIST) *

A–G	The chromosome groups
1–22	The autosome numbers
X, Y	The sex chromosomes
/	Diagonal line indicates mosaicism, e.g., 46/47 designates a mosaic with 46-chromosome and 47-chromosome cell lines
del	Deletion
der	Derivative of chromosome
dup	Duplication
i	Isochromosome
ins	Insertion
inv	Inversion
p	Short arm of chromosome
q	Long arm of chromosome
r	Ring chromosome
s	Satellite
t	Translocation
	rcp Reciprocal translocation
	rob Robertsonian translocation
	tan Tandem translocation
ter	Terminal (may also be written as pter or qter)
→	From → to
+ or −	Placed *before* the chromosome number, these symbols indicate addition (+) or loss (−) of a whole chromosome; e.g., +21 indicates an extra chromosome 21, as in Down syndrome.
	Placed *after* the chromosome number, these symbols indicate increase or decrease in the length of a chromosome part; e.g., 5p− indicates loss of part of the short arm of chromosome 5, as in cri du chat syndrome.

*For further details, see Paris Conference references.

ment are the white cells of the blood. (Red cells have no nucleus and therefore no chromosomes.) To prepare a short-term culture, a sample of peripheral blood is obtained and mixed with heparin to prevent clotting. It is then centrifuged at a carefully regulated speed so that the white cells form a distinct layer. Cells of this layer are collected, placed in a suitable tissue culture medium and stimulated to divide by the addition of a mitogenic (mitosis-producing) agent, phytohemagglutinin, which is an extract of red bean. The culture is then incubated until the cells are dividing well, usually for about 72 hours. When the cultured cells are multiplying rapidly, a very dilute solution of colchicine is added to the medium. This interferes with the action of the spindle by binding specifically to the tubulin of the spindle microtubules, but it also prevents the centromeres from dividing. Because colchicine stops mitosis at metaphase, cells in metaphase accumulate in the culture. A hypotonic solution is then added to swell the cells and to separate the chromatids while leaving the centromeres intact. The cells are fixed, spread on slides and stained by one of several techniques. They are then ready for microscopic examination, photography and karyotype preparation.

Chromosome cultures prepared from peripheral blood have the disadvantage of being short-lived. Long-term cultures can be derived from other tissues, such as skin. A skin biopsy is a minor surgical procedure. The sample, which must include dermis, grows in culture and produces fibroblasts,

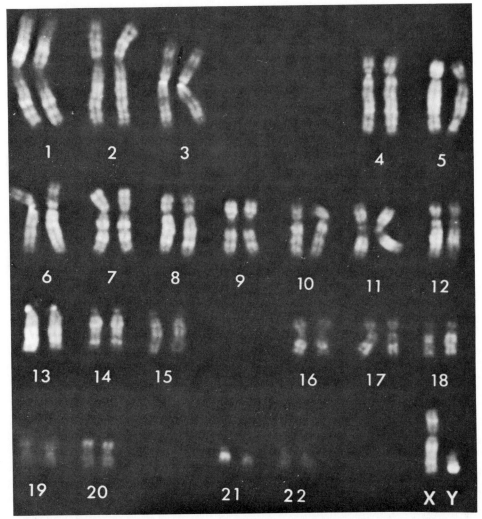

Figure 2–16 A human karyotype stained with quinacrine dihydrochloride and photographed under fluorescent light, to show Q bands. Photomicrograph courtesy of I. Uchida.

which are elongated, spindle-shaped cells capable of continuous multiplication *in vitro* for many cell generations. These cells can be used for a variety of biochemical and histochemical studies as well as for chromosome analysis. Fetal cells from amniotic fluid obtained by the procedure of **amniocentesis** can be cultured by a similar technique.

When the chromosomes of a cell have been karyotyped, it is possible to examine them for abnormalities of number or structure. Though numerical abnormalities are easy to identify, detection of structural ones requires excellent technique and careful observation. Even then, many structural abnormalities may remain beyond the limits of analysis.

STAINING METHODS

Prior to 1970, solid staining (as shown in Fig. 2–12) was the only staining method available, but since that time several special techniques have

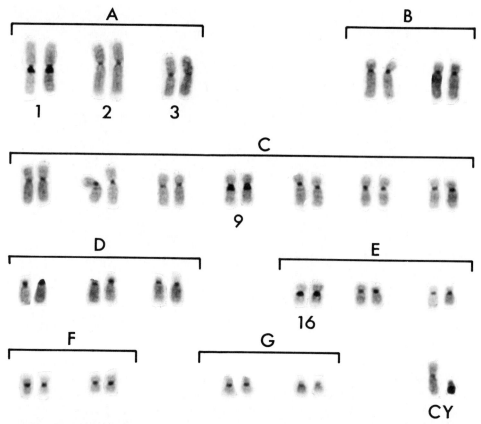

Figure 2–17 C banding. A karyotype stained to show C bands, which are regions containing constitutive heterochromatin. Note that not all the chromosomes can be individually identified. For further discussion, see text. Photomicrograph courtesy of R. G. Worton.

been developed for staining chromosomes in banded patterns. Some but not all of the special staining methods are included in the following list.

Q Banding. Caspersson and his colleagues (1970) found that when chromosomes are stained with quinacrine mustard or related compounds and examined by fluorescence microscopy, each pair stains in a specific pattern of bright and dim bands (Q bands), as shown in Figure 2–16. The Q bands were used as the reference bands for the Paris Conference classification.

G Banding. In this widely used technique chromosomes are treated with trypsin, which denatures chromosomal protein, and are then stained with Giemsa stain. The chromosomes take up stain in a pattern of dark and light staining bands (G bands), with the dark bands corresponding to the bright Q bands (Fig. 2–14).

R Banding. If the chromosomes receive a heat pretreatment and then Giemsa staining, the resulting dark and light stained bands (R bands) are the reverse of those produced by Q and G banding.

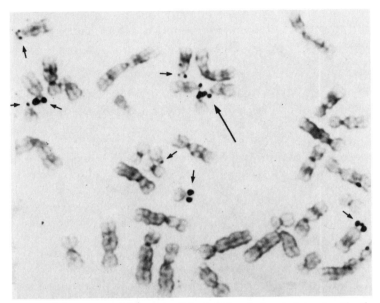

Figure 2–18 NOR staining. A human cell both G banded and stained with silver, to show the nucleolar organizing regions. Arrows show NORs; the large arrow indicates two D chromosomes and a G chromosome in association. For further discussion, see text. From Markovic et al., Hum. Genet. 12:1–17, 1978.

C Banding. This method (Fig. 2–17) specifically stains the centromere regions and other regions containing constitutive heterochromatin, i.e., the secondary constrictions of chromosomes 1, 9 and 16 and the distal segment of the long arm of the Y chromosome. (**Constitutive heterochromatin** is the kind of heterochromatin that forms part of the normal constitution of several chromosomes, as distinguished from **facultative heterochromatin,** which makes up the inactive X chromosome in female somatic cells. Heterochromatin is discussed further in Chapter 3.)

NOR Staining. This is a method that uses ammoniacal silver to stain the nucleolar organizing regions, i.e., the secondary constrictions (stalks) of the satellited chromosomes (Fig. 2–18).

G 11 Banding. This is a modification of Giemsa banding at high pH that preferentially stains the secondary constriction of chromosome 9, the distal segment of the long arm of the Y chromosome and the pericentromeric area of chromosome 20. G 11 banding is particularly useful for demonstrating common normal variants (polymorphisms) of chromosome 9.

MEDICAL APPLICATIONS OF CHROMOSOME ANALYSIS

Apart from their intrinsic genetic interest, chromosome studies are used in medicine in a number of important ways. Though each of the applications noted here is also discussed elsewhere, they are summarized here for convenience. Their significance may not be clear without additional background information, which is supplied in later chapters. These medical applications are as follows:

1. Clinical Diagnosis. Chromosome studies are useful in clinical diagnosis, especially in patients with congenital malformations involving several organ systems, mental retardation, failure to thrive or disorders of sexual development.

2. Linkage and Mapping. Chromosome studies lead to the assignment of specific human genes to their linkage groups and chromosomal positions.

3. Polymorphisms. Minor heritable differences in banding patterns are not unusual, especially for chromosomes 1, 9, 16 and the Y chromosome. These polymorphisms may sometimes be used to trace individual chromosomes through families. Polymorphisms may be used, for example, as markers in family studies or to help in determination of the source of the abnormal gamete in chromosome abnormalities. As described later, a familial chromosome polymorphism (a secondary constriction) allowed Donahue and his colleagues (1968) to map the Duffy blood group locus to chromosome 1, thus making the first assignment of any gene to a specific autosome. In a triploid infant who survived for a few hours after birth, Uchida and Lin (1972) found that the Q banding pattern showed that the extra chromosome set had in all probability come from the father in a diploid sperm. Currently, chromosome polymorphisms are being analyzed in trisomic children and in their parents to determine the relative frequency of nondisjunctional events in the first meiotic division as compared with the second, and in fathers as compared with mothers.

4. Studies of Malignancy. One of the first applications of Q banding was the demonstration that the so-called Philadelphia chromosome (a G chromosome with deletion of part of the long arm) found in the bone marrow cells of most patients with chronic myelogenous leukemia is chromosome 22. More recently Rowley (1973) has shown that the Philadelphia chromosome is actually a translocation of the distal segment of chromosome 22 to the long arm of chromosome 9. In a few patients, the translocation may involve chromosome 22 and some other chromosome or a more complicated rearrangement. The reason for the highly specific chromosome abnormality and its relationship to the leukemic process remain unexplained. Data on chromosome abnormalities in other neoplastic conditions are being collected, and evidence for nonrandom patterns of chromosome change in many of these is being assembled; for example, retinoblastoma is known to be commonly associated with deletion of a specific segment of chromosome 13, and a specific abnormality of chromosome 14 is found in various kinds of lymphoma. Though valuable observations have been made, much more work is needed before the association of chromosome abnormalities with human neoplasia is understood.

5. Reproductive Problems. The very high incidence of chromosome abnormalities found in spontaneous first-trimester abortions is described in a later section. Chromosome analysis may also be useful in determining the cause of infertility or repeated abortion, although only a small proportion of couples with such problems have a chromosome abnormality in one or the other parent that could account for their reproductive difficulties.

6. Prenatal Diagnosis. Because of the comparative ease and safety with which the karyotype of the fetus can be determined and the known association of chromosome abnormalities with late maternal age, many older pregnant women now receive amniocentesis to allow analysis of the chromosomes of the fetus. Familial chromosome abnormalities can also be monitored by this technique.

GENERAL REFERENCES

Hamerton, J. L. 1971. *Human Cytogenetics.* Vol. 1, *General Cytogenetics.* Vol. 2, *Clinical Cytogenetics.* Academic Press, Inc., New York.

Paris Conference (1971): Standardization in human cytogenetics. *Birth Defects: Orig. Art. Ser.* 8(7), 1972.

Paris Conference (1971), Supplement (1975). *Birth Defects: Orig. Art. Ser.* 11(9), 1975, reprinted in *Cytogenet. Cell Genet.* 15:201–238, 1975.

Yunis, J. J., ed. 1977. *Molecular Structure of Human Chromosomes.* Academic Press, New York.

Yunis, J. J., ed. 1977. *New Chromosomal Syndromes.* Academic Press, New York.

PROBLEMS

1. At a certain locus an individual is heterozygous, having the genotype *Aa.*
   a) What are the genotypes of his gametes?
      b) When do *A* and *a* segregate:
         1) if there is no crossing over between the *A* locus and the centromere of the chromosome?
         2) if there is a single crossover between this locus and the centromere?
2. How many different genotypes are possible in the ova of a woman who is:
   a) heterozygous at a single locus?
   b) heterozygous at 5 independent loci?
   c) heterozygous at *n* independent loci?
3. What is the proportion of normal human germ cells that contain chromosome sets in which there has been no random recombination?
4. How do Mendel's laws reflect chromosome behavior?

3

THE STRUCTURE AND FUNCTION OF CHROMOSOMES

Chromosomes are composed of **DNA** (deoxyribonucleic acid) complexed with histones (small basic proteins) and nonhistone proteins, in roughly equal amounts. These constituents together form nucleoprotein fibers or **chromatin.** The central role of DNA in encoding the genetic information, transmitting it to subsequent generations and determining protein structure is accepted as the central dogma of molecular genetics.

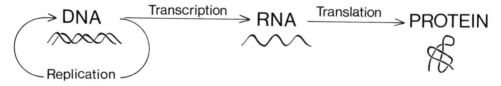

In recent years major improvements in our understanding of chromosomal structure and its functional significance have indicated that the central dogma requires qualification; not all DNA is transcribed into **RNA** (ribonucleic acid), not all RNA is translated into protein, and transcription can even proceed in reverse from RNA to DNA by means of a recently discovered enzyme, reverse transcriptase. **Euchromatin,** the type of chromatin that is transcribed, is normally not distinguishable in interphase nuclei, whereas **heterochromatin,** which is not transcribed, remains highly compacted and may be visible in stained interphase cells. Some of the new findings about chromosome structure have clinical implications apart from their general

biological significance, and these are mentioned later in the appropriate con-
texts.

Man is a **eukaryote**. This ponderous term means only that his cells, like
those of many other organisms including protozoa and fungi, have a genuine
nucleus. In contrast, prokaryotes such as *E. coli* (*Escherichia coli*, the intestinal bacterium widely used in molecular genetic research) do not have their
genetic information enclosed in a nucleus. Eukaryotic cells differ from pro-
karyotic cells in a number of other important ways: eukaryotic cells are larger
in size, have an extensive internal organization of membranes and organelles
in addition to the nuclear membrane, and may be capable of phagocytosis and
independent movement.

Evidence that DNA is the genetic material began to appear half a century
ago when Griffith (1928) in a classic experiment demonstrated the "transformation" of one strain of pneumococcus into another by mixing the first strain
with killed bacteria of the second strain in a host mouse and recovering live
bacteria of the second strain. Avery, MacLeod and McCarty (1944) showed
that the transforming factor was DNA by demonstrating transformation when
only a DNA extract from the second strain was used. Later Hershey and Chase
(1952), working with bacteriophage, showed that only the DNA of the phage
and not its protein was important in allowing the phage to reproduce.

NUCLEIC ACID STRUCTURE

The nucleic acids, DNA and RNA, are macromolecules (polymers) composed of three types of units: a five-carbon sugar (**deoxyribose** in DNA, **ribose**

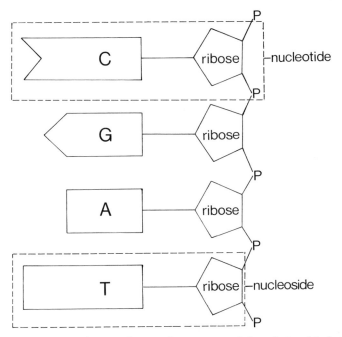

Figure 3–1 Diagram to show nucleic acid structure and the relationship between nucleoside (base and sugar) and nucleotide (base, sugar and phosphate).

in RNA), a nitrogen-containing base and phosphate. The bases are of two types, **purines** and **pyrimidines**. In DNA there are two purine bases, **adenine** (A) and **guanine** (G), and two pyrimidines, **thymine** (T) and **cytosine** (C). In RNA, **uracil** (U) replaces thymine. (There are a few rare exceptions, such as the presence of unusual pyrimidine bases in the type of RNA known as transfer RNA.) The relationship between the basic units, nucleoside (base plus sugar), and nucleotide (base, sugar and phosphate) is shown in Figure 3–1. The nucleotides polymerize into long polynucleotide chains by covalent bonding of one nucleotide to the 3' hydroxyl of the adjacent one.

DNA

In the DNA molecule the number of nucleotides containing A equals the number containing T, and the number containing G equals the number containing C. However, the A+T/G+C ratio in the DNA of various organisms varies within rather wide limits, 0.5 to 2.5; in man it is about 1.4.

A molecule of DNA is composed of two long polynucleotide chains coiled around one another to form a double helix. Figure 3–2 reproduces the illustration that appeared in the paper in which Watson and Crick (1953) proposed the double helix structure. The 1:1 ratio of A:T and G:C was one piece of evidence that led to the Watson-Crick model. Another important observation was made by Wilkins and Franklin, who showed by X-ray diffraction studies

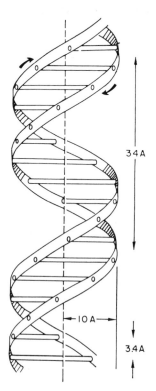

Figure 3–2   Structure of the DNA molecule as proposed by Watson and Crick in their original paper. From Nature 171:737–738, 1953.

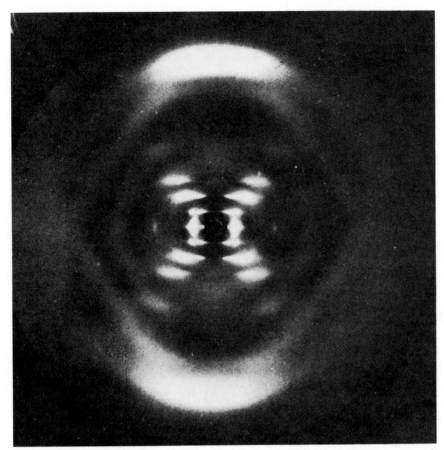

Figure 3–3 X-ray photograph confirming the helical structure of DNA. From Franklin and Gosling, Nature 171:740–741, 1953.

that the DNA molecule had a spiral shape and contained more than one polynucleotide chain (Fig. 3–3).

The two polynucleotide chains of the DNA molecule run in opposite directions and are held together by hydrogen bonds between A (a purine) in one chain and T (a pyrimidine) in the other or between G (a purine) and C (a pyrimidine). For physical and chemical reasons, other combinations of the bases do not fit, and they occur only rarely and by error. Obviously, the specificity of base pairing explains why equal amounts of A and T and of G and C occur in DNA.

Since A:T and G:C pairing is obligatory, the parallel strands must be complementary to one another; thus if one strand reads AGT CCA, the complementary strand must read TCA GGT. An important consequence of complementarity is that a DNA molecule can replicate precisely by separation of the strands followed by formation of two new complementary strands.

Almost infinite variations are theoretically possible in the arrangement of the bases along a polynucleotide chain. In any one position there are four possibilities; thus there are 4^n possible combinations in a sequence of n bases.

For three bases, 64 combinations are possible. It is now known that three bases (a **triplet**) constitute a **codon**, a unit of the genetic code.

THE GENETIC CODE

A major function of DNA is to direct the synthesis of polypeptides, which are molecules built of amino acids in a specific amino acid sequence (primary structure) held together by peptide linkages. A protein molecule comprises one or more polypeptide chains. The properties of the protein depend upon the order of the amino acids in its constituent polypeptides.

The "one gene–one enzyme" hypothesis proposed by Beadle and Tatum in 1941 may be interpreted in more up-to-date terms as "one gene–one polypeptide chain." Biochemically, both the polynucleotide chain that constitutes a gene and the polypeptide are linear molecules. The sequence in which amino acids are incorporated into a polypeptide chain is dictated by the order of the corresponding triplets of bases in *one* of the pair of polynucleotide chains of the DNA. The DNA sequence and the corresponding polypeptide sequence are therefore said to be **colinear.**

The **genetic code** (Table 3–1) was worked out through experiments using synthetic polynucleotides. The first synthetic polyribonucleotide used as a messenger RNA was polyU (polyuracil, a sequence of nucleotides in which all the bases are U). PolyU directs the synthesis of a polypeptide chain composed exclusively of phenylalanine, thus demonstrating that the code for phenylalanine is UUU. Since there are only 20 amino acids and 64 possible codons, most amino acids are specified by more than one codon; hence the code is said to be **degenerate.** For instance, the base in the third position of a triplet can often be either purine or either pyrimidine, or in some cases any one of the four bases, without altering the coded message. Leucine and arginine can each be coded for by six different codons. Three of the 64 codons designate termination of a message. With a few possible exceptions, the code is **universal;** the same amino acids are coded for by the same codons in all organisms studied, from bacteria to man.

DNA REPLICATION

The complementary structure of the DNA molecule allows for a conceptually simple method of replication. The two strands separate and each serves as a template upon which the missing partner can be reconstructed, by base pairing, from nucleotides present in the cell (Fig. 3–4). The nucleotides are bound to one another by the action of an enzyme, DNA polymerase, and are hydrogen bonded to the template strand. Replication begins at several points along the DNA molecule and proceeds in both directions, the "bubbles" merging where they meet (Fig. 3–5).

TABLE 3–1   THE GENETIC CODE*

First Base	Second Base				Third Base
	U	*C*	*A*	*G*	
U	UUU phe UUC phe	UCU ser UCC ser	UAU tyr UAC tyr	UGU cys UGC cys	U C
	UUA leu UUG leu	UCA ser UCG ser	UAA stop UAG stop	UGA stop UGG try	A G
C	CUU leu CUC leu	CCU pro CCC pro	CAU his CAC his	CGU arg CGC arg	U C
	CUA leu CUG leu	CCA pro CCG pro	CAA gln CAG gln	CGA arg CGG arg	A G
A	AUU ile AUC ile	ACU thr ACC thr	AAU asn AAC asn	AGU ser AGC ser	U C
	AUA ile AUG met	ACA thr ACG thr	AAA lys AAG lys	AGA arg AGG arg	A G
G	GUU val GUC val	GCU ala GCC ala	GAU asp GAC asp	GGU gly GGC gly	U C
	GUA val GUG val	GCA ala GCG ala	GAA glu GAG glu	GGA gly GGG gly	A G

Abbreviations for amino acids:

ala	alanine	leu	leucine
arg	arginine	lys	lysine
asn	asparagine	met	methionine
asp	aspartic acid	phe	phenylalanine
cys	cysteine	pro	proline
gln	glutamine	ser	serine
glu	glutamic acid	thr	threonine
gly	glycine	try	tryptophan
his	histidine	tyr	tyrosine
ile	isoleucine	val	valine

Other abbreviation:
     stop   termination of a gene

*Codons are shown in terms of messenger RNA. The corresponding DNA codons are complementary to these.

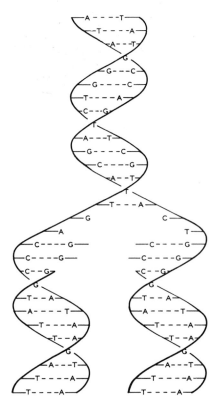

Figure 3–4 Replication of DNA. The original molecule unwinds, the strands separate, and each acts as a template on which a new strand can be formed by base pairing. The process results in two complete molecules each identical to the original. From Sutton, *Genes, Enzymes and Inherited Diseases*. Holt, Rinehart and Winston, Inc., New York, 1961.

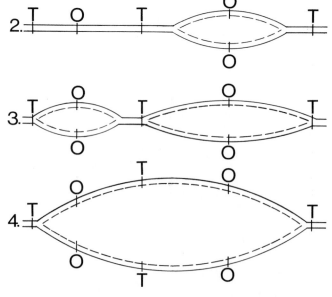

Figure 3–5 Model to show DNA replication occurring in both directions along a chromosome, O, points of origin of DNA replication; T, termination points. Redrawn from Huberman and Riggs, J. Mol. Biol. 32: 327, 1968.

Mutation

Normally replication is highly accurate, but occasionally an incorrect nucleotide is inserted into a new strand during its synthesis. The frequency of errors is very low, probably in the range of 10^{-7} to 10^{-9}. Any such error is copied at subsequent replications, unless a reverse mutation happens to occur. These changes, which alter the genetic code of the triplet within which they originate, may alter the amino acid sequence of the corresponding polypeptide, though the degeneracy of the code prevents many such alterations. Mutations can also originate by insertion or deletion of a nucleotide, and in these cases, since the code is read in triplets, the whole "reading frame" of the gene may be altered (Fig. 3–6). Mutations that result in termination codons stop the biosynthesis of the corresponding polypeptide chain at that point, and mutations of termination codons to other codons can cause elongation of the corresponding polypeptide, with a nonsense sequence at the end. Other kinds of possible changes in the DNA blueprint are described in Chapter 5.

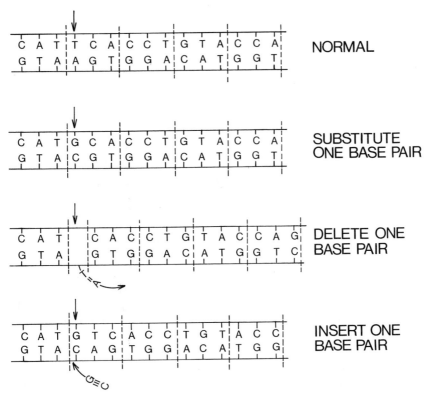

Figure 3–6 Different kinds of mutation. "Mis-sense" mutations involve substitution of a single nucleotide, resulting in an amino acid substitution in the corresponding polypeptide. "Nonsense" mutations involve deletion or addition of a nucleotide, resulting in an altered "reading frame."

THE COMPONENTS OF DNA

The human genome is estimated to contain about 2.7×10^9 nucleotide pairs, enough to code for about three million genes. However, it appears that a very small percentage of the total genome consists of genes that are transcribed into RNA and translated into protein, i.e., **structural genes.** A rough estimate is that man may have 30,000 structural genes. The function of most of the remainder of the DNA is unknown. Two classes of DNA have been recognized: "unique sequence" (nonrepetitive) DNA in 60 to 70 percent of the genome and repetitive DNA in the remaining 30 to 40 percent. Though logically one would expect the genetic information to be encoded in the unique sequence component, there is far too much of it to be accounted for by the structural genes alone. Some of the excess may be transcribed into heterogeneous RNA and into precursors of transfer RNA and ribosomal RNA. About a quarter of the repetitive DNA is classed as "highly repetitive," and most of the remainder is interspersed with unique sequence DNA.

Satellite DNA (not to be confused with chromosomal satellites, but named because of its method of separation by centrifugation in a cesium chloride gradient) is the name given to several kinds of a minor DNA component made up of very short, simple sequences which are highly repetitive, occurring in hundreds of thousands or even millions of copies. Satellite DNA is found in regions of constitutive heterochromatin, such as the C bands adjacent to the centromeres or in areas related to the synthesis of the nucleolus. It is not transcribed, and is thought to have a regulatory function. Satellite-rich heterochromatin tends to interfere with crossing over and thus to preserve linkage groups in the region of the centromere, but the selective advantage of such a mechanism is obscure.

The secondary constrictions (stalks) of the five pairs of acrocentric chromosomes carry the genes for 18S and 28S ribosomal RNA. [In this context S (Svedberg) is a measure of the speed of sedimentation of a given fraction in the centrifuge, i.e., its sedimentation constant.] As previously mentioned, the stalks of the acrocentric chromosomes are the nucleolus organizer regions of the genome. It is probably not coincidence that numerical aberrations of the acrocentric chromosomes are quite frequent causes of abnormality in liveborn children or of first-trimester abortion, but the mechanism by which the risk of nondisjunction is increased in these chromosomes is not known. The genes for 5S ribosomal RNA have not been definitely mapped, but appear to be near the ends of the longer chromosomes.

Moderately repetitive DNA sequences are found chiefly in Q and G bands, and it is suspected that most of the aberrations of large chromosomes in man involve chromosomes with large amounts of such sequences.

DNA REPAIR

If a DNA strand is broken by radiation or other agents, repair can normally be effected by a series of enzymatic reactions. Figure 3–7 illustrates the best understood repair process; thymine dimers produced by exposure of DNA to ultraviolet light are excised, and a new short segment is synthesized by base pairing with the intact strand.

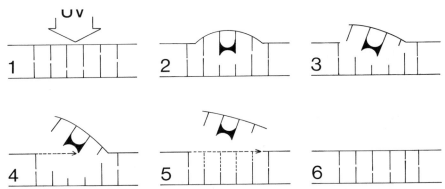

Figure 3–7 A model of DNA repair. 1, damage of DNA strand by ultraviolet light; 2, formation of thymine dimer; 3, breakage of the DNA strand by an endonuclease; 4, synthesis of a new DNA strand begins, using the intact strand as a template; 5, complete excision of the damaged region by an exonuclease; 6, joining of the repaired section to the old strand by a polynucleotide ligase.

In several genetic disorders, cultured fibroblasts of affected individuals show a high frequency of chromosome breaks, indicating defective repair synthesis. One such disorder is xeroderma pigmentosum (XP), a rare autosomal recessive characterized by sensitivity to sunlight, development of pigmented lesions on exposed skin and a high risk of malignancy. Cleaver (1968) showed that the ability to repair DNA is lost or greatly reduced in cells cultured from XP patients. If fibroblasts from two different XP patients, both with defective ability to repair DNA, are fused experimentally, the hybrid cells of certain pairs may be able to effect repair. Fusion experiments have shown that there are at least four groups of mutually correcting forms of XP (complementation groups). In genetic terms this means that mutations at a minimum of four different loci can produce XP, and that enzymes produced by at least four different loci are involved in the repair process. (There is still another form of XP in which DNA repair is not defective.)

Repair synthesis provides a degree of protection from spontaneous chromosome breakage due to natural radiation and hence may serve an evolutionary function.

RECOMBINANT DNA

Recently advances in knowledge and technology concerning DNA and its enzymes have made it possible to combine DNA of different organisms to manufacture new hybrid forms. This type of research holds great promise for the elucidation of the chromosome structure and genome organization of higher forms, since it allows them to be studied within much simpler systems.

Figure 3–8 shows diagrammatically the basic method for making recombinant DNA. Two types of DNA molecule are involved, often called the "vehicle" and the "passenger." The vehicle is usually a plasmid, a small circular DNA molecule of *E. coli* that is capable of replication independently

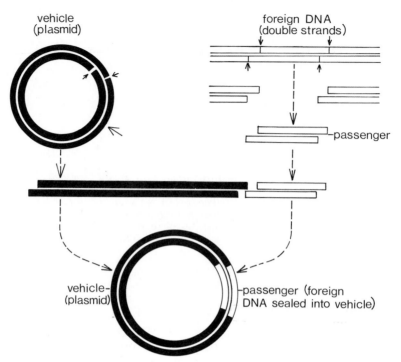

Figure 3–8 Technique for the manufacture of recombinant DNA. Small arrows mark sites of cleavage of DNA by restriction endonuclease. For discussion, see text.

of the *E. coli* chromosome. Methods using other vehicles and cultured mammalian cells rather than *E. coli* as the host are also under development.

The passenger DNA and the vehicle DNA with which it is to be combined are exposed to a type of DNA enzyme known as a restriction endonuclease, which has the special property of cleaving DNA only at specific sequences in which the code reads the same in both directions. Because the cut ends match, a segment of the foreign DNA can be inserted into the plasmid, where it is sealed in place by another enzyme, DNA ligase. The newly formed recombinant molecule is then introduced into host cells, where it can replicate. A host cell thus transformed can then be cloned in order to produce enough material to be analyzed.

An additional experimental step may be required to produce the foreign DNA that is to be hybridized. As an example, hemoglobin DNA can be obtained in quantity only if it is synthesized from hemoglobin messenger RNA, which is present in large amounts in reticulocytes. Complementary DNA (cDNA) can be made on an mRNA template by means of the enzyme reverse transcriptase and can then be introduced into plasmids. It is not yet known whether globin DNA can be transcribed and translated under these experimental conditions.

The possibilities for useful application of recombinant DNA research are manifold. Among the benefits proposed are the addition of nitrogen-fixing genes to nonleguminous plants, genetic engineering to "cure" genetic disorders caused by lack of a specific enzyme and industrial manufacture of hormones for medical use.

Nevertheless, there is considerable concern about the possible dangers of such research. *E. coli* is a ubiquitous resident of the human intestine. If, for example, genetic information that could cause malignancy were incorporated into *E. coli* that later escaped from the laboratory, the consequences might be very serious, though the probability that this might happen seems remote. Because of concern outside and to some extent within the scientific community, guidelines for recombinant DNA research have now been drawn up in many countries.

The debate about the safety and significance of recombinant DNA research has been an example of the kind of difficulty that can arise when a scientific problem of great public importance has to be resolved by the very people who have an investment in continuing the work, the scientists themselves, or by members of the lay public who all too often have had no opportunity to obtain enough background information to objectively assess its risks and benefits. There is a great need for members of the public, especially physicians with their responsibility for the health of the population, to become sufficiently well informed about the issues to be knowledgeable participants in decision-making about such problems.

HISTONES

As noted previously, about one-third of chromatin consists of histones. Five different histones are known, differing chiefly in their content of lysine and arginine. Unlike most other proteins, they have remained virtually unchanged throughout evolution. They are involved with DNA in the composition of **nucleosomes**, the units that give chromatin its beaded appearance in electron micrographs. Their function and the reason for their immutable structure remain unknown, though they are thought to play a regulatory role.

CHROMOSOMAL EVIDENCE OF GENE ACTIVITY

Evidence of gene activity and inactivity can be seen microscopically in the lampbrush chromosomes of amphibian oocytes (see later) and in the giant salivary gland chromosomes of the larvae of *Drosophila* and some other flies.

The salivary gland chromosomes consist of many parallel chromosome strands produced by a series of replications without separation. Homologous loci remain in apposition, so the whole of a chromosome pair shows a banded appearance (Fig. 3–9). In these chromosomes the loci actively synthesizing RNA are seen as lightly staining puffs and the inactive loci as densely staining bands that contain arginine-rich histone. Different patterns of puffing are seen in different tissues or in the same tissue at different periods of differentiation.

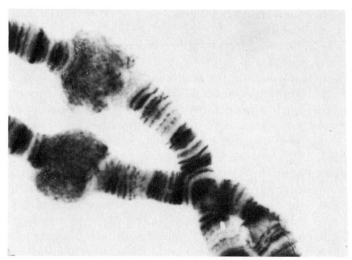

Figure 3–9 Chromosome puffs in the black fly. Note the two puffs on the strands at left and the banded appearance of the unpuffed regions. For details, see text. Photomicrograph courtesy of J. Pasternak.

THE ORGANIZATION OF DNA IN CHROMOSOMES

The duplicated DNA molecule of the G2 stage of the cell cycle becomes the very precisely folded, condensed chromosome seen at metaphase. The shape and consistent staining pattern of the chromosomes show that the DNA is not haphazardly contracted, but instead undergoes a highly organized condensation process.

Figure 3–10 shows a small acrocentric chromosome (21 or 22) as it appears under the electron microscope after critical-point drying and whole-mounting. This figure may be compared with the idiograms of chromosomes 21 and 22 from the Paris Conference publication (Fig. 2–16). It is generally accepted that each chromatid of a chromosome is a single long DNA molecule, a mononeme (or unineme). The processes that condense and pack the extended interphase chromatid into the complex structure shown, which has regular regional organization in spite of its irregular appearance, remain unexplained.

RNA

The genetic code is contained in DNA in the chromosomes within the cell nucleus, but polypeptide synthesis takes place in the cytoplasm in association with cytoplasmic organelles known as **ribosomes**. The link between DNA and polypeptide is RNA.

As already mentioned, the primary structure of RNA is similar to that of DNA, except that it contains a different sugar, ribose, and the base uracil in place of thymine. In its secondary structure it is single-stranded rather than double-stranded, though in special circumstances a single-stranded RNA mol-

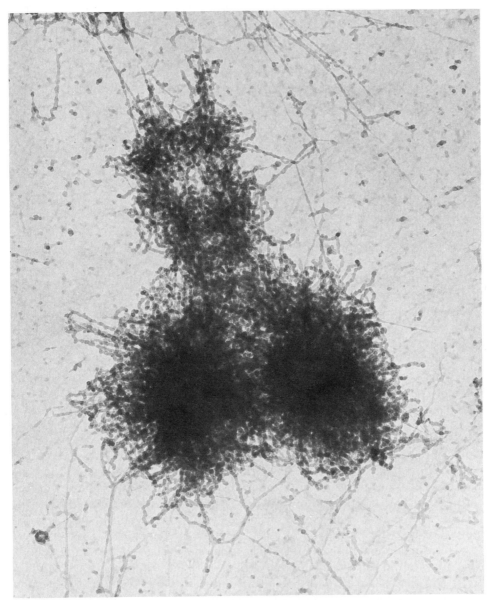

Figure 3–10 Electron micrograph of a G-group metaphase chromosome after critical point drying and whole mounting. From Bahr, Chromosomes and chromatin structure. In: *Molecular Structure of Human Chromosomes.* Yunis, J. J., ed. Academic Press, New York, 1977, p. 183, by permission.

ecule can form a double helix with another part of its own structure. RNA molecules vary in molecular weight from about 25,000 to 2 million.

Transcription of RNA depends upon **RNA polymerase**, a complicated enzyme that is one of the components of the nonhistone portion of chromatin. It is spoken of as "DNA-dependent" because it functions only in association with DNA and follows the base sequence of the corresponding DNA. RNA polymerase recognizes the start signal of a gene (a specific nucleotide sequence) and opens up a short section of the DNA helix, thus allowing synthesis of RNA along the DNA template. It then moves along the template, allowing the DNA double strands to become rebonded behind it and allowing the RNA to peel off and move out of the nucleus to a ribosome. Synthesis stops when a specific termination codon is reached. Figure 3–11 shows electron micrographs of this process in an oocyte of the newt *Triturus*.

TYPES OF RNA

Four types of RNA have been identified: messenger RNA, transfer RNA, ribosomal RNA and heterogeneous RNA.

Messenger RNA (**mRNA**) forms the template for protein synthesis, as described above. It is variable in molecular size, the length being determined by the corresponding DNA message.

Transfer RNA (**tRNA**), also known as adaptor RNA, is not translated into protein. Its function in protein synthesis is to transfer amino acids from the cytoplasm to their specific positions along the mRNA template. The different forms of tRNA are each able to recognize and transfer a specific amino acid, which must first be activated by a particular amino acid activating enzyme. The tRNA molecule is relatively small, having only 80 nucleotides and a molecular weight of about 25,000. It contains several unusual bases as well as the usual four. Its three-dimensional structure is complex, with hairpin folds that allow base pairing within their length (Fig. 3–12). One unpaired site on the molecule is an **anticodon** that is complementary to a specific codon on the mRNA chain and bonds to it, thus bringing the amino acid into position in the growing polypeptide.

Ribosomal RNA (**rRNA**) is associated with protein in the ribosomes and makes up a large proportion of the total RNA (80 percent in bacteria). It does not (normally) carry genetic information, but its specific function in the ribosome is still poorly understood. Figure 3–13 shows a cluster of ribosomes (polyribosomes) in a preparation of rabbit reticulocytes.

Heterogeneous RNA (**HnRNA**) is a category of high molecular weight RNA that, unlike other forms of RNA, never leaves the nucleus. It contains both an "informative" region and a "noninformative" region; the latter is presumably involved in regulation. Though the role of HnRNA is not precisely known, it ~~seems likely that it~~ is the precursor from which mRNA is processed.

m RNA

PROTEIN BIOSYNTHESIS

The steps in the biosynthesis of a polypeptide chain, most of which have already been mentioned, can now be summarized.

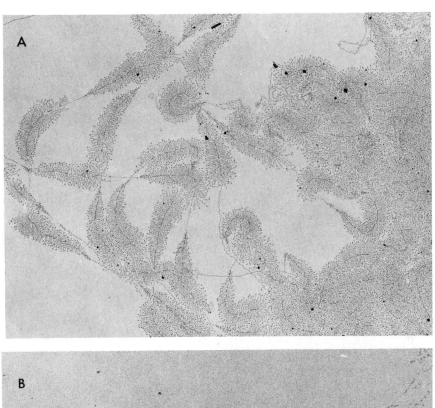

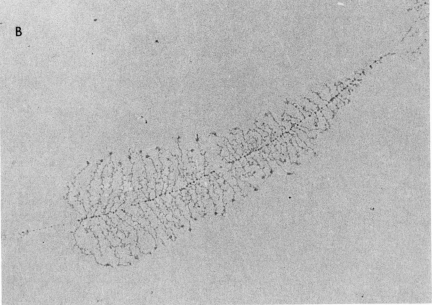

Figure 3–11 Electron micrograph of synthesis of repetitive RNA on a DNA template in a lampbrush chromosome of an oocyte of the newt *Triturus*. A, DNA molecules (long strands) each with many RNA molecules in the process of transcription projecting from it. B, a DNA strand enlarged to show RNA polymerase (black dots) with RNA molecules projecting laterally from each polymerase unit. Note increase in length of RNA molecules as transcription proceeds along the DNA template. From Miller and Beatty, J. Cell Physiol. 74, Suppl. 1:225–232, 1969, by permission.

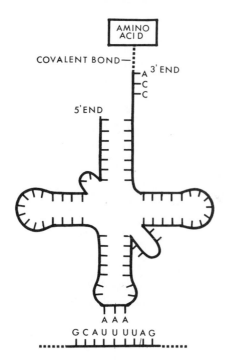

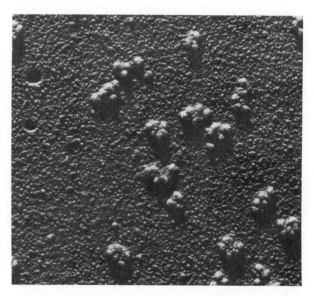

Figure 3–12 Schematic representation of a molecule of transfer RNA. For details, see text.

Figure 3–13 Polyribosomes as seen in a rabbit reticulocyte preparation. Electron photomicrograph courtesy of A. Rich.

Proteins, whether structural components or enzymes, are encoded in DNA. Their synthesis, however, is not dictated directly by the DNA template, but requires the participation of three types of RNA. Usually protein synthesis takes place in the cytoplasm.

Only one of the two DNA strands functions as a template. The two strands of DNA dissociate in the area of the gene that is to be transcribed, and mRNA forms on the DNA template under the influence of RNA polymerase. The mRNA molecule diffuses to the cytoplasm, where it associates with a group of ribosomes (polyribosome).

Amino acids (AA) in the cytoplasm are recognized and bonded to specific tRNA molecules by specific activating enzymes. After activation, the AA-tRNA complexes move to the ribosomes, where they are lined up in the correct order by base-pairing between a codon of the mRNA molecule and a corresponding anticodon of tRNA.

Ribosomes are composed of approximately equal amounts of protein and nonspecific rRNA. Structurally, each ribosome consists of a large and a small subunit that dissociate after completion of chain synthesis and reunite only after a complex of AA-tRNA and mRNA has combined with the smaller subunit.

Initiation of chain synthesis is followed by chain elongation, with peptide bonding between the successive amino acids. Chain termination occurs when a particular chain-termination codon is reached. A schematic interpretation of protein biosynthesis is shown in Figure 3–14.

THE CONTROL OF PROTEIN SYNTHESIS

At a given time, only some of the genes of a cell are active in protein synthesis, and those that are active may be synthesizing their different proteins in widely different amounts. The factors that determine whether a

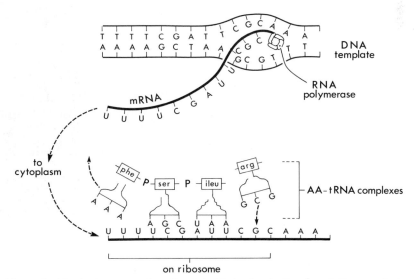

Figure 3–14   An interpretation of protein biosynthesis. For description, see text.

certain protein will be synthesized and in what amounts are still poorly understood, especially in mammalian systems. So far, efforts to induce cells in culture to make products other than those characteristic of that particular cell type have been disappointing. The problem is important, not only in connection with understanding cell function in molecular detail, but also with respect to the control of differentiation and development.

The regulation of protein synthesis may be effected at any of several levels: transcription of DNA to RNA, translation of RNA to protein or by cellular regulatory mechanisms acting on the completed protein. One model, first proposed by Jacob and Monod (1961) for control of the synthesis of enzymes in *E. coli,* has been substantiated for several bacterial systems, but may not apply to man and other higher organisms. The model (Fig. 3–15) involves the interaction of two kinds of units, **regulatory genes** and **operators,** as distinct from the structural genes.

According to the Jacob-Monod model, regulatory genes exert a negative control over protein synthesis by coding for **repressors,** specific protein molecules that act in the cytoplasm to repress the synthesis of specific proteins or groups of related proteins. Repressors inhibit protein synthesis by binding to corresponding operators, thus forestalling messenger RNA transcription and consequently preventing protein synthesis.

Because a repressor inhibits synthesis of a specific protein, it must be inactivated (the operator must be derepressed) if synthesis of the protein is to proceed. A variety of substances can act as inducers, combining with the repressor molecules to prevent them from binding with their specific operators.

An operator is a site at one extremity of a gene or of a sequence of related genes that when repressed can inhibit transcription by the gene(s) related to

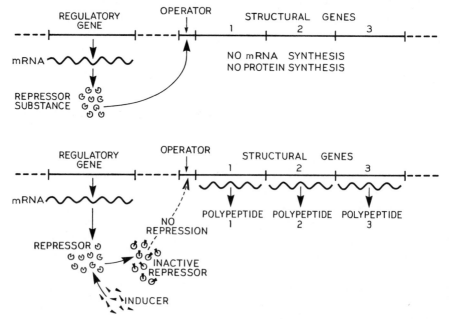

Figure 3–15   An interpretation of the regulation of protein synthesis. For description, see text.

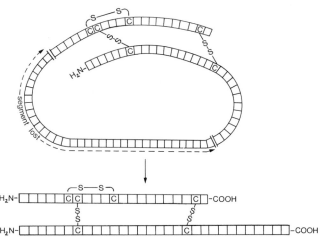

Figure 3–16 Synthesis of the insulin molecule. For description, see text.

it. It is a relatively short sequence of nucleotides, not regarded as a "whole" gene, but rather as a site within a gene or gene complex. The operator and the adjacent structural gene or genes controlled by it constitute an **operon**. Unlike regulatory genes, the operator affects only the genes in its own operon, not those on the homologous chromosome, i.e., it has an effect in *cis* but not in *trans*.

POSTSYNTHETIC MODIFICATION OF PROTEINS

Many proteins undergo extensive modification after synthesis. The polypeptide chain that is the primary gene product is folded and bonded in a very specific way that appears to be determined by the amino acid sequence itself. Two or more chains, alike or different, may combine to form a single protein; when this happens, the molecular size and amino acid sequence of the component polypeptides usually suggest that the corresponding loci have arisen by duplication of a single locus in the distant past (Hopkinson et al., 1976).

Proteins may also be modified by removal of a portion of the molecule. Insulin is a case in point. This protein is a hormone which is responsible for reducing the level of glucose in the blood; when it is present in insufficient amounts, diabetes mellitus results. Insulin is made up of two chains, one with 21 and the other with 30 amino acids, held together by disulphide bonds. By analogy with other proteins it was suspected that each chain was coded by a separate gene, but this proved not to be the case. Instead, the primary gene product, proinsulin, is a single polypeptide 82 amino acids long, which after folding and bonding loses a large interstitial section (Fig. 3–16). Though insulin production is defective in diabetes mellitus, the insulin molecule itself does not have abnormal structure; in other words, diabetes is not caused by a mutation in the structural gene for insulin.

GENERAL REFERENCES

Comings, D. E. 1972. The structure and function of chromatin. *In: Advances in Human Genetics* 3:237–431. Harris, H., and Hirschhorn, K., eds. Plenum Publishing Corporation, New York
Watson, J. D. 1976. *Molecular Biology of the Gene.* 3rd ed. W. A. Benjamin, Inc., Menlo Park.
Yunis, J. J., ed. 1977. *Molecular Structure of Human Chromosones*, Academic Press, New York.

PROBLEMS

1. What kind of mutation is illustrated by each of the following amino acid sequences?

   Wild type   – – lys arg his his tyr leu – – –
   Mutant I   – – lys arg his his cys leu – – –
   Mutant II   – – lys arg ile ile ile – – – – –
   Mutant III   – – lys glu thr ser leu ser – – –

2. Match the following:

   A
   – DNA molecule
   – cis
   – point mutation
   – trans
   – satellite of acrocentric chromosomes
   – deletion of a nucleotide
   – polynucleotide ligase
   – operator
   – satellite DNA
   – restriction endonuclease

   B
   1) cleaves DNA where code reads the same in both directions
   2) on the homologous chromosome
   3) single amino acid substitution
   4) found in constitutive heterochromatin
   5) chromatid
   6) contains genes for ribosomal RNA
   7) reunites broken DNA strand
   8) produces nonsense mutation
   9) on the same chromosome
   10) controls transcription of structural gene or genes

4

PATTERNS OF TRANSMISSION OF SINGLE-GENE TRAITS

In Chapter 1 the three main types of genetic disorders were named and briefly defined. This chapter is concerned with a more complete description of the patterns of transmission shown by traits determined by genes at a single locus. Single-gene phenotypes are sometimes referred to as "**Mendelian**" or "Mendelizing" traits because they segregate sharply within families and, on the average, occur in fixed proportions as did the characteristics studied by Mendel in garden peas. As noted earlier, the pedigree patterns shown by such traits depend upon two factors: (1) whether the gene responsible is on an autosome or on the X chromosome and (2) whether it is **dominant,** i.e., expressed even when present on only one chromosome of a pair, or **recessive,** i.e., expressed only when present on both chromosomes. Thus there are only four basic patterns:

$$
\text{Autosomal}
\begin{cases}
\text{dominant} \\
\\
\text{recessive}
\end{cases}
$$

$$
\text{X-linked}
\begin{cases}
\text{dominant} \\
\\
\text{recessive}
\end{cases}
$$

An individual pedigree pattern is also determined by the chance distribution of genes from parents to children through the gametes. Typical patterns can be altered or obscured in a number of ways, and it is necessary to take into account complications arising from heterogeneity, pleiotropy, reduced penetrance, variable expressivity, onset age, sex limitation, gene interaction or environmental effects. In this chapter the patterns of single-gene inheritance are described and some of the ways these patterns may be altered are discussed.

Some terms with special connotations in genetics must first be introduced. The family member who first brings a family to the attention of an investigator is the **propositus** (proband, index case). **Sibs** (or siblings) are brothers or sisters, of unspecified sex. The parent generation is designated the P_1, and the first generation offspring of two parents the F_1, but these terms are used more frequently in experimental genetics with inbred lines of plants or animals than in human genetics.

Recall that genes at the same locus on a pair of homologous chromosomes are **alleles.** In more general terms, alleles are alternative forms of a gene. When both members of a pair of alleles are identical, the individual is **homozygous** (a homozygote); when they are different, the individual is **heterozygous** (a heterozygote or carrier). The term **compound** is used to describe a genotype in which two different *mutant* alleles are present, rather than one normal and one mutant. These terms (homozygous, heterozygous and compound) can be applied either to an individual or to a genotype.

An allele which is expressed whether homozygous or heterozygous is **dominant;** an allele which is expressed only when it is homozygous is **recessive.** Strictly speaking, it is the **trait** (phenotypic expression of a gene) rather than the gene itself which is dominant or recessive, but the terms **dominant gene** and **recessive gene** are in common use.

The distinction between dominant and recessive genes is not absolute. In heterozygotes, each member of the gene pair forms a gene product, whether or not the heterozygous phenotype is distinguishable from the homozygous form. By definition a recessive has no detectable expression in heterozygotes, but some genes are defined as recessive simply because when heterozygous they are not phenotypically evident *under the conditions of analysis.* If the phenotype is examined in a different way, the expression of the gene may be detectable. For example, if its anticipated phenotypic expression is a manifest disease, a gene may be regarded as recessive even if on the biochemical level the "recessive" gene is identifiable when heterozygous. Later we will mention a number of examples of so-called recessive genes that are detectable in heterozygotes.

As a general but by no means rigid rule, disorders in structural (nonenzymatic) proteins are inherited as dominants, whereas changes in enzyme proteins are usually recessive. The explanation is that in heterozygotes an abnormal structural protein is formed in all cells and consequently leads to an abnormal phenotype, whereas the gene for an abnormal enzyme usually produces no obvious phenotypic effect in heterozygotes, because the margin of safety in enzyme systems allows normal function even if only one of a pair of alleles is present to produce the normal enzyme.

Family data can be summarized in a **pedigree chart,** which is merely a shorthand method of classifying the data for ready reference. The symbols used in drawing up a pedigree chart are shown in Figure 4–1. Variants of these symbols are acceptable, but the ones shown are all in common use. Special symbols may be invented to demonstrate special situations. By convention, gene symbols are always in italics. Usually a capital letter is used to indicate a dominant gene and the same letter in lower case for a corresponding recessive allele; alternatively, a plus sign (+) may indicate the normal allele, which is often referred to as the "wild type." A genotype is often shown

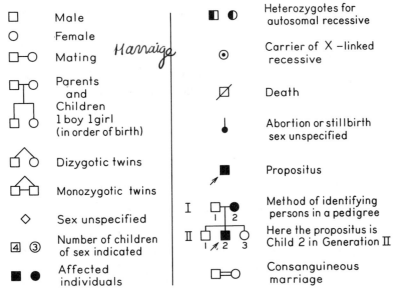

Figure 4–1 Symbols commonly used in pedigree charts.

with a slash (to symbolize the chromosome pair) between the two gene symbols: thus, *A/a*. In common usage, the slash is often omitted and will be omitted throughout this book.

Many genetic traits are determined by genes at a single locus, in either homozygous or heterozygous state. There are 2786 such conditions, most of which are abnormalities rather than normal variants, catalogued in a most useful reference: *Mendelian Inheritance in Man* by V. A. McKusick (5th edition, 1978). This is an appreciable proportion of the estimated 30,000 human structural genes. Of the total, 1473 are autosomal dominant, 1108 autosomal recessive and 205 X-linked.

GENES, GENOTYPES AND MATING TYPES

A simple example of a common hereditary trait governed mainly by a single pair of autosomal alleles is the ability to taste phenylthiourea, also called phenylthiocarbamide (PTC). (The genetics and biochemistry of this trait are discussed in additional detail in Chapter 5.) The ability to taste PTC is dominant to the inability to taste it. If *T* is the taster gene and *t* the nontaster gene, then the phenotype of both the *TT* (homozygous dominant) and *Tt* (heterozygous) genotypes is taster, and that of the *tt* (homozygous recessive) genotype is nontaster. In summary:

Genes	Genotypes	Phenotypes
T and *t*	*TT*	taster
	Tt	taster
	tt	nontaster

Since there are three possible genotypes and since either male or female may have any one of the three, there are six possible mating types, as shown in Table 4–1. Each gamete will have just one allele of the pair, and the possible gametes of each genotype are as shown below.

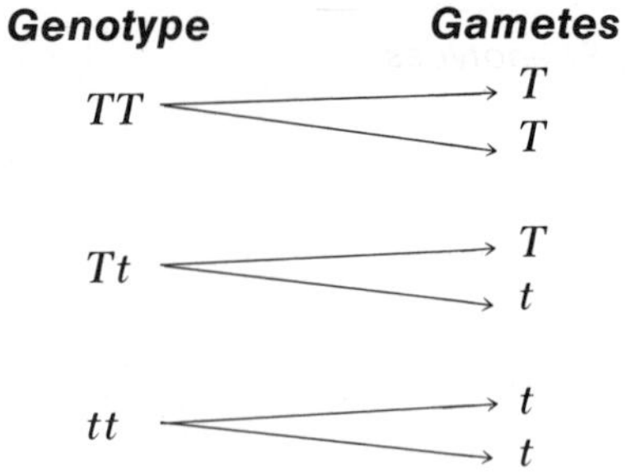

Genotype	Gametes
TT	T, T
Tt	T, t
tt	t, t

To illustrate how the information contained in Table 4–1 is derived, let us consider the progeny of a heterozygous *(Tt)* male and female. A "checkerboard" is sketched to show the gametes of each parent (in the margins) and their possible progeny (in the closed squares). A checkerboard is sometimes called a Punnett square. It was invented by R. C. Punnett, who is better known as the codiscoverer, with Wm. Bateson, of genetic linkage in 1906.

PROGENY OF *Tt* × *Tt* MATING

	Ova	
	T	*t*
T	TT	Tt
t	Tt	tt

Sperm (row labels)

Each of the offspring has a 25 percent chance of being a dominant homozygote *(TT)*, and a 50 percent chance of being a heterozygote; either of these genotypes causes the taster phenotype. There is a remaining 25 percent chance that the child will be a recessive homozygote, with the nontaster phenotype. (See Table 4–1, fourth line.)

Another way of working out the possible combinations of gametes in the progeny of a cross is by a mating diagram, as follows:

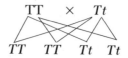

TT × Tt

TT   TT   Tt   Tt

All the possible mating types involving a single pair of autosomal alleles are set out in Table 4–1, but two of these are of particular importance: the *Tt* × *Tt* mating, the common one for autosomal inheritance of a rare recessive gene *t*, and the *Tt* × *tt* mating, typical of autosomal inheritance of a rare dominant gene *T*.

TABLE 4-1   AUTOSOMAL INHERITANCE: MATING TYPES AND
EXPECTED PROPORTIONS OF PROGENY FOR A PAIR OF
AUTOSOMAL ALLELES *T* AND *t*.

Mating Types		Progeny	
Genotypes	Phenotypes	Genotypes	Phenotypes
TT × *TT*	taster × taster	All *TT*	All taster
TT × *Tt*	taster × taster	1/2 *TT* 1/2 *Tt*	All taster
TT × *tt*	taster × nontaster	All *Tt*	All taster
Tt × *Tt**	taster × taster	1/4 *TT* 1/2 *Tt* 1/4 *tt*	3/4 taster 1/4 nontaster
Tt × tt‡	taster × nontaster	1/2 *Tt* 1/2 *tt*	1/2 taster 1/2 nontaster
tt × *tt*	nontaster × nontaster	All *tt*	All nontaster

*This is the usual pattern of inheritance for a rare autosomal recessive trait.
‡This is the usual pattern of inheritance for a rare autosomal dominant trait.

AUTOSOMAL INHERITANCE

AUTOSOMAL DOMINANT INHERITANCE

Table 4–1 shows five patterns of inheritance involving autosomal dominant genes in both parent and child, but if the gene is rare, only the *Tt* × *tt*
mating is frequent enough to be of practical significance. In this mating, one
parent is heterozygous for a rare autosomal dominant gene and the other
parent is homozygous for the normal allele. Each child of the affected parent
has a 50 percent chance of receiving the abnormal allele *T* and thus being
affected, and a 50 percent chance of receiving the normal allele *t* and thus
being normal. (The normal parent will give a normal allele to each child.)
Though *on the average* half the children will have the trait, the formation of
each zygote is statistically an independent event, so that in any single family
the ratio of affected to normal may be quite different from 1:1. The distribution of the trait in the family is not affected by sex. Figure 4–2 is a stereotype
pedigree illustrating these characteristics.

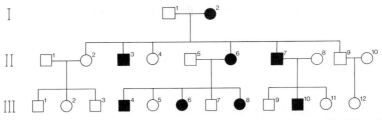

Figure 4–2   Stereotype pedigree of autosomal dominant inheritance. Half of the offspring of
affected persons (7 out of 14) are affected. The condition is transmitted only by affected family
members, never by unaffected ones. Equal numbers of males and females are affected. Male-to-
male transmission is seen.

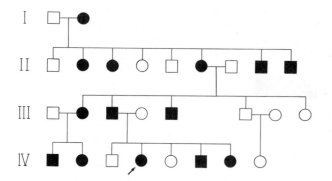

Figure 4–3 Pedigree of dentinogenesis imperfecta, an autosomal dominant disorder of dentine formation.

An Example of Autosomal Dominant Inheritance: Dentinogenesis Imperfecta

Dentinogenesis imperfecta is a rare, dominantly inherited tooth defect with an incidence of about one in 8000. The teeth have a peculiar opalescent brown color and their crowns wear down readily; their abnormality is so obvious that the condition is quite easy to ascertain in members of a kindred in which the gene is segregating. Figure 4–3 is a pedigree of dentinogenesis imperfecta in four generations of a family. The condition of the teeth and gums of one of the affected children is shown in Figure 4–4.

Let us use the symbol D for the dominant gene for dentinogenesis imperfecta and d for its normal allele. Then an affected person has the genotype Dd and a normal person has the genotype dd. In the general population, 50 percent of the progeny of $Dd \times dd$ matings are Dd and 50 percent dd. Stated differently, each child has a 1/2 chance of being affected and a 1/2 chance of being normal. The Dd parent forms D and d gametes in equal numbers. The dd parent forms d gametes only:

PROGENY OF $Dd \times dd$ MATING

Normal Parent

		d	d
Affected Parent	D	Dd Affected	Dd Affected
	d	dd Normal	dd Normal

The Homozygous Dominant Genotype

A dominant gene, by formal definition, has the same expression in the heterozygous state as in the homozygous state, but in actual practice homozygotes for rare dominants are seldom encountered. This is because homozy-

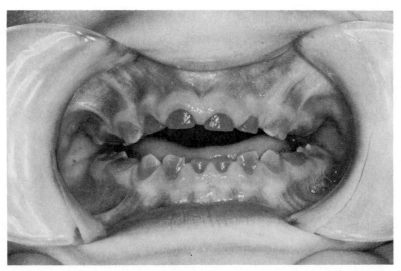

Figure 4–4 Dentinogenesis imperfecta in the propositus (IV–4) of the family shown in Figure 4–3. Photograph courtesy of N. Levine.

gotes are produced only by the mating of two heterozygotes, and a mating of two persons, both affected by the same rare dominant trait, is statistically improbable. Any offspring of two heterozygous parents has a 25 percent risk of being homozygous affected.

PROGENY OF *Dd* × *Dd* MATING

	D	d
D	*DD* Homozygous affected (usually rare)	*Dd* Heterozygous affected
d	*Dd* Heterozygous affected	*dd* Homozygous normal

In experimental animals, with large numbers of progeny, homozygotes can be distinguished from heterozygotes because, when mated with a homozygous recessive, a homozygote has only affected offspring. However, in human families, by chance a *Dd* person with a *dd* spouse might have only *Dd* children even though the chance of a *dd* child is 50 percent in each pregnancy. Thus it may be difficult to identify a homozygote positively, even if she or he is the offspring of two heterozygotes. As it happens, many rare dominants

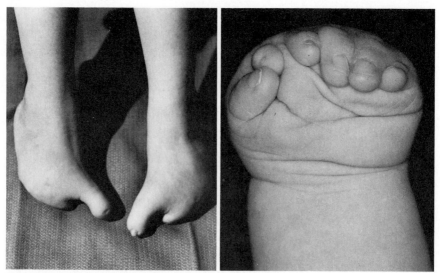

Figure 4–5 A probable example of homozygosity for an autosomal dominant trait. Shown are the feet and hands of a child whose first-cousin parents had minor foot and hand malformations. From Edwards and Gale, Am. J. Hum. Genet. *24*:464–474, 1972, by permission.

are more severe in homozygotes and so can be identified phenotypically. Figure 4–5 illustrates such a condition.

In this family both parents, who are first cousins, have brachycamptodactyly, a mild skeletal anomaly affecting the hands and feet. The metacarpals and metatarsals are short, and the middle phalanges are short and bent. Many women in the kindred have urinary incontinence and a longitudinal vaginal septum. The presumably homozygous children have very severe limb malformations (brachydactyly, polydactyly, syndactyly) as shown, and are retarded and deaf; the female sib also has a vaginal septum.

Achondroplasia is an example of an autosomal dominant phenotype in which the homozygous form is lethal in early infancy. This is a skeletal disorder of short-limbed dwarfism and large head size, with bulging forehead and "scooped out" nose bridge (Fig. 4–6). True achondroplasia can be distinguished radiologically from the many other forms of short-limbed dwarfism because the interpeduncular distance of the vertebrae narrows, rather than widens, toward the caudal end of the vertebral column. Most achondroplastics have normal intelligence and wish to lead normal lives within their physical capabilities. Understandably, marriages between two achondroplastics are not uncommon, and in these marriages achondroplastic offspring may even be preferred to those of normal height. The homozygous achondroplastic offspring have a severe skeletal disorder and do not survive.

Since a heterozygote has one normal allele and a homozygote has none, it is not surprising to see a more severely abnormal phenotype in homozygotes; the observation demonstrates that the classification of genes as "dominant" and "recessive," though both alleles are functional, is a convenient label rather than an essential distinction between dominance and recessivity.

New Mutation in Genetic Disorders

When a disorder has a deleterious effect on the ability of the affected individual to reproduce (his or her "reproductive fitness"), an appreciable proportion of patients with the disorder may be new mutants who have received the defective gene as a fresh mutation in a germ cell from a genetically normal parent. We will refer to this topic again in Chapter 15. New mutation is not an unusual observation in autosomal dominant or X-linked pedigrees of severe disorders. It is much less common in autosomal recessives, in which it is much more likely that the normal parents are heterozygotes. In achondroplasia, at least 80 percent of the patients are new mutants.

The significance of new mutation for genetic counseling is that when a particular patient can confidently be classified as a new mutant, the risk that the parents (or other relatives) will produce another such child is no greater than the general population risk. The risk for the offspring of the patient, however, is the usual Mendelian risk (50 percent for each child of an achondroplastic, for example).

Criteria for Autosomal Dominant Inheritance

The criteria for autosomal dominant inheritance may be summarized as follows: *In a pedigree :*

1. The trait appears in every generation, with no skipping.
2. The trait is transmitted by an affected person to half his children on the

Figure 4–6 Achondroplasia, an autosomal dominant disorder. From Tachdjian, M. O., 1972. Pediatric Orthopedics, Vol. I. W. B. Saunders Company, Philadelphia, p. 284.

average, though in an individual family wide discrepancy from the 1:1 ratio may be seen.

3. Unaffected persons do not transmit the trait to their children.

4. The occurrence and transmission of the trait are not influenced by sex; i.e., males and females are equally likely to have the trait and equally likely to transmit it.

AUTOSOMAL RECESSIVE INHERITANCE

A trait transmitted as an autosomal recessive is expressed only in homozygotes, i.e., in persons who have received the recessive gene *from both parents*. The trait may or may not appear among the sibs of the propositus but typically does not occur in relatives other than sibs. Exceptionally, other relatives may also be affected, especially in large inbred kindreds. Table 4–1 shows three mating types which can produce recessively affected offspring, but only one of these crosses (heterozygote × heterozygote) is at all common.

The most frequent autosomal recessive disorder in white ("Caucasian") children is **cystic fibrosis,** a condition in which there are abnormalities of several exocrine secretions, including pancreatic and duodenal enzymes, sweat chlorides and bronchial secretions. The thick, viscid mucus produced by the bronchi is particularly serious, since it makes the affected children highly susceptible to pneumonia. Males are infertile as a secondary consequence of abnormal mucous secretions in the vas deferens. The loss of chlorides in the sweat may be severe enough to cause heat prostration in warm weather.

In whites, cystic fibrosis affects perhaps one child in 2000 births. About one person in 22 is a heterozygous carrier. Both homozygotes and heterozygotes are much less common in Orientals. (The mathematical relationship between gene frequency and genotype frequency is discussed in Chapter 15.)

In a hypothetical family of four children produced by carrier parents (as shown in Figure 4–7), one child will be homozygous affected, two will be heterozygous and phenotypically normal, and one will be homozygous for the normal allele and also phenotypically normal. These proportions are rarely observed in actual families but are true for the general population.

Because children with recessive traits usually have phenotypically normal (but heterozygous) parents, the only families that can be recognized and studied are those in which there is already at least one affected child. Families in which no child is affected merge with the general population and are not ascertained. This **bias of ascertainment** creates a statistical problem in that the proportion of affected children in the sibships that can be ascertained will be well above the theoretical one-fourth, unless the sibship size is very large (about 12). We will return to this problem and ways of dealing with it in Chapter 14.

Consanguinity and Recessive Inheritance

A carrier of an autosomal recessive gene can have affected children only if the other parent is also a carrier. The risk that a carrier of cystic fibrosis will

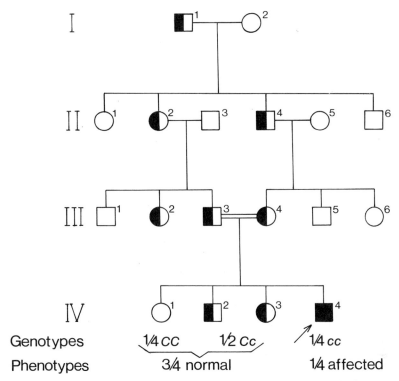

Figure 4–7   Stereotype pedigree of autosomal recessive inheritance, including a cousin marriage. A gene from a common ancestor I–1 has been transmitted down two lines of descent to "meet itself" in IV–4.

mate with another carrier is one in 22 (the incidence of carriers in the population). However, since rare recessive genes are passed down in families and so are concentrated in family groups, the risk that one carrier will marry another is usually higher by at least one order of magnitude if he marries a near relative than if he marries at random (i.e., without regard for the genotype of the spouse).

For example, consider the chance that a carrier of cystic fibrosis who marries a first cousin will have children homozygous for the gene. Figure 4–8 sets out the various probabilities involved. In brief, the risk that a first cousin of a carrier is also a carrier because of inheritance of the gene from a common ancestor is one in eight. The mathematical aspects of consanguinity are treated in additional detail in Chapter 15.

The effect of **inbreeding** in producing homozygous recessives is the biological basis of the prohibition of cousin marriage in many societies. For a condition as common as cystic fibrosis, the chance that a carrier who marries a cousin will have an affected child is only a little greater (about three times as high, on the basis of the incidence figure used here) than the general population risk. For less common conditions, the risk is higher. Some examples are given in Table 4–2, which compares the incidence of recessive traits of three different frequencies in children of cousin matings with the incidence in the general population.

In clinical medicine, parental consanguinity is often a strong clue that a

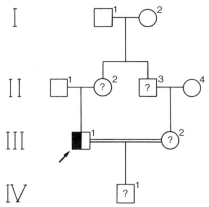

Figure 4–8 The probability that a carrier of cystic fibrosis (CF) who marries a first cousin will have an affected child.

1. If III–1 is a carrier, the chance that II–2 is a carrier is $\frac{1}{2}$.
2. If II–2 is a carrier, the chance that II–3 is a carrier is $\frac{1}{2}$.
3. If II–3 is a carrier, the chance that III–2 is a carrier is $\frac{1}{2}$.

In summary, the chance that III–2 carries the same CF gene as III–1 is $\frac{1}{2} \times \frac{1}{2} \times \frac{1}{2} = \frac{1}{8}$, and the chance that their child will be a CF homozygote is $\frac{1}{8} \times \frac{1}{4} = \frac{1}{32}$.

For simplicity, this discussion ignores the additional small chance that the mother will inherit the CF gene from another line of descent. This probability is the same as that of the frequency in the population of the CF allele, which is about $\frac{1}{44}$ in white North Americans.

disorder is autosomal recessive. Even if the parents consider themselves unrelated, they may have common ancestry within the last few generations if they are of closely similar geographic origin.

Rare Recessives in Genetic Isolates

There are many small groups in which the frequency of certain rare recessive genes is quite different from that in the general population. Such groups, **genetic isolates,** may have become separated from their neighbors by geographic, religious or linguistic barriers.

In Ashkenazi Jews in North America, the gene for **Tay-Sachs disease** (GM$_2$ gangliosidosis) is very common. Tay-Sachs disease is an autosomal recessive neurological degenerative disorder which develops at about six months of age. Affected children become blind and regress mentally and

TABLE 4–2 EFFECT OF COUSIN MARRIAGE ON THE INCIDENCE OF A RECESSIVE CONDITION*

	Affected Individuals		
Frequency of Recessive Gene	A. In General Population	B. Among Children of Cousin Marriages	Ratio B/A
0.2	0.04	0.05	1.25
0.02	0.0004	0.0016	4.00
0.002	0.000004	0.00013	32.25

*Data of Li, Amer. J. Med. 34:702, 1963.

physically. A "cherry-red spot" in the fundus of the eye is a striking diagnostic sign. The disease is usually fatal in early childhood. The frequency is 100 times as high in Ashkenazi Jews (one in 3600) as in other populations (one in 360,000).

When a recessive reaches such a high frequency, cousin marriage is no longer a striking feature of pedigrees of the trait. This is because the gene frequency is so high that a carrier who marries another member of the same group is almost as likely to marry another carrier as if he married a close relative. Consequently, among Ashkenazi Jews the parents of affected chil-dren are usually not closely consanguineous, whereas in other North Ameri-cans the consanguinity rate in the parents of Tay-Sachs patients is high.

There are many other examples of rare recessives in genetic isolates. **Tyrosinemia,** a very rare and lethal hepatic disease of early infancy, has been recognized in 51 French-Canadian children in 29 families in the isolated Lac St. Jean-Chicoutimi region of Quebec in recent years. It is virtually unknown elsewhere. Laberge (1969) has been able to trace both parents of each affected child to a couple who were in Quebec City by 1644. Probably, one member of this couple was a carrier of the tyrosinemia gene, which has been passed down 10 or more generations to "meet itself" in double dose in some of the remote descendants. The carrier frequency in the isolate is of the order of one in 30, and, as expected, the parents of the affected children are not closely consanguineous.

Genetic Counseling in Consanguineous Marriages

Are cousin marriages unwise? There is considerable disagreement among geneticists as to whether the extra risk of defective offspring is significant. A commonly quoted statistic from Japan is that the proportion of abnormal progeny was about 2 percent for random marriages and about 3 percent for consanguineous marriages (Neel and Schull, 1962). From the population standpoint this may be an inconsequential increase, but from the standpoint of the individual family it does mean that the risk of a defective child is increased by half. The high frequency of cousin marriage observed among the parents of congenitally deaf children requiring special educational facilities (Sank, 1963) is one measure of the personal and social cost of consanguineous marriages. (Incidentally, life insurance companies do not consider the off-spring of cousins to be different in any way from normal insurance risks.)

If cousin parents have a child who suffers from a recessively inherited disease, they are thereby proven to be carriers, with a one-quarter chance of having a similarly affected child at any later pregnancy. Furthermore, first cousins may share more than one deleterious recessive gene; in fact, they have one-eighth of their genes in common, and their progeny will, on the average, be homozygous at 1/16 of their gene loci.

Consanguineous marriages more distant than those of first cousins have a correspondingly lower risk of producing affected offspring. For example, second cousins have only 1/32 of their genes in common and third cousins only 1/128. For couples less closely related than second cousins, the risk of defective offspring is near enough to the general population risk to be of little or no genetic consequence. Nevertheless, if a child has a rare or undiagnosed defect and his parents are even remotely consanguineous, it is a reasonable hypothesis that his defect has autosomal recessive inheritance, with a one in four recurrence risk for later-born sibs.

Matings closer than first-cousin marriages are not legal anywhere in North America (with the exception of marriages of double first cousins, which are not proscribed although the relationship of double first cousins is as close as the uncle-niece relationship). However, parent-child and brother-sister matings do take place and do produce children who may become public wards and need to be assessed as candidates for adoption. Children produced by such close inbreeding are homozygous at one-fourth of their gene loci and have a correspondingly high risk of being homozygous for some deleterious gene. Very few studies of children of incest have been reported. Carter (1967) in a series of 13 cases found four to have been severely abnormal, definitely or probably because of homozygosity for recessives.

Not only do cousin marriages have an increased risk of producing children homozygous for deleterious recessive genes, but they also have a higher than normal likelihood of producing children with common congenital malformations or other conditions whose manifestation depends upon several genes. This aspect is discussed under multifactorial inheritance (Chapter 12).

Carrier Identification

Though by definition recessive genes are not expressed in heterozygotes, many recessives that are not clinically significant in heterozygotes may nevertheless have a detectable phenotype if an appropriate carrier test is used. Carrier identification is helpful in genetic counseling. An unaffected sib of a recessively affected child has a two-thirds chance of being heterozygous; uncles and aunts have a one-half chance; and first cousins have a one-fourth chance. Because only carrier × carrier matings can produce affected offspring, these individuals often wish to know definitely whether they do or do not have the abnormal gene, and, if so, whether their mates or prospective mates are also carriers. Identification and investigation of carriers can also add to our understanding of the mechanism of specific diseases, as exemplified by the hemoglobinopathies (see Chapter 5).

The Offspring of Homozygous Recessives

To indicate the types of offspring a homozygous recessive can have, we can return to our original example, the nontaster trait. Table 4–1 shows three matings, involving at least one nontaster parent: a *tt* person may have a *TT*, *Tt* or *tt* spouse. In brief:

Parents	Progeny	
	Genotypes	Phenotypes
tt × *TT*	All *Tt*	All taster
tt × *Tt*	¹/₂ *Tt*, ¹/₂ *tt*	¹/₂ taster, ¹/₂ nontaster
tt × *tt*	All *tt*	All nontaster

Note that the *tt* × *Tt* mating, in which an affected individual has an apparently normal spouse but half of the children are affected, mimics the pattern of inheritance of a rare dominant trait. Recessive inheritance which resembles dominant inheritance because an affected person marries a carrier is termed **quasidominant inheritance.** This pattern cannot always be firmly distinguished from the ordinary autosomal dominant pattern, but a pedigree of such a trait over more than two generations will probably reveal that in each generation the characteristics of recessive inheritance are present; in particular, the trait will usually appear only in the parent sibship and the offspring sibship, not elsewhere in the kindred, and consanguinity may be present in the parents and/or in one set of grandparents. A typical quasidominant pedigree is shown in Figure 4–9.

Criteria for Autosomal Recessive Inheritance

1. The trait characteristically appears only in sibs, not in their parents, offspring or other relatives.
2. On the average, one-fourth of the sibs of the propositus are affected; in other words, the recurrence risk is one in four for each birth.
3. The parents of the affected child may be consanguineous.
4. Males and females are equally likely to be affected.

CODOMINANCE AND INTERMEDIATE INHERITANCE

Up to this point we have discussed only the pedigree patterns produced by genes that are fully dominant or fully recessive in their expression. These are not the only relationships possible between pairs of alleles; it has already been pointed out that dominance and recessivity are relative concepts, and that when a gene is said to be recessive in heterozygotes, the reason may be our inability to recognize an effect, rather than the total absence of an effect.

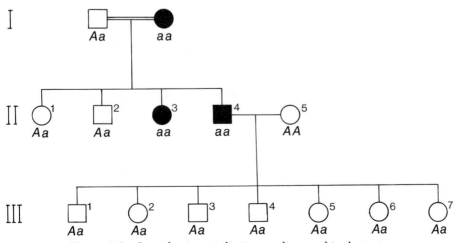

Figure 4–9 Quasidominant inheritance, discussed in the text.

If both alleles of a pair are fully expressed in the heterozygote, the genes are said to be **codominant.** Many examples of codominance are provided by the various blood group and enzyme systems. For example, a person of blood group AB has both A and B antigens on his red cells. The allelic genes *A* and *B* are therefore codominant.

If the heterozygote is different from both homozygotes, the genes concerned are said to show **intermediate inheritance.** When red and white snapdragons are crossed, the pink progeny are clearly intermediate. In humans, an example is provided by sickle cell anemia. Heterozygotes for the abnormal allele do not have the severe sickle cell anemia found in homozygotes, but a proportion of their red cells do show the sickling phenomenon; in other words, heterozygotes are intermediate between normal homozygotes and sickle cell homozygotes and are said to have sickle cell trait. (The use of the term *trait* to indicate a heterozygous phenotype, as in *sickle cell trait,* is a specialized clinical usage. In human genetics, *trait* is used interchangeably with *phenotype* to indicate the visible or detectable expression of a gene.)

The difference between codominance and intermediate inheritance is often indistinct. If we look at the hemoglobin of normal individuals and that of heterozygotes and homozygotes for the sickle cell anemia gene, we see that heterozygotes do not have an "intermediate" type of hemoglobin but have both normal hemoglobin A and sickle cell hemoglobin (hemoglobin S). Therefore, these genes are more properly codominant.

MULTIPLE ALLELES

All the examples used so far in this chapter have involved only a single pair of alleles, usually one "normal" and the other "abnormal." This simple situation is not the only possibility, for at many loci more than two different alleles are known. When for a single locus more than two alternative alleles exist in the population, the alleles are called **multiple alleles.**

The classic example of multiple allelism is provided by the series of alleles that determines the ABO blood groups. The best known alleles of the series are designated *O, A¹, A²* and *B.* For simplicity we will disregard A^2 and consider only the three main alleles. On this basis, the relationships of genes, genotypes and phenotypes are as follows:

Genes	Genotypes	Phenotypes (Blood Groups)
O, A, B	OO	O
	AA AO	A
	BB BO	B
	AB	AB

Note that the gene O is recessive to genes A and B, which are codominant. If A and B are present, both corresponding antigens are formed. O is an **amorph,** that is, a gene which has no effect and leaves the substrate (H substance) unaltered.

It was originally believed that the ABO blood groups were determined by genes at two independent loci, i.e., by two pairs of nonallelic genes. Alleles can be distinguished from nonalleles by analysis of family data, since alleles segregate in the progeny, whereas nonalleles assort randomly. The progeny of an AB × O mating is always A or B, never AB or O. In other words, gene A and gene B are never both transmitted from the parent to the child but always segregate; thus they must be allelic.

It is not always easy in human genetics to prove whether two rare genes are alleles or whether they are at independent loci. One reason is that families suitable for analysis are not often seen. Furthermore, even if two genes do not assort independently, they may be *linked* rather than *allelic*. The critical test for distinguishing linkage from allelism is that recombination can occur between the loci of linked genes (by crossing over in meiotic prophase), but not within a locus, i.e., not between alleles. However, as we shall see, there are exceptions even to this rule.

X-LINKED INHERITANCE

Genes on the sex chromosomes are distributed unequally to males and females within kindreds. This inequality produces characteristic and readily recognized patterns of inheritance, and has led to the identification of many "sex-linked" conditions in man.

"Sex-linked" genes may be X-linked or Y-linked, but for all practical purposes only X linkage has any clinical significance. Apart from genes essential for male sex determination, the Y chromosome appears to have few loci, or at least few loci at which segregation has been recognized. Genes on the Y show **holandric** inheritance; that is, they are passed down rather like the family surname, in the male line exclusively, by an affected man to all his sons and to none of his daughters. From time to time pedigrees showing holandric inheritance have been reported, but few of these have withstood searching examination (Stern, 1957). At present there is only one well-substantiated Y-linked anomaly, the "hairy pinna" trait (Dronamraju, 1960). Apart from this one example, the terms sex linkage and X linkage may be used synonymously, but most medical geneticists prefer to use X linkage because it is more specific.

The distribution of X-linked traits in families follows the course of the X chromosome carrying the abnormal gene. Since females have a pair of X chromosomes but males have only one, there are three possible genotypes in females but only two in males. One convenient way to indicate that a gene is X-linked is this: let X_H represent a dominant gene H on the X, and X_h its recessive allele h. Expressed in symbols, the following are the possible combinations in males and females:

MALES	FEMALES
$X_H Y$	$X_H X_H$
$X_h Y$	$X_H X_h$
	$X_h X_h$

A male is said to be **hemizygous** with respect to X-linked genes. A female is homozygous or heterozygous.

An important difference between autosomal and X-linked inheritance is that whereas for autosomal alleles both members of a pair are genetically active, in females only one member of the pair of X chromosomes is active; the second X remains compacted and nonfunctional, appearing in interphase cells as a heterochromatic body, the sex chromatin (Barr body). Thus in females as in males there is only one functional X. In heterozygous females it is a chance matter whether the paternal or maternal X is the functional one in a given cell. Consequently, though we distinguish X-linked "dominant" and "recessive" patterns of inheritance, it is important to keep in mind that a female heterozygous for either a dominant or a recessive X-linked mutant gene has the mutant as the only functional allele at that locus in about half her body cells.

X-LINKED RECESSIVE INHERITANCE

The inheritance of recessive genes on the X chromosome follows a well-defined pattern. A trait inherited as an X-linked recessive is expressed by all males who carry the gene: but females are affected only if they are homozygous. Consequently, X-linked recessive diseases are practically restricted to males and are rarely if ever seen in females.

Hemophilia (classical hemophilia, hemophilia A) is an X-linked recessive disease in which the blood fails to clot normally because of an abnormality in antihemophilic globulin. The clinical features, which include severe arthritis as a consequence of internal hemorrhages into the joints, are secondary to the clotting defect. The incidence is about one in 10,000 male births. The hereditary nature of hemophilia was recognized in ancient times, and it has since achieved notoriety by its occurrence among descendants of Queen Victoria, who was a carrier.

We have used the symbol X_h to represent a recessive gene h on the X chromosome, and X_H to represent its dominant allele. To demonstrate the pedigree patterns of X-linked recessive inheritance, these symbols will now be used to denote the hemophilia gene and its normal counterpart.

An affected male has the genotype $X_h Y$, and a normal female has the genotype $X_H X_H$. The offspring of these parents can be demonstrated by the checkerboard method.

OVA

		X_H	X_H
SPERM	X_h	$X_H X_h$	$X_H X_h$
	Y	$X_H Y$	$X_H Y$

Daughters: 100 percent heterozygotes
Sons: 100 percent normal

As the checkerboard shows, an affected male does not transmit the gene to any of his sons, but he gives it to all of his daughters, who are therefore carriers. If a carrier daughter marries a normal male, four genotypes are possible in the offspring, with equal probabilities.

OVA

		X_H	X_h
SPERM	X_H	$X_H X_H$	$X_H X_h$
	Y	$X_H Y$	$X_h Y$

Daughters: 50 percent normal, 50 percent carriers
Sons: 50 percent normal, 50 percent affected

Note that the X-linked recessive trait of the maternal grandfather, which did not appear in any of his own progeny, now reappears among his grandchildren. Figure 4–10 is a stereotype pedigree to show the characteristics of X-linked recessive inheritance.

Note also that half the daughters of carriers are carriers. By chance, an X-linked trait may be transmitted through a series of carrier women before it makes its appearance in an affected male. **Duchenne muscular dystrophy** (pseudohypertrophic muscular dystrophy) is an X-linked disease of muscle which affects young boys (Fig. 4–11). It is usually apparent by the time the child begins to walk, and progresses inexorably, so that the child is confined to a wheel chair by about the age of 10, and is unlikely to survive his teens. This disorder is a **genetic lethal,** in that its nature prevents its transmission by affected males; but it is spread by carrier females, who themselves rarely show any clinical manifestation of muscular dystrophy. A representative pedigree of Duchenne muscular dystrophy is shown in Figure 4–12. (For further discussion of the genetics of this disorder, see Chapter 18.)

For convenience, the mating types, the gametes produced and the progeny expected on the basis of X-linked inheritance are shown in Table 4–3.

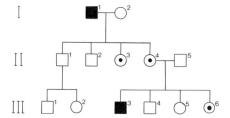

Figure 4–10 Stereotype pedigree of X-linked recessive inheritance. The affected male I–1 transmits the gene to each daughter, who therefore is a carrier. Each son of a carrier has a 50 percent chance of being affected, and each daughter has a 50 percent risk of being a carrier. There is no male-to-male transmission.

Criteria for X-Linked Recessive Inheritance

1. The incidence of the trait is much higher in males than in females.
2. The trait is passed from an affected man through all his daughters to half their sons.
3. The trait is never transmitted directly from father to son.
4. The trait may be transmitted through a series of carrier females; if so, the affected males in a kindred are related to one another through females.

Figure 4–11 Duchenne muscular dystrophy, illustrating the Gowers sign, the characteristic "climbing up himself" maneuvers by which the child rises from the prone position. From Dubowitz, *Muscle Disorders in Childhood.* W. B. Saunders Company, Philadelphia, 1978.

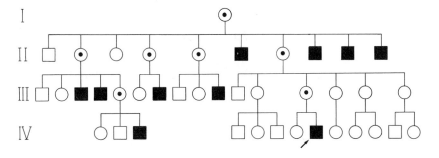

Figure 4–12 A pedigree of Duchenne muscular dystrophy, an X-linked recessive disorder in which affected males almost never reproduce.

TABLE 4–3 X-LINKED INHERITANCE: MATING TYPES, GAMETES AND EXPECTED PROPORTIONS OF PROGENY FOR A PAIR OF X-LINKED ALLELES X_H AND X_h

Mating Types	Gametes		Progeny	
	Ova	Sperm	Genotypes	Phenotypes
$X_H X_H \times X_H Y$	X_H,X_H	X_H,Y	$X_H X_H$ $X_H Y$	all normal
$X_H X_h \times X_H Y$	X_H,X_h	X_H,Y	$X_H X_H$ $X_H X_h$	daughters 1/2 normal, 1/2 carriers
			$X_H Y$ $X_h Y$	sons 1/2 normal, 1/2 affected
$X_h X_h \times X_H Y$	X_h,X_h	X_H,Y	$X_H X_h$	daughters all carriers
			$X_h Y$	sons all affected
$X_H X_H \times X_h Y$	X_H,X_H	X_h,Y	$X_H X_h$	daughters all carriers
			$X_H Y$	sons all normal
$X_H X_h \times X_h Y$	X_H,X_h	X_h,Y	$X_H X_h$ $X_h X_h$	daughters 1/2 carriers, 1/2 affected
			$X_H Y$ $X_h Y$	sons 1/2 normal, 1/2 affected
$X_h X_h \times X_h Y$	X_h,X_h	X_h,Y	$X_h X_h$ $X_h Y$	all affected

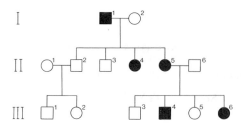

Figure 4–13 Stereotype pedigree of X-linked dominant inheritance. Affected males have no affected sons and no normal daughters.

X-LINKED DOMINANT INHERITANCE

Whereas X-linked recessive traits typically occur in males only, X-linked dominants are approximately twice as common in females as in males. (For further discussion of this point, see Chapter 15.) The chief characteristic of X-linked dominant inheritance is that an affected male transmits the gene (and the trait) to *all* his daughters and to *none* of his sons (Fig. 4–13). If any daughter is normal or any son affected, the gene concerned must be autosomal, not X-linked. Because females have a pair of X chromosomes just as they have pairs of autosomes, it is not possible to distinguish between autosomal and X-linked inheritance of a dominant trait by observing its transmission to the progeny of affected females; whether autosomal or X-linked, it is transmitted to half the progeny of each sex.

An example of X-linked dominant inheritance is provided by the X-linked blood group system Xg. The Xg system is governed by a pair of alleles, Xg^a and Xg, which produce two phenotypes, Xg(a+) and Xg(a−). (For discussion of blood group terminology and symbols, see Chapter 9.) The possible genotypes and phenotypes in the two sexes are as follows:

Males		Females	
"Genotypes"*	Phenotypes	Genotypes	Phenotypes
Xg^aY	Xg(a+)	Xg^aXg^a Xg^aXg	Xg(a+)
XgY	Xg(a−)	$XgXg$	Xg(a−)

*This is not really a genotype, because the Y chromosome is shown as well as the X-linked gene, but it is easier to visualize the pattern of transmission when both members of the chromosome pair are shown.

The transmission of the Xg groups in families demonstrates the general pattern of X-linked dominant inheritance.

HOMOZYGOUS Xg(a−) FEMALE AND Xg(a+) MALE

$XgXg \times Xg^aY$

	Ova			Genotypes	Phenotypes
		Xg	Daughters:	Xg^aXg	Xg(a+) (like father)
			Sons:	XgY	Xg(a−) (like mother)
Sperm	Xg^a	Xg^aXg			
	Y	XgY			

HETEROZYGOUS Xg(a+) FEMALE AND Xg(a−) MALE

$Xg^aXg \times XgY$

	Ova			Genotypes	Phenotypes
	Xg^a	Xg	Daughters: $Xg^aXg, XgXg$	½ Xg(a+), ½ Xg(a−)	
			Sons: Xg^aY, XgY	½ Xg(a+), ½ Xg(a−)	
Sperm Xg	Xg^aXg	$XgXg$			
Y	Xg^aY	XgY			

Males and females each have a 50 percent chance of receiving the dominant allele.

Only a few genetic disorders exhibit the X-linked dominant pattern. An example is hypophosphatemia, also called vitamin D−resistant rickets. In rare X-linked dominants, affected females (heterozygotes) are twice as common as affected males (hemizygotes) but usually have a milder form of the defect. Hypophosphatemia fits this criterion in that although both sexes are affected, the serum phosphorus is less depressed and the rickets less severe in heterozygous females than in males.

Criteria for X-Linked Dominant Inheritance

1. Affected males have no normal daughters and no affected sons.
2. Affected females who are heterozygous transmit the condition to half

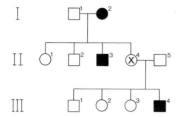

Figure 4-14 Reduced penetrance of an autosomal dominant trait. For further details, see text.

their children of either sex. Affected females who are homozygous transmit the trait to all their children. Transmission by females follows the same pattern as an autosomal dominant. In other words, X-linked dominant inheritance cannot be distinguished from autosomal dominant inheritance by the progeny of affected females, but only by the progeny of affected males.

3. Affected females are more common than affected males (twice as common if the disorder is rare).

VARIATION IN THE EXPRESSION OF GENES

The pattern of inheritance shown by any trait that is determined by a rare gene at a single locus can often be readily recognized if the trait segregates sharply, i.e., if the normal and abnormal phenotypes can easily be distinguished. In ordinary experience, however, the abnormal persons in a kindred may not show an obvious pattern in their relationships, even though the incidence of the abnormality is clearly higher within the kindred than in the general population. We will now consider some of the factors that can affect gene expression and lead to confusion in interpretation of pedigree data.

PENETRANCE AND EXPRESSIVITY

A mutant gene may not always be phenotypically expressed or, if it is expressed, the degree of expression of the trait may vary widely in different individuals. **Penetrance** applies to a gene's ability to be expressed at all; **expressivity** refers to the degree of expression, i.e., whether mild, moderate or severe.

When the frequency of expression of a trait is below 100 percent, that is, when some individuals who have the appropriate genotype fail to express it, the trait is said to exhibit **reduced penetrance.** For example, in Figure 4–14, the mutant gene must be carried by II-4, since she transmits it, though she does not express it; in her case, it is therefore said to be **nonpenetrant.** Penetrance is an all-or-none concept. In mathematical terms, it is the percentage of genetically susceptible individuals who actually show the trait.

If, on the other hand, a trait takes somewhat different forms in different members of a kindred, it is said to exhibit **variable expressivity.** Expressivity may range from severe to mild, and members of the same kindred may express the same gene in different ways and with different severity. Expressivity may be roughly equated with clinical severity.

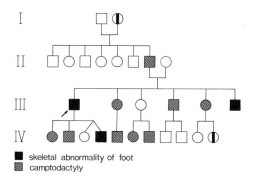

Figure 4–15 A pedigree showing both re-duced penetrance and variable expressivity. Dark symbols represent full expression of foot abnormality; half-filled symbols represent par-tial expression; and hatched symbols represent camptodactyly (flexion deformity of interphalan-geal joints). II–7, father of the propositus, has only camptodactyly, or virtual nonpenetrance of the defect.

■ skeletal abnormality of foot
▨ camptodactyly

Failure of penetrance and variable expressivity may both appear in the pedigree of a single trait. Figure 4–15 is a pedigree of a family segregating for an unusual form of polydactyly, involving a bifid second metatarsal bone. Figure 4–16 shows the abnormality as expressed in the propositus. The trait is transmitted as an autosomal dominant, but its expression differs in various members of the family. The accessory spur of the second metatarsal may be large or small, polydactyly may or may not be present and a variety of other foot deformities may appear. Furthermore, in some individuals the gene is nonpenetrant; that is, those individuals transmit the gene but have no obvious deformity. In these people, penetrance might actually be revealed by radio-logical examination; if so, they would be classified as examples of mild expression rather than of failure of penetrance. It is not unusual for the concepts of penetrance and expressivity thus to blur into one another at the normal end of the spectrum of expression.

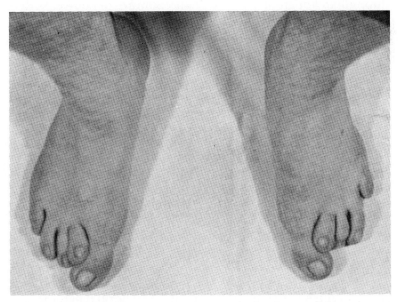

Figure 4–16 The feet of the propositus (III–1) of the family recorded in Figure 4–15. From Neel and Rusk, Am. J. Hum. Genet. 15:288–291, 1963, by permission.

Variability of expression is usually more marked in members of different family groups than in members of a single kindred. This suggests that expression depends at least partly upon **modifying genes,** which would differ more between than within families.

Reduced penetrance and variable expressivity are more often seen in autosomal dominant than in autosomal recessive pedigrees. Perhaps this observation is not surprising when one considers that an autosomal dominant disorder is usually the expression of a heterozygous genotype; the individual has one normal allele as well as an abnormal one, and both alleles are active. At least two other factors are important: (1) The *primary* action of the gene may be many developmental or biochemical steps removed from the observed effect; (2) the expression of the gene or genes at that locus is modified by the "background" genes of the organism.

Though failure of penetrance and variability of expression are more commonly observed for autosomal dominants than for autosomal recessives, many recessives are also variable in their expression; for example, albinism may be complete or partial in different members of the same sibship, and cystic fibrosis may be much more severe in one sib than in another.

Forme Fruste

The expression of a clinical syndrome may be very mild (subclinical, or of little clinical significance), and so may be difficult to distinguish from the normal range of variation. An extremely mild and clinically insignificant expression of an abnormality, disease or syndrome is called a **forme fruste.** For example, in **Marfan syndrome** (Fig. 4–17), which is autosomal dominant, the full clinical picture includes elongated extremities, dislocation of the lens of the eye and cardiovascular abnormalities; but a pedigree of this syndrome might show a grandfather and grandson severely affected, related to one another through an apparently normal man. Stature, of course, is determined by many genes and also influenced by environment; accordingly, the Marfan gene in a person whose height would otherwise be below average might only raise it to about normal. If in this person the ocular and cardiovascular effects were minimal, he might well be recorded as completely normal and the trait would appear to "skip a generation." Clinically, this is an example of a *forme fruste.*

Tuberous Sclerosis. Tuberous sclerosis (epiloia) is an autosomal dominant condition in which the chief clinical signs are sebaceous adenomata of the skin, a "butterfly rash" on the face, muscle tumors (rhabdomyomata) and epileptic seizures. Phakomata (retinal tumors) are present in some genetically susceptible persons who are otherwise normal, as well as in some patients. White leaf-shaped macules on the skin, present from birth, are often useful in diagnosis, especially in otherwise unaffected persons who nevertheless have the causative gene. Though the majority of cases are new mutants, there are also cases in which the gene has been inherited from a very mildly affected parent in whom the disorder has not been recognized. It is important to attempt to distinguish between new mutants with normal parents and cases in which one parent is heterozygous but demonstrates no overt signs of the

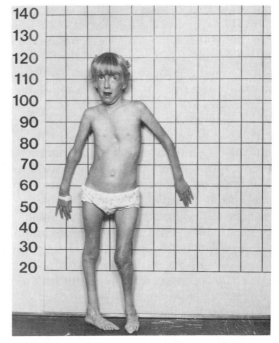

Figure 4-17 Marfan syndrome in a girl aged 5.50 years. Her height is above the 97th percentile for her age. Note limb length with long extremities and arachnodactyly, genu valgum ("knock knees"), long facies and pectus excavatum. She also has a high, narrow palate, bilateral ectopia lentis and myopia. Photograph courtesy of V. A. McKusick.

disease. This is because the risk that the disorder will recur in the family is zero if both parents are genetically normal, but if either parent has the gene any child has a 50 percent chance of inheriting it.

PLEIOTROPY: ONE GENE, SEVERAL EFFECTS

Each gene has only one primary effect in that it directs the synthesis of a polypeptide chain. From this primary effect, however, many different consequences may arise. It is probable that most genes, even those which have a clear-cut major effect, produce quite diverse secondary effects. Multiple phenotypic effects produced by a single mutant gene or gene pair are examples of the principle of **pleiotropy** or pleiotropism.

In any sequence of events, interference with one early step may have ramifying effects. Thus, Gruneberg (1947) speaks of a "pedigree of causes," implying that a single defect occurring early in development can lead through branching pathways to various abnormalities in fully differentiated structures. In some cases, perhaps a fan might be a more accurate simile than a family tree; a primary gene product might be envisaged as participating in a number of unrelated biosynthetic pathways, possibly at different times.

Clinical syndromes offer many examples of pleiotropy. The literal meaning of the word *syndrome* is "running together," and the term is used to describe the set of signs and symptoms characteristic of a specific disorder. In some syndromes, there is a clear or at least a plausible mechanism by which the primary effect of the gene could produce the diverse features of the syndrome. For example, in phenylketonuria, a metabolic disease inherited as an autosomal recessive, the enzyme phenylalanine hydroxylase is lacking. The primary effect in homozygotes is the specific enzyme deficiency. There are multiple secondary effects, notably severe mental retardation, excretion of phenylketones in the urine and dilution of pigmentation. Though the pathways by which these secondary effects are produced are not known in full detail, the steps by which they are related to the primary defect can at least be conjectured. Similarly, in galactosemia a lack of the enzyme galactose-1-phosphate uridyl transferase is the primary effect of homozygosity for the recessive gene concerned. The array of secondary effects, which include cirrhosis of the liver, cataracts, galactosuria and mental retardation, can be produced in experimental animals if they are fed a diet high in galactose. Furthermore, the secondary effects can be prevented in many genetically susceptible infants if the condition is recognized and the child is placed on a galactose-free diet at birth. Thus, the characteristic features of the disease are clearly secondary to the enzyme deficiency.

In other syndromes, the various clinical signs and symptoms may be less obviously related to a single underlying defect. In Marfan syndrome, mentioned in the preceding section, the pleiotropic effects (skeletal, ocular and cardiovascular anomalies) may have as a common basis a defect in the elastic fibers of connective tissue. By contrast, in the rare syndrome of hypogonadism, polydactyly, deafness, obesity, retinitis pigmentosa and mental retardation known as the Laurence-Moon-Bardet-Biedl syndrome (Fig. 4–18), there is certainly no one obvious common basis for the assortment of abnormalities.

Pleiotropy and Linkage

Pleiotropy must not be confused with linkage, in which two or more traits may be transmitted together for a number of generations simply because they are determined by linked genes. For example, the connective tissue disorder osteogenesis imperfecta (OI), shown in Figure 4–19, is nearly always autosomal dominant, though there may be a rare recessive form. The three main signs of OI are brittle bones, blue sclerae and otosclerosis. The nature of the basic connective tissue defect that leads to these different manifestations is still unknown. OI is highly variable in expression; some severely affected children have multiple fractures at birth, but in other heterozygotes only one or two stigmata may be present, and the expression may be so mild that the diagnosis of OI is never made. If a parent with all three stigmata had a child with blue sclerae but without deafness or brittle bones, one might suspect a trio of linked genes, one of which had become separated from the other two by crossing over. But recurrence of brittle bones and deafness in a later generation would demonstrate that the mildly affected child was genetically competent to transmit all three stigmata.

In contrast, an example of linkage as the mechanism underlying the transmission of a group of characteristics is provided by the HLA complex. The HLA (human lymphocyte antigen) complex is the region on chromosome 6 that includes the genes determining the major transplantation antigens as well as many other genes important in immune processes. The various gene loci of the HLA complex are very closely linked, so that within a kindred the

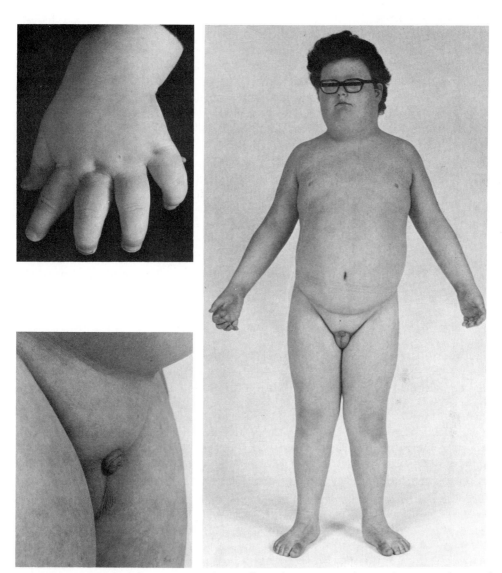

Figure 4–18   Laurence-Moon-Bardet-Biedl syndrome. Note obesity, hypogenitalism and polydactyly (expressed as a small tag of tissue on the fifth finger).

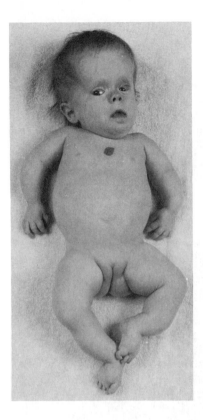

Figure 4-19 Osteogenesis imperfecta in an eight-month-old female. Note triangular face, pointed nose, broad forehead, short chest and bowing of extremities secondary to fractures. The nevus on the chest is not part of the syndrome.

combinations of alleles at these loci (haplotypes) are transmitted as units. When recombination by crossing over occurs, the new combinations formed are permanent, until a further recombination event takes place in some future generation. The HLA complex is discussed in more detail in Chapter 8.

GENETIC HETEROGENEITY: SEVERAL GENES, ONE EFFECT

When a genetic disorder that at first glance appears to be uniform is carefully analyzed, not infrequently it is found to comprise a number of separate conditions that are only superficially alike. If mutations at different loci can independently produce the same trait, or traits that are difficult to distinguish clinically, that trait is said to be genetically **heterogeneous.** Recognition of genetic heterogeneity is an important aspect of clinical diagnosis and counseling for genetic defects.

Profound childhood deafness is a condition that shows a remarkable degree of genetic heterogeneity. In almost any community that has a high degree of consanguinity, congenital deafness is one of the autosomal recessive disorders found. Fraser (1976) estimates that there are about 16–18 types of autosomal recessive deafness alone and suggests that 10 percent or more of the general population are heterozygous carriers of one or more of these

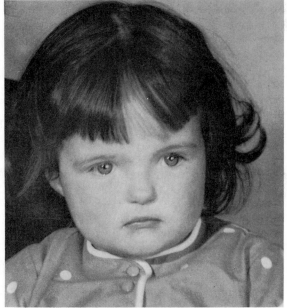

Figure 4–20 Waardenburg syndrome. Both mother and child have a white forelock, and both have pale irides. From Partington, M. W., Arch. Dis. Child. 34:154–157, 1959.

forms. There are also autosomal dominant types, such as Waardenburg syndrome, which combines deafness with pigmentary disturbances (Fig. 4–20); X-linked forms; malformation syndromes, probably not genetic, involving faulty embryological development of the first and second branchial arches; acquired deafness, such as the types resulting from prenatal rubella or postnatal otitis media; and many cases of unknown etiology.

Profound childhood deafness produces difficulties in communication that may be so severe as to isolate affected persons from the rest of society. An understandable social consequence is that the deaf usually marry within the

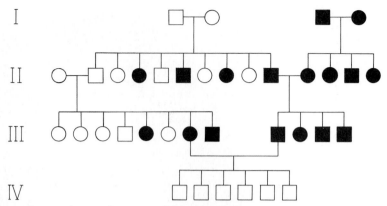

Figure 4-21   Pedigree of congenital deafness showing genetic heterogeneity, discussed in the text. Adapted from Figure HD–2, p. 219, "Hereditary deaf mutism with particular reference to Northern Ireland" by Stevenson and Cheeseman, Ann. Hum. Genet. *20*:177–231, 1956, with the permission of Cambridge University Press.

deaf community. Three such marriages are shown in Figure 4–21. In this pedigree, the marriage of I–3 and I–4 produces only deaf children, as does the marriage of II–10 and II–11. However, the marriage of III–7 and III–9 produces only hearing children. The probable explanation is that in the first two of these marriages the parents were homozygous for the same autosomal recessive gene, but in the third marriage the parents' deafness was caused by different recessive genes at different loci, each parent having normal alleles at the locus for which the partner had only abnormal alleles. Thus if *d* and *e* represent the two recessive genes concerned, the father could be *dd EE* (deaf because he is homozygous *dd*) and the mother *DD ee* (deaf because she is homozygous *ee*). All the children would then be *Dd Ee,* and since there is a dominant normal allele at each locus, all have normal hearing.

Though genetic methods can be helpful in analyzing genetic heterogeneity, as in this example, often heterogeneity is recognized by clinical and biochemical observations.

Heterogeneity in the Muscular Dystrophies

Clinical heterogeneity is well shown by diseases of voluntary muscle, for example, the progressive muscular dystrophies and myotonic disorders.

The three main types of muscular dystrophy differ both in clinical features and in genetics. The Duchenne type, discussed earlier, is X-linked, expressed in early childhood and marked by involvement of the muscles of the pelvic girdle followed by those of the shoulder girdle. Hypertrophy of the calf muscles is an early and striking sign. The course is severe, with inability to walk by about age 10 and death usually by about age 20. Limb-girdle muscular dystrophy is autosomal recessive, typically has its onset in the second or third decade and, though milder than the Duchenne type, usually leads to severe disability by middle life. Facioscapulohumeral muscular dystrophy is autosomal dominant. The expression is irregular, and formes frustes may occur. Onset is usually in the teens but may be much later. The incapacity usually progresses slowly and does not shorten life.

There are many other forms of muscular dystrophy, some well-defined and others still unclear. In particular, a second X-linked type, the Becker type, has been separated from the Duchenne type by its later onset (usually in the teens) and generally milder course.

Myotonic dystrophy is one of several myotonic disorders and is noted here because it is used as an example later. Myotonia is the continued active contraction of a muscle after cessation of voluntary effort or stimulation, which is best seen in the difficulty affected persons have in relaxing the grip. Other features of myotonic dystrophy include cataracts, gonadal atrophy and frontal baldness especially in males. Inheritance is autosomal dominant, with some irregularity of expression, to be described later.

Resolution of the heterogeneity shown by muscle disorders is clearly important both for prognosis in individual cases and for genetic counseling in families.

Biochemical Heterogeneity

Heterogeneity may be at the biochemical level. In Chapter 5 many examples are given of heterogeneity in human biochemical disorders. Many of these variant forms are determined by different mutations at a single locus (glucose-6-phosphate dehydrogenase [G6PD] variants, hypoxanthine-guanine phosphoribosyl transferase [HPRT] variants, and so on). In the case of hemoglobin there are several different loci, at least two of which, α and β, have many different known mutations. Among the mucopolysaccharidoses and probably in numerous other disorders, it is known that compound genotypes, with two different mutant alleles, can determine a phenotype different from that of either homozygote.

SEX-LIMITED AND SEX-INFLUENCED TRAITS

If a trait is determined by a gene on the X chromosome (or, for that matter, on the Y), the sex ratio of the affected individuals is not 1:1. An abnormal sex ratio may provide the first hint that a trait is caused by an X-linked gene. However, there are many traits in which the sex ratio is abnormal even though the determining gene is autosomal. This should not be surprising, because the milieu in which any gene acts is determined in part by the sexual constitution of the individual.

Sex-Limited Traits

A trait which is autosomally transmitted but expressed in only one sex is said to be sex-limited. An example is **precocious puberty**, in which heterozygous males develop secondary sexual characteristics and undergo an adolescent growth spurt at about four years of age, sometimes even younger (Fig. 4–22). There is no expression in heterozygous females. Affected boys are much taller than their contemporaries at first, but because of early fusion of the epiphyses of the long bones they soon stop growing and end up as short men. Figure 4–23 is part of a large pedigree of precocious puberty, showing male-to-male transmission.

For traits in which males do not reproduce, sex-limited autosomal dominant inheritance cannot always be readily distinguished from X-linkage, because the most critical evidence of X-linkage, namely absence of male-to-male transmission, cannot be provided. Duchenne muscular dystrophy and testicular feminization are examples. In testicular feminization, the affected males are XY and have testes, but have female external genitalia and are raised as females. At puberty there is female breast development, but usually pubic and axillary hair is absent. There are other forms including an "incomplete" form in which the affected males may have either male, ambiguous or female external genitalia but virilize at puberty. The basic defect is believed to be deficient response of the target tissues to androgen, possibly through deficiency of the enzyme that converts testosterone to the form active within cells, dihydrotestosterone. A comparable gene in the mouse, *tfm,* is known to be X-linked (Lyon and Hawkes, 1970). By analogy to other genes known to be

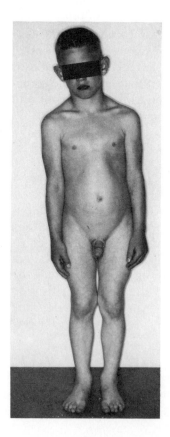

Figure 4–22 Precocious puberty. At age 4.75 years, the patient's height is 120 cm., far above the normal range.

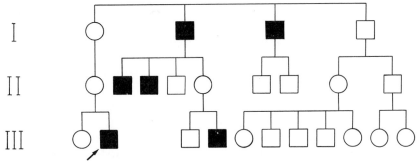

Figure 4–23   Pedigree of precocious puberty in the family of the patient shown in Figure 4–22.

X-linked throughout the mammals, such as G6PD and HPRT, it can be concluded that testicular feminization is probably X-linked.

Sex-Influenced Traits

Traits are said to be sex-influenced when they are expressed in both sexes, but with widely different frequencies. Table 4–4 lists the sex ratios in a number of different autosomal disorders, to illustrate the effect of sex on gene expression. Many multifactorial traits are also sex-influenced (see Chapter 12). Aberrations of the autosomes may also be more common in one sex than the other, e.g., trisomy 18, in which four-fifths of the patients are female. This may however be due to differential viability *in utero*.

TABLE 4–4   EXAMPLES OF AUTOSOMAL PHENOTYPES WITH UNEQUAL EXPRESSION IN MALES AND FEMALES*

Phenotype	Genetics	Sex Difference
Baldness	Autosomal dominant in males, recessive in females?	Great excess of males
Congenital adrenal hyperplasia	Autosomal recessive	More often recognized in females
Hemochromatosis	Autosomal dominant	Excess of males; loss of blood in menses may protect females
Hypogonadism, male	Autosomal recessive types	Limited to males
Legg-Perthés disease	Autosomal dominant type	Excess of males
Precocious puberty	Autosomal dominant	Limited to males

*Data chiefly from McKusick, 1978.

ONSET AGE

Many genetic disorders are not present at birth but become manifest later in life, some at a characteristic age and others at variable ages throughout the life span. Genetic disorders are, of course, not necessarily congenital, nor are congenital disorders necessarily genetic. To classify a disorder as genetic means that genes are plainly implicated in its etiology; but to say that a disorder is congenital means only that it is manifest at birth.

Some genetic disorders have a prenatal onset. Chromosomal aberrations, as described later, are found in many spontaneous first-trimester abortuses, indicating that their effects are lethal in prenatal life. Dysmorphic conditions of many kinds originate during embryological development. Some disorders are not expressed until the infant begins its independent life, e.g., phenylketonuria and galactosemia. Later, other defects of genetic origin may make their appearance, some at characteristic ages, others at less restricted times. Some characteristic onset ages of genetic disorders are shown in Table 4–5.

That a genetic disorder need not be present at birth should cause no surprise, since many genes are known to be expressed at specific times during development. Actually, remarkably little is known about the action of genes that govern normal development. Why does puberty occur when it does? What causes the menopause? Why do people become senile? These questions are just as puzzling as the problem of why the Huntington chorea gene lies

TABLE 4–5 CHARACTERISTIC ONSET AGES OF SOME
GENETIC DISEASES

Typical Onset Age	Condition
Lethal during prenatal life	Some chromosomal aberrations Some gross malformations
Present at birth	Neural tube malformation Chromosomal aberrations Some forms of adrenogenital syndrome Some forms of deafness
Soon after birth	Phenylketonuria Galactosemia Cystic fibrosis
During the first years of life	Tay-Sachs disease Duchenne muscular dystrophy
Near puberty	Limb-girdle muscular dystrophy
Young adulthood	Acute intermittent porphyria Hereditary juvenile glaucoma
Variable onset age	Diabetes mellitus (0 to 80 years) Facioscapulohumeral muscular dystrophy (2 to 45 years) Huntington chorea (15 to 65 years) Myotonic dystrophy (birth to old age)

dormant for many years before the disease becomes manifest and runs its progressive and eventually fatal course.

Variability of Onset Age in Huntington Chorea

Huntington chorea, which was first described by Huntington in 1872 in an American kindred of English descent, is an autosomal dominant disease characterized by choreic movements and progressive mental deterioration. Its incidence is only about one in 25,000, but because of its genetic pattern and variable onset age it is seen frequently in genetic counseling centers.

The onset age distribution is shown in Figure 4–24. If H is the dominant gene for the disease and h the recessive normal allele, then anyone of the genotype Hh will develop the disease if he or she lives long enough. But until the disease actually makes its appearance there is no sure way to detect those who have the Hh genotype. Moreover, there would be a major ethical problem in applying such a test if it existed, since many at-risk family members would prefer not to know the result. As Figure 4–24 shows, it is not until about age 40 that even half of those who are Hh develop the disease.

If no affected persons reproduced, an autosomal dominant disorder could be virtually eradicated in a single generation; only those cases caused by new mutation would prevent it from disappearing completely. Unfortunately, in Huntington chorea many heterozygotes do not know they carry the H gene until they have already completed their families. In biological terms, selection against the mutant gene is not strong, since it does not greatly impair the biological fitness of persons who carry it. In fact, some studies of Huntington chorea have actually shown that choreics have larger families than their nonchoreic sibs. If this is generally true, and remains true for a very long time, the ultimate consequence is that the mutant H allele gradually replaces h and becomes the "normal" allele at the locus.

The variability of onset age in this disease complicates genetic counseling. Figure 4–25 shows how to allow for it in counseling phenotypically normal relatives of choreics, especially offspring. (We are grateful to the medical student who corrected our discussion of this point in an earlier edition.) Of 100 babies each with an Hh parent, 50 will be Hh (genetically susceptible to the disease) and 50 hh (normal). By age 40, half the Hh individuals (25) will already manifest the disease. Of the remaining 75, 25 (one in three) are still at risk. Thus, the chance that any child of a choreic who is still normal at age 40 will eventually develop the disease is one in three.

ANTICIPATION

Anticipation is the phenomenon of the apparent onset of a disease at successively earlier ages in succeeding generations. Since earlier onset may be associated with more severe signs and symptoms, as in diabetes mellitus, progressive worsening from generation to generation used to be regarded as the normal course for many inherited diseases. Now most geneticists believe that anticipation is a statistical artifact rather than a biological reality.

In myotonic dystrophy, described earlier, there is still controversy about

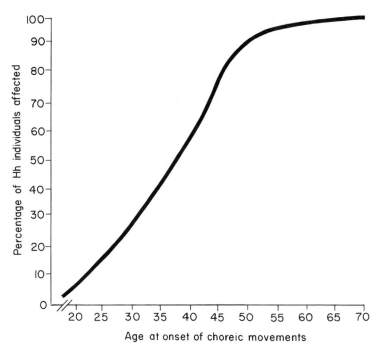

Figure 4–24   Onset age distribution in Huntington chorea. From data of Reed and Chandler, Am. J. Hum. Genet. *10*:201–225, 1958.

anticipation. The average age at onset of this disorder may be widely different in parent and child. According to Penrose (1948) the average onset age is 38 in parents and 15 in children, a difference of 23 years.

The difference has been explained on the basis of ascertainment bias. If an affected parent is ascertained through his affected child (the propositus), the parent-child pairs most often seen will be those in which the disease coincides in time in parent and child. If both are affected simultaneously, the onset age difference may be 20 to 30 years.

The difficulty of observing certain combinations of onset age in successive generations can be shown by comparison of typical dates of onset for a parent born in 1940 and a child born in 1965.

| Onset Age | | Year of Onset | | Number of Years |
Parent	Child	Parent	Child	Apart in Time
40	10	1980	1975	5 (child earlier)
40	40	1980	2005	25 (child later)
40	70	1980	2035	55 (child later)

Clearly, the situation with relatively late onset in the parent and relatively early onset in the child is the one most likely to be recognized as an example of heredity. There are also other complications to be considered.

Early-onset persons are, on the average, less likely to transmit the defect than are late-onset persons. This is a general rule that applies to many other heritable conditions as well. In fact, childhood myotonic dystrophy is virtually always inherited from the mother (Harper and Dyken, 1972), though it is not clear if this is caused only by the gradual loss of fertility in affected males. It may be that the combination of the gene in the fetus and in the mother is responsbile in some unknown way for the phenomenon. For the present, therefore, the problem of apparent anticipation in myotonic dystrophy must remain open.

GENE INTERACTION

Although so far in this chapter the various patterns of genetic transmission have been discussed as though genes at any one locus were expressed independently of the remainder of the genome, this is an oversimplification. The expression of a gene may be affected by its own allele, by the presence of specific genes at other loci involved in the same biosynthetic pathway or by the remainder of the genome, the "genetic background."

Interaction of Allelic Genes. The expression of a gene may be affected by its allele. The most obvious examples are the different phenotypes resulting from homozygous, heterozygous and compound genotypic combinations, mentioned earlier. It also appears that isoallelism (the presence of more than one kind of normal allele at a locus) can be a significant cause of variation in expression of mutant genes. Renwick (1956) noted that in nail-patella syndrome (a rare autosomal dominant malformation characterized by variable degrees of dystrophy of the nails, absence or reduction of the patella, other bone dysplasias and, occasionally, nephropathy), the severity of the defect was closely correlated in sibs but not in parent-child pairs. The presence of isoalleles could explain this observation, because each of the affected sibs has inherited the affected parent's deleterious allele and not its normal counterpart; instead, each has a normal allele from the unaffected parent. Thus the sibs have a 50 percent chance of having identical genotypes at the locus concerned, whereas if there are several possible normal alleles it is unlikely that parent and child would match exactly. The possible significance of isoalleles in the expression of other disorders has hardly been investigated.

Figure 4–25 Risk of Huntington chorea after age 40 in offspring of choreics. For discussion, see text.

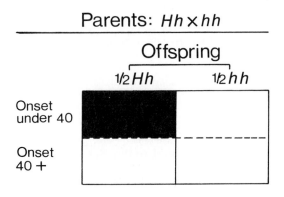

Nonallelic Interaction.   There are many examples of genes at two or more different loci acting together to produce a phenotype. Many such instances involve biosynthetic pathways, in which the product of one reaction acts as the substrate for the next. An example is the formation of the A, B and H blood group substances by the sequential action of genes at the H locus and the ABO locus (see Chapter 9). In prokaryotes genes concerned with the same pathway have often been found to be genetically as well as functionally linked, but in higher organisms including man there is no evidence of genetic linkage of functionally related genes. Other examples involve nonallelic genes affecting the same process in different ways, such as the combined effects of structural hemoglobin genes and thalassemia genes in producing hematological disorders, mentioned in Chapter 5.

Genetic Background.   In inbred animals such as laboratory mice there are many examples of mutant genes expressed with different degrees of severity against different genetic backgrounds. Analysis of gene action against different backgrounds is not practical in man, but it is logical to expect that the effect of genetic background would be important. For example, normal stature is believed to have multifactorial inheritance, i.e., to be determined by multiple genes of minor effect, together with environmental factors such as nutrition. On a genetic background for tall stature, a gene for Marfan syndrome (described earlier) may have a striking effect, whereas against a genetic background for short stature its expression may not be apparent. Similarly, though mental retardation is a standard observation in Down syndrome, the degree of retardation varies, and one factor in the variation appears to be the genetic background of the patient with respect to intelligence.

GENERAL REFERENCES

See the General References to Chapter One, and the following more specialized texts.

Dubowitz, V. 1978. *Muscle Disorders in Childhood.* W. B. Saunders Company, Philadelphia.

Gorlin, R. J., Pindborg, J. J., and Cohen, M. M. 1976. *Syndromes of the Head and Neck.* 2nd ed. McGraw-Hill Book Company, New York.

McKusick, V. A. 1972. *Heritable Disorders of Connective Tissue.* 4th ed. C. V. Mosby Company, Saint Louis.

Smith, D. W. 1976. *Recognizable Patterns of Human Malformation.* 2nd ed. W. B. Saunders Company, Philadelphia.

The Birth Defects: Original Article Series, published for the National Foundation-March of Dimes by The Williams & Wilkins Company, Baltimore, is also an excellent source of clinical and genetic information.

PROBLEMS

1. Prepare a complete three-generation pedigree of your own family, showing dates of births, dates and causes of deaths, miscarriages, stillbirths and any genetic disorders. Use standard symbols.

2. A woman (A) and her paternal aunt (B) marry related men; B's husband is the maternal uncle of A's husband. Each couple produces a child with cystic fibrosis (autosomal recessive).
   a) Draw the pedigree.
   b) If A and her husband had sought genetic advice before having a child, what risk of producing a child with cystic fibrosis would you have cited?

3. A man (C) with classic hemophilia marries his mother's sister's daughter (D). His maternal grandfather had the same disease. C and D have one affected son, two affected daughters and two normal daughters.
   a) Draw the pedigree.
   b) Why are two of the daughters affected?
   c) What is the risk that a son of one of the affected daughters will be affected?
   d) What is the risk that a son of one of the unaffected daughters will be affected?

4. A man (E) with muscular dystrophy married his first cousin (F). Their mothers are sisters. The type of muscular dystrophy is unknown; onset was at age 16, and he was disabled at age 55. There was no previous family history, but the children of E and F include two similarly affected sons, a normal son and four normal daughters.
   a) Draw the pedigree.
   b) On the basis of this pedigree, what patterns of single-gene inheritance can be ruled out?
      You now learn that F has two brothers with the same disease.
   c) What is now the most likely pattern of inheritance?
   d) On this basis, what is the risk that an affected son of E and F will have an affected child?
   e) What type of muscular dystrophy described in this chapter would fit this pattern of inheritance?

5. A young woman (G) comes to Genetics Clinic with the following problems. Her father has retinitis pigmentosa (a disease of the eye). Her paternal grandmother (H) had the same disease. Her paternal grandparents are first cousins.
      Retinitis pigmentosa is genetically heterogeneous. About 5 percent of cases are autosomal dominant, the great majority are autosomal recessive and there are X-linked forms.
   a) Draw the pedigree.
   d) What is the risk for a child of G?
   c) You investigate further and learn that a paternal aunt of G is also affected, as is one of her three sons. What is now the probable risk for a child of G?
   d) On examination, G is found to have early changes characteristic of retinitis pigmentosa. How does this finding affect your estimate of risk for her child?

6. A woman (H) who is the normal sister of a girl with the Laurence-Moon-Biedl syndrome marries her father's sister's son.
   a) What is the risk that their first child will have the same syndrome?
   b) If the first child is affected, what is the risk that the second child will also be affected?

7. A color-blind albino man and a woman with normal color vision and pigmentation produce an albino daughter who is color blind. State the genotypes of the parents and the child.

8. Figure 4–21 is a pedigree of congenital deafness, discussed in the text.
   a) State the probable genotypes of the children in generation IV.
   b) If one of these boys marries a daughter of III–1, what is the probability that they will have a deaf child?

9. Figure 4–23 is a pedigree of precocious puberty, discussed in the text.
   a) What is the probability that II–9 has the gene for this condition?
   b) What is the probability that a son of III–1 would be affected?

10. a) From Figure 4–24, what percentage of persons heterozygous for the Hunting-
    ton gene are still phenotypically normal at age 30?
    b) What genetic advice would you give to a phenotypically normal man after age
    30 whose mother had Huntington chorea? (Use Figure 4–25.)

5

HUMAN BIOCHEMICAL GENETICS

Many different proteins are synthesized in body cells. These proteins, which may be either enzymes or structural components or even serve both functions, are responsible for all the developmental and metabolic processes of the organism.

As discussed in an earlier chapter (Chapter 3), the fundamental relationship between genes and proteins is that the sequence of bases in the DNA of a given gene codes for the sequence of amino acids in the corresponding polypeptide chain. Alteration of the base sequence in the gene results in the synthesis of a variant polypeptide with a correspondingly altered amino acid sequence. Proteins are composed of one or more polypeptide chains. Hence, a gene mutation results in the formation of a variant protein, which may have altered properties as a consequence of its altered structure.

The phenotypic changes produced by gene mutations are numerous, varied and frequently unexpected. If a mutation results in an amino acid substitution in a so-called structural protein, such as hemoglobin, the phenotypic effect will depend on how the alteration in amino acid sequence affects such properties of the hemoglobin molecule as its affinity for oxygen or its tendency to sickle. If the amino acid sequence of an enzyme polypeptide is altered, the enzyme synthesized by the mutant gene may have altered enzymatic activity. Most of the variant enzymes known have less activity than the normal "wild type" allele, and some are completely inactive. Rarely, a mutation may produce excessively high activity. Occasionally a mutation leads to synthesis of an unstable polypeptide chain that is rapidly destroyed *in vivo*, or to other types of change which are described in the subsequent pages.

Not all genetic changes lead to clinically abnormal phenotypes. On the contrary, many proteins are known to exist in two or more relatively common, genetically distinct and structurally different "normal" forms. Such a situation is known as a **polymorphism** (see later discussion).

95

The subject matter of this chapter falls into three main areas: (1) **hemoglobins,** normal and abnormal, which have done more than any other type of protein to clarify the relationship between gene, protein and disease; (2) **human biochemical disorders** caused by metabolic blocks that in turn are caused by gene mutations; and (3) some examples of genetic variations in response to drugs (**pharmacogenetics**).

THE HEMOGLOBINS

The human hemoglobins occupy a unique position in medical genetics for many reasons. They have taught us more about the molecular basis of human and medical genetics than any other system. They are historically important for their part in the demonstration of the relationship between genetic information and protein structure. They also illustrate mechanisms of forming new genes other than by point mutation, cast light on the process of evolution at both the molecular and the population level, and provide a model of gene action during development. Furthermore, hemoglobin variants are clinically important as causes of a variety of genetic disorders of blood. Consequently, the genetic background of the human hemoglobins merits examination in some detail.

The first step in understanding the genetics of the hemoglobinopathies was taken by Neel (1949), who showed that patients with the blood disorder known as sickle cell disease were homozygous for a gene that produced a similar but much milder abnormality, sickle cell trait, in both parents, who were heterozygous. Shortly afterwards Pauling et al. (1949) designated sickle cell disease as the prototype of the "molecular diseases," in which an abnormal hemoglobin molecule was the basic defect (see later). Ingram (1956) then discovered that the abnormality in the hemoglobin present in sickle cell disease constituted a replacement of only one of the 287 amino acids present in the hemoglobin half-molecule; this was the first demonstration *in any organism* that a mutation in a structural gene could cause an amino acid substitution in the corresponding protein.

THE STRUCTURE AND FUNCTION OF HEMOGLOBIN

Hemoglobin is the respiratory carrier in vertebrate red blood cells and is also found in some invertebrates and in the root nodules of legumes. The molecule is a tetramer of four subunits, each of which has two parts: a polypeptide chain, globin, and a prosthetic group, heme, which is an iron-containing pigment that combines with oxygen and gives the molecule its oxygen-transporting ability. The heme portion is alike in all forms of hemoglobin, genetic variation being restricted to the structure of the globin portion only. Even within the globin polypeptide there is some restriction on the kinds of amino acid substitutions that are acceptable in terms of natural selection, since impairment of the oxygen-carrying function of the molecule cannot be tolerated.

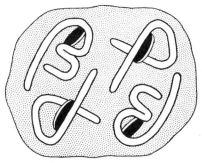

Figure 5–1 Representation of a molecule of normal adult hemoglobin. There are two α and two β chains, each associated with a heme (black disc). After Ingram, Nature 183:1795–1798, 1959.

The hemoglobin molecule typically consists of two each of two different types of polypeptide chain. In normal adult hemoglobin (Hb A) these chains are designated α and β. The four chains are folded and fitted together to form a roughly globular molecule with a molecular weight of about 64,500 (Fig. 5–1). The "formula" for the composition is $\alpha_2^A\beta_2^A$, which translates to "two α chains typical of those in Hb A plus two β chains typical of those in Hb A." The formula is often written $\alpha_2\beta_2$ omitting the superscript. The two kinds of chains are almost equal in length, the α chain having 141 amino acids and the β chain 146. The α and β chains are coded by genes at two separate loci (also designated α and β), so a mutation affects one chain or the other but not both. The chains resemble one another markedly both in amino acid sequence (primary structure) and in three-dimensional configuration (tertiary structure). They also resemble myoglobin, the respiratory pigment of muscle, though less closely; the myoglobin molecule has only a single polypeptide chain, but similarities in amino acid sequence and tertiary structure indicate that the hemoglobin and myoglobin molecules have evolved from a common precursor.

The details of the structure of hemoglobin are shown in Figure 5–2. Some points of the structure should be emphasized. The heme and the amino acid to which the iron is linked (a histidine) are the same in all hemoglobins. In all hemoglobins, the porphyrin is "wedged into its pocket" by a phenylalanine, and about 35 other sites along the chain are occupied by nonpolar residues. Otherwise, the sequence of amino acids along the chain is not rigidly specified but shows considerable variation between species, a fact that has allowed the hemoglobins to be used in studies of evolution.

THE HEMOGLOBIN LOCI

Adult Hemoglobins

As noted above, Hb A is a tetramer coded by two gene loci, designated α and β. There is also a minor hemoglobin component, Hb A$_2$, which constitutes only 2 percent of the total. Hb A$_2$ has the same tetrameric structure and the same two α chains as Hb A, but its non-α chains are not β chains. They are designated δ chains, coded by a δ locus. The close relationship of the β and δ loci is shown by the close similarity of their products, which differ, so far as we know, at only 10 sites. The composition of Hb A$_2$ may be expressed as $\alpha_2\delta_2$.

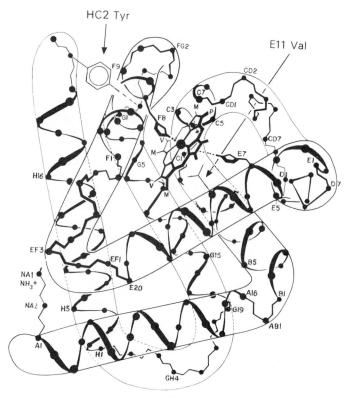

Figure 5–2 The hemoglobin molecule. The various hemoglobins and myoglobin differ only slightly. The molecule has eight helical regions (designated A to H); the amino acids are numbered from the N-terminal end of each segment. The letters M, V and P refer to methyl, vinyl and propionate side chains of the heme. Note the bond between F8 (histidine) and the iron atom of the heme, which is the only covalent link between heme and globin. From Perutz, Structure and function of haemoglobin, Brit. Med. Bull. 32:195–208, 1976. Reproduced by permission of The Medical Department, The British Council.

The complexity of the genetic coding for hemoglobins has been deepened by the finding that there are probably, in most people, two separate α loci producing identical products. The evidence comes from individuals homozygous for α-chain variants who are found to have some normal Hb A. There are two chief examples. (1) Hb Buda and Hb Pest are both α chain variants. In a Hungarian family reported by Brimhall et al. (1974), three members had both mutant genes, but produced 25 percent Hb Buda, 25 percent Hb Pest, and 50 percent Hb A. (2) Hb Constant Spring, an α-chain termination mutant described later, has been found in a homozygote who has Hb A as well as Hb Constant Spring. These and other examples provide good evidence that there are two separate α loci, thus a total of four α genes. Even when two of the genes are mutant, the other two can produce Hb A.

Fetal and Embryonic Hemoglobin

Additional gene loci are active during early development, coding for hemoglobins characteristic of prenatal and early postnatal life.

Fetal hemoglobin (Hb F), the main hemoglobin of fetal and early postna-

TABLE 5–1   THE HUMAN HEMOGLOBIN LOCI

Gene Locus	Number Per Haploid Set	Hemoglobin(s) Formed
α	Probably 2	All except embryonic hemoglobins
β	1	Hb A
δ	1	Hb A$_2$
$^{G}\gamma$	1 ?	Hb F
$^{A}\gamma$	1 ?	Hb F
ϵ	–	Embryonic Hb (Hb Gower 1 and Hb Gower 2)
ζ	–	Embryonic Hb (Hb Portland, Hb Gower 1 ?)

tal life, has the same α chain as the adult hemoglobins. (Consequently it is abnormal when mutant α genes are present.) Its non-α chain is designated the γ chain, and the composition of fetal hemoglobin may be written $\alpha_2\gamma_2$. Two different γ chains are known, differing only at a single site, $^{G}\gamma$ having glycine and $^{A}\gamma$ having alanine at position 136 of the chain. Gamma chains are 146 amino acids long, like β chains, and differ from β chains in 39 positions. It is possible that there are actually four γ chain loci per chromosome rather than two.

There are two other embryonic hemoglobin chains, designated epsilon (ϵ) and zeta (ζ). They are not fully characterized as yet, because of the difficulty of obtaining enough material for analysis. In the first two months of development, the embryo forms a hemoglobin that has two α and two ϵ chains ($\alpha_2\epsilon_2$); it is known as Hb Gower II. Finally, the ζ chain, which may be an embryonic α chain, forms Hb Portland ($\zeta_2\gamma_2$) and possibly Hb Gower I ($\zeta_2\epsilon_2$).

Table 5–1 summarizes the hemoglobin loci and the chief hemoglobins for which they code.

HEMOGLOBIN VARIANTS

Most of the hemoglobin variants result from point mutations in the structural genes that code for the amino acid sequence of the globin chains of the molecule, but some, especially the hemoglobins Lepore and Wayne, arise by a different mechanism. More than 250 abnormal hemoglobins have been described. Their clinical manifestations, if any, result from altered oxygen affinity (either higher or lower than normal), instability of the molecule or, in a few cases, the presence of iron in the oxidized state, resulting in methemoglobinemia.

In **methemoglobinemia,** the iron of the heme group of the molecule is in the ferric state, unable to combine reversibly with oxygen. Cyanosis is the most obvious clinical sign of the condition. All the hemoglobin variants resulting in methemoglobinemia have an amino acid substitution in either the α or the β chain very close to the attached heme and are able to bond to it so that the normal reduction by methemoglobin reductase cannot take place. Hb

TABLE 5-2   SUMMARY OF NORMAL AND VARIANT HEMOGLOBINS
MENTIONED IN CHAPTER 5

Normal Hemoglobins	Composition	
Hb A	$\alpha_2\beta_2$	
Hb A$_2$	$\alpha_2\,\gamma_2$	
Hb F	$\alpha_2\,\gamma_2^{136\,gly}\,\,\alpha_2\,\gamma_2^{136\,ala}$	
Hb Gower I	$\zeta_2\,\epsilon_2$	
Hb Gower II	$\alpha_2\,\epsilon_2$	
Hb Portland	$\zeta_2\,\gamma_2$	

Variant Hemoglobins	Composition	Principle
Hb S Hb C	$\alpha_2\,\beta_2^{S}\,(\alpha_2\,\beta_2^{6\,glu\rightarrow val})$ $\alpha_2\,\beta_2^{C}\,(\alpha_2\,\beta_2^{6\,glu\rightarrow lys})$	Point mutation producing amino acid substitution

Hb Lepore Hollandia
      Baltimore
      Boston (Washington)
Hb Kenya
	$\alpha_2\,\gamma\beta_2$ (See text for details.)
	
	$\alpha_2\,\gamma\beta_2$
		Intragenic unequal recombination giving rise to a fusion product

Hb M Boston Hb M Saskatoon	$\alpha_2\,\beta_2^{58\,his\rightarrow tyr}$ $\alpha_2\,\beta_2^{63\,his\rightarrow tyr}$	Because site of substitution is near heme, oxygen transport is affected and methemologinemia results.
Hb Ho-2	$\alpha_2^{Ho\text{-}2}\,\beta_2\,(\alpha_2^{112\,his\rightarrow asp}\,\beta_2)$	Non-allelism of α and β loci
Hb Bart's Hb H	γ_4 β_4	Molecule of four identical subunits
Hb C Georgetown Hb C Harlem	$\alpha_2\,\beta_2^{6\,glu\rightarrow val\,+\,?}$ $\alpha_2\,\beta_2^{6\,glu\rightarrow val.\,73\,asp\rightarrow asn}$	Double substitution
Hb Gun Hill Hb Freiburg Hb McKees Rocks	$\alpha_2\,\beta_2^{93\text{-}97\,del}$ $\alpha_2\,\beta_2^{23\,del}$ $\alpha_2\,\beta_2^{145\text{-}6\,del}$	Deletion
Hb Constant Spring	$\alpha_2\,\beta_2\,(\alpha_2^{+141\text{-}171}\,\beta_2)$	Chain termination mutation
Hb Buda Hb Pest	$\alpha_2^{61\,lys\rightarrow asn}\,\beta_2$ $\alpha_2^{74\,asp\rightarrow asn}\,\beta_2$	Evidence for two α loci

M Boston and Hb M Saskatoon are two of the examples. Methemoglobinemia can also originate by quite a different mechanism, through deficiency of the enzyme methemoglobin reductase itself, and in this case the inheritance is autosomal recessive rather than autosomal dominant.

The hemoglobin variants mentioned in this section are listed in Table 5-2 for convenient reference. The first to be detected and still the most important clinically is sickle cell hemoglobin (Hb S).

Sickle Cell Disease

Sickle cell disease is a severe hemolytic disease characterized by a tendency of the red cells to become grossly abnormal in shape ("sickled cells") under conditions of low oxygen tension (Fig. 5-3). The clinical manifestations include anemia, jaundice and "sickle cell crises" marked by impaction of sickled cells, vascular obstruction and painful infarcts in various tissues such as the bones, spleen and lungs. The disease has a characteristic geographic distribution, occurring most frequently in equatorial Africa, less commonly in

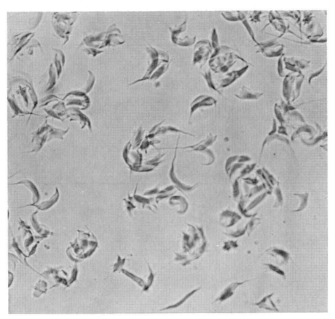

Figure 5–3   The red cell phenotype in sickle cell anemia. Homozygotes for the sickle cell hemoglobin gene form only sickle cell hemoglobin, no normal adult hemoglobin. Under conditions of reduced oxygen tension, the abnormal hemoglobin aggregates to form long projections that distort the cell. Photomicrograph courtesy of J. H. Crookston.

the Mediterranean area and India and in countries to which people from these regions have migrated. About 0.25 percent of American blacks are born with this disease, which is often fatal in early childhood though longer survival is becoming more common.

The parents of affected children, though usually clinically normal, have red cells that sickle when subjected to very low oxygen pressure *in vitro* (that is, they show a positive sickling test). Occasions when this might happen *in vivo* are very unusual. The heterozygous state, known as **sickle cell trait,** is present in approximately 8 percent of American blacks. As noted earlier, sickle cell disease results from homozygosity for the mutant gene.

The physicochemical abnormality of sickle cell hemoglobin was demonstrated in 1949 by Pauling et al., who studied Hb A and Hb S by electrophoresis. They found that Hb A and Hb S were readily distinguishable from one another by their mobility in an electrical field and concluded that their globin molecules were different. They also found that the hemoglobin of persons with sickle cell trait behaved like a mixture of normal and sickle cell hemoglobin. The structure of sickle cell hemoglobin was, in their words, "a clear case of a change produced in a protein by an allelic change in a single gene involved in synthesis."

The precise nature of the change in the protein molecule predicted by these workers was identified by Ingram, by means of "fingerprinting." This is a technique developed by Sanger for determining the structure of proteins by breaking them down with trypsin and then separating the resulting small peptides by electrophoresis in one direction and by chromatography at right angles to it (Fig. 5–4). Ingram demonstrated that the difference between normal and sickle cell hemoglobin lay in the β chain and involved only *one* of the 146 amino acids in the chain, the amino acid sixth in position from the

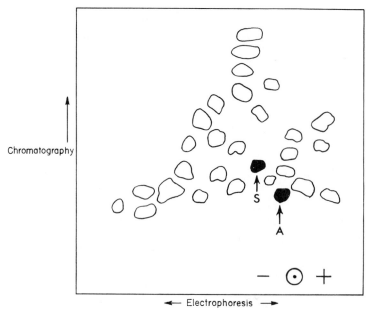

Figure 5-4 Diagram to compare the "fingerprints" of normal adult hemoglobin (Hb A) and sickle cell hemoglobin (Hb S). Each spot represents a tryptic peptide. Hb A and Hb S differ in only a single peptide. A is present in Hb A but not in Hb S; conversely, S is present in Hb S but not in Hb A.

N-terminal end of the β chain. At this position, the amino acid valine has replaced the glutamic acid of normal hemoglobin. The sequences in the corresponding sections of the two hemoglobins are as follows:

Hb A: val-his-leu-thr-pro-*glu*-glu-lys
Hb S: val-his-leu-thr-pro-*val*-glu-lys

The substitution of a valine for a glutamic acid in each of the two β chains of the hemoglobin molecule is the only biochemical difference between Hb A and Hb S. This difference, which results in an altered electrical charge, explains the difference in electrophoretic mobility of the two hemoglobins. All the clinical manifestations of sickle cell hemoglobin are consequences of this relatively minor change. The physical basis of the sickling phenomenon is the tendency of the abnormal hemoglobin molecules to aggregate, forming rod-like masses that distort the red cells into sickle shapes under conditions of low oxygen tension.

Because the abnormality of Hb S is localized in the β chain, the formula for sickle cell hemoglobin may be written $\alpha_2^A\beta_2^S$, or simply $\alpha_2\beta_2^S$. This symbol indicates that the α chain is normal, but that the β chain is of the sickle cell hemoglobin type. Note that the abnormality affects both β chains of the molecule. A heterozygote has a mixture of the two types of hemoglobin, A and S, not a compound type. The relationships of clinical status, hemoglobin types and genes can be summarized as follows:

Clinical Status	Hemoglobin	Hemoglobin Composition	Genotype
Normal	Hb A	$\alpha_2{}^A\beta_2{}^A$	$\alpha\alpha\ \beta\beta$
Sickle cell trait	Hb A + Hb S	$\alpha_2{}^A\beta_2{}^A$ and $\alpha_2{}^A\beta_2{}^S$	$\alpha\alpha\ \beta\beta^S$
Sickle cell disease	Hb S	$\alpha_2{}^A\beta_2{}^S$	$\alpha\alpha\ \beta^S\beta^S$

Other Beta Chain Variants

The first several abnormal hemoglobins to be discovered all involved mutations of the β locus and thus were allelic. In fact, the second variant to be found, Hb C, also has a substitution at position 6 of the β chain, a change from glutamic acid to lysine. The sequence is:

Hb C: val his leu thr pro *lys* glu lys

Note that only a single substitution in the triplet coding for glutamic acid alters it to code for valine or for lysine:

Amino Acid	mRNA Code
Glutamic acid	GA(A or G)
Valine	GU−
Lysine	AA (A or G)

In addition to Hb S and Hb C, over 150 other β chain variants have been discovered, almost all of which are <u>point mutations</u>. Substitutions are known for many positions other than position 6. Though all the mutations that produce β chain variants are allelic in that they affect the amino acid sequence of the same chain, the mutation may be at the same or a different site within the gene, affecting the same or a different codon. The term **heteroalleles** is used to refer to allelic mutations in which different codons are affected, as distinct from **eualleles,** in which the same codon is involved (Fig. 5–5).

The allelism of the β chain variants can be demonstrated both genetically and biochemically. For example, Figure 5–6 shows a family in which a person with two different mutations at the β locus has parents with one abnormality or the other, but not both. The genotypes and corresponding phenotypes are as follows:

	Genotype	Hemoglobin Composition	Comments
Parent II 1	$\alpha^A\alpha^A\ \beta^S\beta^C$	$\alpha_2{}^A\beta_2{}^S$ and $\alpha_2{}^A\beta_2{}^C$	Hb S and Hb C, no Hb A
Parent II 2	$\alpha^A\alpha^A\ \beta^A\beta^A$	$\alpha_2{}^A\beta_2{}^A$	Hb A only
Child III 1	$\alpha^A\alpha^A\ \beta^A\beta^S$	$\alpha_2{}^A\beta_2{}^A$ and $\alpha_2{}^A\beta_2{}^S$	Hb A and Hb S
Child III 2	$\alpha^A\alpha^A\ \beta^A\beta^C$	$\alpha_2{}^A\beta_2{}^A$ and $\alpha_2{}^A\beta_2{}^C$	Hb A and Hb C

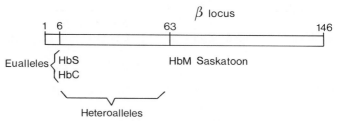

Eualleles { HbS / HbC   HbM Saskatoon

Heteroalleles

Figure 5–5 Point mutations in the β gene, which may be either eualleles (at the same site) or heteroalleles (at different sites). Intragenic crossing over between heteroalleles can produce a gene with two different amino acid substitutions (see also Fig. 5–8).

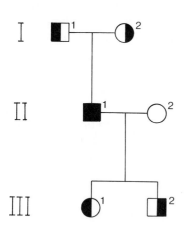

Figure 5–6 A pedigree to demonstrate allelism of the β chain variants Hb S and Hb C. For details, see text.

■ HbAS heterozygote
□ HbAC heterozygote

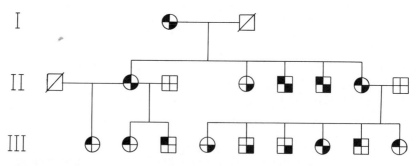

⊕ Tested, normal
⬙ Heterozygote for Hopkins-2 hemoglobin
⬗ Heterozygote for sickle hemoglobin
◪ Deceased

Figure 5–7 A pedigree to demonstrate independent assortment of Hb S and Hb Hopkins-2. Redrawn with abbreviations from Smith and Torbert, Bull. Johns Hopkins Hosp. 102:38–45, 1958.

Alpha Chain Variants

Since Hb A, Hb A_2, Hb F and Hb Gower II all have α chains, all are affected by an α chain mutation. More than 70 α chain variants are known. The first to be discovered is known as Hb Hopkins-2. In a family with both Hb Hopkins-2 and Hb S, it was found that the members with both abnormal hemoglobins also had large quantities of Hb A and had children with either one or both of the abnormal hemoglobins as well as Hb A (Fig. 5–7). Thus it was clear that the two mutations were not allelic. As all three main hemoglobins were affected, the mutation was obviously in the α chain.

If the α chain is completely missing, as in α thalassemia 1 (see later), tetramers of four identical chains may form. One such type of hemoglobin is Hb Bart's (so named because it was first identified at St. Bartholomew's Hospital, London), composed of four γ chains. The other, Hb H, has four β chains.

Delta Chain Abnormalities

A mutation in the gene coding for the δ chain leads to production of an abnormal Hb A_2. A few persons with both an abnormal Hb A_2 and sickle cell trait have been identified. By marriage to a normal homozygote, these double heterozygotes have produced children with neither abnormality and children with both, but none with only one or the other, thus providing genetic evidence that the β and δ genes are closely linked. Even better evidence for β-δ linkage comes from the Lepore hemoglobins, described later.

Double Mutations

At least two abnormal hemoglobins, Hb C Georgetown and Hb C Harlem, possess two separate amino acid substitutions in the same polypeptide chain. They are identified as Hb C because they migrate electrophoretically to the same position as Hb C, not because they have the same amino acid substitution. On the contrary, each has the same mutation as Hb S, plus a second substitution at a different site. The mechanism may be either a second mutation in a β^s allele or intragenic crossing over during meiosis between two β genes carrying mutations at different sites (Fig. 5–8).

Deletions

As previously mentioned, the α hemoglobin chain is five amino acids shorter than the β chain. From comparison of amino acid sequences, it seems likely that amino acids 54 to 58 of the β chain are missing in the α chain. The variant hemoglobin Gun Hill appears to have lost five amino acid residues from the β chain, and Hb Freiburg has lost one. All these deletions could have arisen as a result of abnormal pairing and intragenic crossing over during meiosis (Fig. 5–9).

Linkage Relations of the Hemoglobin Genes

The hemoglobin genes have not yet been definitely assigned to chromosomes, but it seems clear that the non-α genes (i.e., β, δ, $^G\gamma$, $^A\gamma$ and possibly

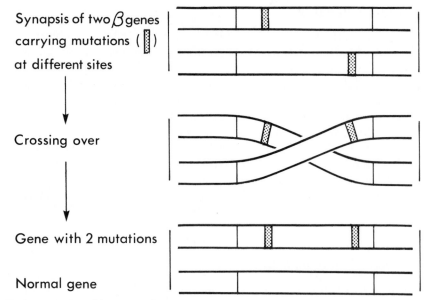

Synapsis of two β genes
carrying mutations ( ) at different sites

Crossing over

Gene with 2 mutations

Normal gene

Figure 5–8 Possible origin of a β gene carrying mutations at two different sites by intragenic crossing over. For discussion, see text.

also ε) are very closely linked, probably in tandem on the same chromosome, and that the two α loci are not close to the non-α loci, though they are probably closely linked to one another.

Lepore Hemoglobins and the Linkage of the β and δ Loci. In the Lepore hemoglobins, the non-α chain has a segment homologous to the N-terminal end of a normal δ chain followed by a segment homologous to the C-terminal

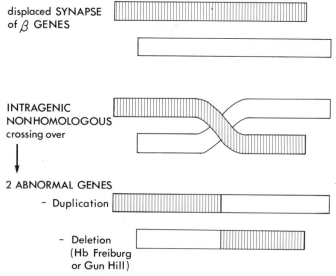

displaced SYNAPSE of β GENES

INTRAGENIC NONHOMOLOGOUS crossing over

2 ABNORMAL GENES

– Duplication

– Deletion (Hb Freiburg or Gun Hill)

Figure 5–9 Possible origin of a β gene carrying a deletion by intragenic nonhomologous crossing over. For discussion, see text. Adapted from Giblett, *Genetic Markers in Human Blood.* Blackwell Scientific Publications, Oxford, 1969, p. 562, by permission.

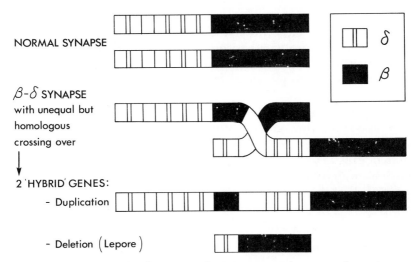

NORMAL SYNAPSE

δ

β

β-δ SYNAPSE
with unequal but
homologous
crossing over

2 'HYBRID' GENES:

- Duplication

- Deletion (Lepore)

Figure 5–10 Model for the formation of a Lepore gene by unequal crossing over. The δ and β loci are very closely linked and their sequences differ at only 10 sites (shown on the δ gene as vertical lines). If mispairing occurs at meiosis, followed by intragenic crossing over, two hybrid genes can result, one with a deletion of a portion of the two loci (Lepore), the other with a duplication (anti-Lepore). Adapted from Giblett, *Genetic Markers in Human Blood.* Blackwell Scientific Publications, Oxford, 1969, p. 564, by permission.

end of a normal β chain, which together form a new $\delta\beta$ fusion chain. The explanation of the origin of these variants is shown in Figure 5–10. The δ and β genes have extensive homology; as previously mentioned, they may differ at only 10 of their 146 sites. Misalignment between the δ gene of one chromatid and the β gene of the homologous chromatid, occurring during meiosis, could happen as a rare accident. Crossing over would then be possible between the two mismatched genes, resulting in the formation of two products: (1) a new gene beginning as a δ and ending as a β and (2) a reciprocal product with a normal δ and a normal β gene but also a fusion product coded for a β segment followed by a δ segment. The first is a Lepore gene, the second an anti-Lepore. This mechanism is known as homologous but unequal crossing over.

If a crossover between misaligned β and δ genes did not include at least one of the codon differences between β and δ, the crossover products could not be distinguished. As there are (at least) 10 differences between β and δ, there are (at least) nine possible Lepores and anti-Lepores. So far, three different Lepore hemoglobins have been identified, differing from one another in the relative lengths of their δ and β portions:

Lepore Hemoglobin	*Site of Crossover*
Hb Lepore Hollandia	Between 22 and 50
Hb Lepore Baltimore	Between 50 and 86
Hb Lepore Boston (Washington)	Between 87 and 116

Three anti-Lepores have also been found so far, though only two have been well characterized. The anti-Lepore gene is inevitably associated with normal β and δ genes, so there is no deleterious effect on Hb A synthesis. However, Lepore hemoglobin does have a deleterious effect, with clinical and hematological consequences resembling β thalassemia.

The observation that a Lepore chain begins as a δ chain and ends as a β chain indicates that the sequence of the genes on the chromosome is $\delta\beta$.

Hb Kenya and the Linkage of the γ and β Loci. Hb Kenya provides information about the linkage and order of the γ and β genes comparable to the information provided by Lepore for the β and δ genes. The Hb Kenya non-α chain begins as γ and ends as β (Fig. 5–11). Heterozygotes for Hb Kenya ($\alpha_2\gamma\beta_2$) have hereditary persistence of fetal hemoglobin (HPFH), and the fetal hemoglobin is all of the $^G\gamma$ type. Probably, then, the order of the genes on the chromosome is $^G\gamma^A\gamma\delta\beta$.

Chain-Termination Mutations

Two types of chain termination mutation can occur: mutation *to* a termination codon (UAA, UAG, or UGA) at some position within a gene, or mutation *of* the termination codon itself to a codon specifying an amino acid. The first of these types, which would presumably code for a shorter than normal polypeptide, is thought to be represented by Hb McKees Rocks, which lacks the two C-terminal amino acids of the β chain.

Four examples of mutation of chain termination codons have been found so far, all involving the α chain. The first and best known is Hb Constant Spring. Each variant chain is 171 amino acids long, and each has an amino acid at position 142 that could be derived by point mutation from a termination codon. The sequence between position 142 and the end of the chain is identical in all four, and does not resemble the amino acid sequence of any known protein. Thus, one of the intriguing aspects of Hb Constant Spring is that it helps to elucidate the composition of the DNA *between* the structural genes.

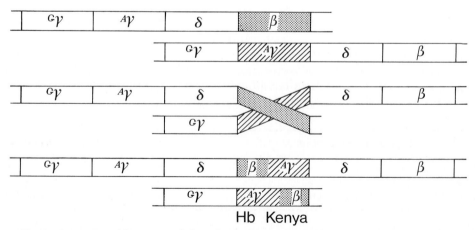

Hb Kenya

Figure 5–11 Possible origin of the Hb Kenya non-α gene by unequal crossing over. For discussion, see text.

THE THALASSEMIAS

The thalassemias are a group of disorders of hemoglobin synthesis in which the basic defect is not in molecular structure, but in a reduced rate of synthesis of either the α or the β chains. They occur in two forms: a severe form, thalassemia major (homozygous) and a mild form, thalassemia minor (heterozygous, or in some cases homozygous for a milder defect). Thalassemia minor is sufficiently common in Canada and the United States to pose the important problem of differential diagnosis from iron deficiency anemia. There is a characteristic distribution of the disorder in a band around the Old World — in the Mediterranean, the Middle East, parts of Africa, India and the Orient. The name is derived from the Greek word for sea, *thalassa*, and signifies that the disease was first discovered in persons of Mediterranean stock. The erythrocytes typically have a target cell appearance.

In recent years much progress has been made in clarifying the molecular and genetic basis of the thalassemias, though many puzzling problems still remain. Some types are so frequent that they are important public health problems in certain areas; this is especially true of the form known as β^+ thalassemia in the Mediterranean, particularly Cyprus and Greece. Like sickle cell anemia and G6PD deficiency, the thalassemias are most common in former malarial areas and are thought to have been maintained in high frequency by their interaction with malaria (see Chapter 15), but they are by no means unknown in other parts of the world.

Two main groups of thalassemias are defined: the α thalassemias, in which α chain synthesis is reduced or absent, and the β thalassemias, in which β chain synthesis is impaired. Both types are genetically heterogeneous, and both are much more severe in homozygotes than in heterozygotes. A further complication is that it is not unusual for genes for both types of thalassemia as well as for structural hemoglobin abnormalities to coexist in an individual and to interact. Since there are many such genes, they account for a wide variety of complex hematological problems. Not all the thalassemias are well characterized, and this brief discussion risks oversimplification by mentioning only some of the most common and best-known types.

Alpha Thalassemia. As noted earlier, there is reason to believe that in man the α chains are duplicated, so that there are two per chromosome and four per diploid set. This fact has significance for understanding the α thalassemias. Alpha thalassemia 1 (α thal 1) results from an actual deletion of both α genes on one chromosome, with abolition of α chain synthesis. If the deletion is homozygous, no α chains at all are formed; if heterozygous, α chain synthesis is reduced to half the normal level. In either case, the non-α chains (especially γ and β chains) are present in excess, and may form tetramers of four identical chains, i.e., γ_4 (Hb Bart's) and β_4 (Hb H). Homozygosity for α thal 1 results in fetal hydrops (gross edema due to heart failure secondary to severe anemia), which causes intrauterine death and abortion at an advanced stage of gestation. In heterozygotes α chain synthesis is sufficiently abnormal to cause thalassemia minor, with tissue anoxia (because both Hb Bart's and Hb H have high oxygen affinity), some impairment of erythropoiesis and mild hemolytic anemia.

Alpha thalassemia 2 (α thal 2) is the result of deletion of only one of the two α genes and is consequently a milder defect than α thal 1, producing thalassemia minor when homozygous but virtually insignificant in heterozygotes.

The structural mutation Hb Constant Spring, an α chain termination mutant described earlier, results in an α chain which is synthesized in only small amounts and thus can lead to a form of thalassemia clinically very similar to the α thal 2 type.

Beta Thalassemia. Whereas in α thalassemia the molecular error is known, in β thalassemia the nature of the defect is less obvious. Genetic studies of families segregating for an abnormal structural allele at the β locus as well as for β thal strongly suggest that the defect is either allelic or very closely linked to the β locus. The β^+ form apparently results from reduced production of β chain mRNA, but the precise reason for the reduced production is not clear. Thalassemia major (Cooley's anemia) due to homozygosity for β^+ thalassemia ranks with G6PD deficiency as one of the most common and important genetic disorders in the world.

Clinically thalassemia major is characterized by severe anemia from infancy, skeletal deformities secondary to compensatory expansion of the erythron, growth retardation, increased gastrointestinal iron absorption and hepatosplenomegaly. Treatment, since it requires repeated transfusion, exacerbates the problem of iron overload. Death usually results from iron toxicity, most often in the age range of 16 to 24 years. Thalassemia minor causes mild anemia and typical red cell abnormalities including target cells.

Prenatal Diagnosis. Although β^+ thalassemia is a disorder of β chain synthesis and β chains are produced only in relatively small amounts in the early stages of fetal development, there is about 10 percent Hb A in normal fetuses at the sixteenth to eighteenth week of gestation, and this is sufficient to allow detection of impairment of β chain synthesis and thus to enable diagnosis of β^+ thalassemia in the fetus. The technique requires sampling of fetal blood and biochemical examination of the fetal red cells for β chain synthesis. Because both the obstetrical and the biochemical aspects are sophisticated procedures not yet available except in a few specialized centers, prenatal diagnosis of β^+ thalassemia is not yet widely available.

Other Forms of Beta Thalassemia. A different class of disorders is represented by β^0 thalassemia, in which β chain synthesis is absent, though mRNA for β chains appears to be present. The Lepore ($\delta\beta$) hemoglobins also lead to β thalassemia, apparently because in some way, as yet unclear, the presence of a message that begins as the code for a δ chain lowers the rate of production of the Lepore chain to the same level as that of a normal δ chain.

Hereditary Persistence of Fetal Hemoglobin (HPFH). A number of conditions are characterized by persistent fetal hemoglobin associated with defective β chain synthesis, without any clinical abnormality. These disorders behave like very mild β thalassemias in which chain imbalance is minimal and is compensated for by persistence of γ chain production. The molecular defect in most of the HPFH syndromes is unknown, but in one form it might represent

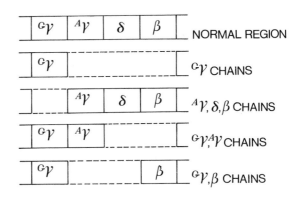

Figure 5–12 Various possible deletions in the γ, δ and β loci. Deletion involving all or part of the β locus might lead to hereditary persistence of high fetal hemoglobin (HPFH).

deletion of various portions of the group of loci that determine the $^G\gamma$, $^A\gamma$, δ and β globin chains in one of the patterns shown in Figure 5–12.

THE EVOLUTION OF HEMOGLOBIN

Because of the technical feasibility of determining the amino acid sequences of polypeptides and the development of the concept that the amino acid sequence of a given polypeptide is encoded in a specific gene, some evolutionary events at the molecular level can now be reconstructed.

The amino acid sequences of the different globin chains of man have extensive homologies. The α chain is composed of 141 amino acids; the β, γ and δ chains each consist of 146. After allowances are made for the difference in length of the chains, the α chain differs from the γ chain in 83 amino acids. The β chain differs from the α chain in 77 amino acids and from the γ chain in 38. The β and δ differ in only 10 amino acids. Forty-one sites are invariant in all four chains.

The muscle protein myoglobin has a single polypeptide chain of 151 amino acids rather than a four-subunit molecule, but in amino acid sequence and three-dimensional configuration it resembles the hemoglobin chains.

According to an evolutionary model proposed by Ingram (1963), and by Epstein and Motulsky (1965), the α gene arose by duplication of a primitive gene for a myoglobin-like molecule. Mutation gradually brought about divergence of the two primitive genes. Further gene duplication led to the formation of "new" genes that in turn diverged from one another (Fig. 5–13) to result eventually in the amino acid sequences of the globin chains of present-day man. Meanwhile, development of the tetrameric configuration allowed much more efficient oxygenation.

There is a limit to the kind of amino acid substitution that is acceptable in terms of natural selection. Only those substitutions that do not impair function can be maintained, and it is of interest that the sites of heme attachment are invariant in the different human globin chains (and in those of other vertebrate species that have been examined; Perutz, 1965). Another constraint is that those amino acids which are classified as polar (hydrophilic) tend to occupy the surface of the hemoglobin molecule, where they are exposed to water, whereas those which are nonpolar (hydrophobic) are in the interior, where they play a

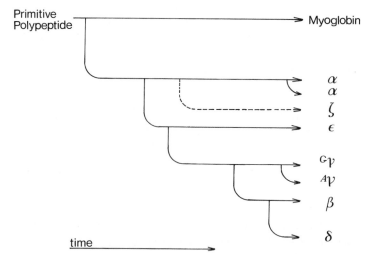

Figure 5–13 Evolution of hemoglobin. By duplication a primitive gene gave rise to the genes for myoglobin and the α chain of hemoglobin, which then diverged by accumulation of mutations. Successive gene duplications gave rise to the other globin genes. Adapted from Epstein and Motulsky, Progr. Med. Genet. 4:85–127, 1965, and from Giblett, *Genetic Markers in Human Blood.* Blackwell Scientific Publications, Oxford, 1969, p. 402, by permission.

role in maintaining the structure of the molecule. Consequently, substitutions are not likely to be preserved if they do not have the right type of polarity. The fact that many sites are alike in all four chains indicates that selection places rigid constraints on some types of changes in specific proteins.

As a generalization, many proteins have "sensitive areas," in which mutations cannot occur without affecting function, and "insensitive areas," where mutations are tolerated more freely.

A NOTE ON HEMOGLOBIN NOMENCLATURE

New hemoglobins are being discovered too rapidly for their nomenclature to follow any firm rules. The normal hemoglobins, Hb A, Hb A_2 and Hb F, and the first abnormal hemoglobin to be discovered, Hb S, are obviously named for their associations with adult, fetal, and sickle cell forms of hemoglobin. Other hemoglobins were at first named alphabetically in order of their discovery, but this system soon broke down in practice. At present, newly identified hemoglobins are usually named for the place where they were discovered (e.g., Hb Bart's, Hb Hopkins-2), or by their electrophoretic resemblance to some previously discovered hemoglobin (e.g., Hb C Georgetown, Hb C Harlem). From time to time, international conferences are held to bring the nomenclature up to date in the light of recent findings and to eliminate obvious duplications.

HUMAN BIOCHEMICAL DISORDERS

Human biochemical disorders, often called inborn errors of metabolism, are conditions in which a specific genetically determined enzyme defect produces a metabolic block that may have pathological consequences. The concept of inborn errors of metabolism as causes of disease was first proposed by the eminent physician Garrod in 1902, shortly after the rediscovery of Mendel's laws. Garrod noted that in alcaptonuria and several other disorders the patients appeared to lack the ability to perform one specific metabolic step. (In the case of alcaptonuria, this step is the breakdown of the benzene ring of the amino acid tyrosine. Failure to complete this reaction leads to the excretion in urine of an abnormal constituent, homogentisic acid or alcapton.) Garrod also noted that these disorders showed a striking familial pattern of distribution, in that two or more sibs might be affected though the parents and other relatives were usually normal, and that the parents themselves were often consanguineous. Through discussions with Bateson, a leading biologist and pioneer geneticist, Garrod came to interpret this pattern as exactly what would be expected on the basis of Mendelian recessive inheritance, and thus described in alcaptonuria the first example of autosomal recessive inheritance in man.

Garrod's concept, that an inborn error of metabolism was a genetically determined enzyme defect leading to interruption of a metabolic pathway at a specific point, was far in advance of its time. The first actual demonstration of a specific enzyme defect in an inborn error of metabolism was not provided until a half century later (1952), when Gerty Cori showed that von Gierke's disease (glycogen storage disease Type I) is due to loss of activity of the enzyme glucose-6-phosphatase. The specific deficiency of the enzyme homogentisic oxidase, which is the cause of alcaptonuria, was not directly demonstrated until 1958 (by La Du et al.).

The majority of inborn errors of metabolism are inherited as autosomal recessives. Several, including some of great clinical or genetic interest, are X-linked, for example, glucose-6-phosphate dehydrogenase (G6PD) variants and hypoxanthine-guanine phosphoribosyl transferase (HPRT) deficiency (Lesch-Nyhan syndrome). Only a few inborn errors of metabolism are autosomal dominants; for example, acute intermittent porphyria and congenital spherocytosis are expressed in heterozygotes.

A large number of biochemical defects that produce inborn errors of metabolism are now known. Table 5–3 lists those mentioned in this chapter. In the following sections we will illustrate some of the mechanisms involved and discuss a few examples of the variety of clinical and genetic problems related to specific enzyme deficiencies.

MECHANISMS AND CONSEQUENCES OF ENZYME DEFECTS

Metabolism is performed as a stepwise series of reactions, each step catalyzed by a specific enzyme. At any step the pathway may be blocked if the enzyme required to catalyze the reaction is mutant, i.e., determined by a mutant gene. Most of the mutational changes are probably substitutions of single amino acids, though by analogy with the hemoglobins we assume that other mechanisms such as double substitution or deletion may sometimes be involved.

The various ways in which a mutation in a gene coding for an enzyme might bring about a metabolic defect can be clarified by considering how enzymes perform their functions. An enzyme catalyzes the conversion of a

Text continued on page 118

TABLE 5–3 SUMMARY OF HUMAN BIOCHEMICAL DISORDERS AND POLYMORPHISMS MENTIONED IN CHAPTER 5

Disorder	Chief Clinical Manifestations	Defective Protein	Genetics	Heterozygote Detection	Prenatal Diagnosis in Cultured Amniotic Fluid Cells
Albinism	Absence of pigmentation in skin and eyes; visual disorders	Tyrosinase (in one common type)	Heterogeneous; usually autosomal recessive; autosomal dominant and X-linked also	—	—
Anemia, pyridoxin-responsive	Anemia, hemochromatosis, splenomegaly	Unknown; related to pathway of porphyrin synthesis	X-linked	Minor hematological changes	—
Alcaptonuria	Pigmentation of cartilage, arthritis	Homogentisic acid oxidase	Autosomal recessive	None	—
Cystinuria	Aminoaciduria, renal lithiasis	Not known; affects transport of cystine, lysine, arginine and ornithine	Three forms, probably allelic; I autosomal recessive, II and III incompletely recessive	In incompletely recessive forms, by urinary excretion of abnormal amounts of the affected amino acids	—
Galactosemia	Hepatosplenomegaly, cirrhosis of liver, cataracts, mental retardation	Galactose-1-phosphate uridyl transferase	Autosomal recessive; Duarte and other allelic variants known	Intermediate enzyme activity	Yes
Glucose-6-phosphate dehydrogenase variants	Hemolytic anemia in response to certain foods and drugs	Glucose-6-phosphate dehydrogenase (G6PD)	X-linked; numerous alleles	Variants may have altered enzymatic activity or electrophoretic mobility, demonstrable in heterozygotes	Yes
Glycogen storage disease Type I (von Gierke's disease)	Hepatomegaly, enlarged kidneys, growth retardation, hypoglycemia, acidosis, xanthomata	Glucose-6-phosphatase	Autosomal recessive	Intermediate enzyme levels in intestine and platelets	—
Type II (Pompe's d...)	Varied; in infants ...	Acid maltase	Autosomal recessive; ...	Reduced enzyme activ-...	Yes

Disease	Clinical manifestations	Basic defect	Inheritance	Heterozygote detection	Prenatal diagnosis
Hartnup disease	Skin rash, neurological changes	Not known; affects tryptophane transport	Autosomal recessive	–	–
Hemoglobinopathies	Anemia; other manifestations depend upon site and nature of hemoglobin abnormality	Globin chain of Hb molecule	Autosomal, clinical manifestations usually recessive; numerous variants	Usually by electrophoresis	Possible (in fetal blood) but not widely available
Homocystinuria	Arachnodactyly, dislocated lenses, mental retardation	Cystathionine synthetase	Autosomal recessive; genetically heterogenous	Not well-established	Theoretically possible
Hypercholesterolemia (Type II hyperlipoproteinemia)	Premature coronary heart diseases, xanthomas, arcus corneae	Unknown; affects ability of LDL receptors to bind LDL	Autosomal; severe childhood disease in homozygotes, later onset and milder expression in heterozygotes	Yes	Yes
Hyperthermia, malignant	Hyperpyrexia and muscle rigidity in response to anesthesia	Unknown	Heterogeneous; most pedigrees autosomal dominant	Susceptible persons may have elevated serum creatine kinase and mild myopathy	–
Hypophosphatemic rickets (vitamin D-resistant rickets)	Rickets, short stature	Not known; affects renal tubular reabsorption of phosphate	X-linked dominant	Affected females are usually if not always heterozygotes.	–
Isoniazid inactivation slow	Neurological problems	Acetyltransferase	Autosomal recessive	–	–
Lesch-Nyhan syndrome	Uric aciduria, cerebral palsy, self-mutilation, mental retardation	Hypoxanthine-guanine phosphoribosyl transferase (HPRT)	X-linked; milder allelic mutations known	Several methods	Yes
Methylmalonicaciduria (cobalamin-responsive type)	Failure to thrive, ketoacidosis, retardation	Step in synthesis of cobalamin (vitamin B_{12}), the cofactor for methylmalonyl CoA carbonylmutase	Autosomal recessive; same clinical condition can also result from defect in synthesis of the enzyme but is not cobalamin-responsive	–	Yes
Mucopolysaccharidoses Hurler syndrome (MPS I)	Gargoyle facies, corneal clouding, mental retardation, dwarfism	α-L-iduronidase	Autosomal recessive	Yes, by storage of certain MPS in cultured fibroblasts	Yes

TABLE 5-3 SUMMARY OF HUMAN BIOCHEMICAL DISORDERS AND POLYMORPHISMS MENTIONED IN CHAPTER 5 *(Continued)*

Disorder	Chief Clinical Manifestations	Defective Protein	Genetics	Heterozygote Detection	Prenatal Diagnosis in Cultured Amniotic Fluid Cells
Scheie syndrome (MPS I)	Corneal clouding, coarse facies, stiff joints, normal intelligence, normal height	α-L-iduronidase	Autosomal recessive, mutation allelic to Hurler mutation	Yes (see above)	Yes
Hunter syndrome (MPS II)	Similar to Hurler syndrome except that corneal clouding is rare	Iduronidate sulphatase	X-linked; probably at least two variants	Yes (see above)	—
Sanfilippo syndrome (MPS III)	Severe mental retardation, mild dysmorphic features	Form A: heparan N-sulphatase Form B: N-acetyl α.glucosaminidase	Autosomal recessive; forms A and B are not allelic	—	—
Morquio syndrome (MPS IV)	Skeletal deformities, growth retardation, usually with normal intelligence	Not clear	Autosomal recessive; probably more than one mutant allele	—	—
Maroteaux-Lamy syndrome (MPS VI)	Skeletal changes, growth retardation, usually with normal intelligence	Arylsulphatase B	Autosomal recessive; probably at least two mutant alleles	—	—
Phenylketonuria	Mental retardation, microcephaly, diluted pigmentation	Phenylalanine hydroxylase	Autosomal recessive; several allelic mutations known	Yes, by measurement of serum phenylalanine and phenylalanine/tyrosine ratio	—
Porphyria, acute intermittent	Episodes of abdominal pain, neurological problems, excessive urinary excretion of δ-amino levulinic acid (ALA)	Deficiency of uroporphyrinogen I synthetase, secondary rise of hepatic ALA synthetase	Autosomal dominant	Patients are usually if not always heterozygotes.	—

Spherocytosis, congenital	Episodes of hemolytic anemia and jaundice, spherocytes	Not known; causes red cell membrane defect	Autosomal dominant	Patients are usually if not always heterozygotes.	—
Succinylcholine sensitivity	Prolonged apnea in response to succinylcholine	Serum cholinesterase	Clinically significant only in subjects with homozygous or compound genotypes; i.e., autosomal recessive	Yes, by reduced serum cholinesterase activity	—
Taster/nontaster trait	Ability to taste phenylthiocarbamide (PTC)	Not known	Nontaster trait is autosomal recessive	No	—
Tay-Sachs disease (GM_2 gangliosidosis)	Severe physical and mental degeneration, cherry-red spot on macula	Hexosaminidase A	Autosomal recessive; at least two loci and two or more allelic mutations at the hex A locus produce different GM_2 gangliosidoses	Yes; mass screening methods exist	—
Thalassemia	Severe anemia, splenomegaly, bone changes	Defect in rate of synthesis of a globin chain	Clinically significant in homozygotes, some types also in heterozygotes; highly heterogeneous; also interacts with genes for abnormal hemoglobins	Yes	Possible (in fetal blood) but not widely available
Tyrosinemia	Acute liver disease, usually fatal in early childhood	Unknown	Autosomal recessive; at least two forms known	—	—

substrate to a product. Most enzymes (holoenzymes) are composed of a coenzyme (which is often a vitamin) specifically bound to an apoenzyme (the protein portion of the molecule). Usually the coenzyme is the chief component of the active site, which is the part of the enzyme directly involved in the catalytic reaction. Mutational changes could alter the stability of the enzyme by distorting its normal three-dimensional configuration or could change its affinity for either coenzyme or substrate, thus altering the kinetics of the reaction. Other mutations could result in a reduced rate of synthesis, or even failure of synthesis. Whatever the mechanism, the consequence is a metabolic block.

The metabolic block may be in either a biosynthetic or a catabolic pathway, and may be either partial or complete. The immediate biochemical consequences are of two main types:

1. Accumulation of a precursor prior to the block
   a. The accumulated precursor may itself be harmful.
   b. Alternate minor pathways may open, with overproduction of toxic metabolites.
2. Deficiency of product
   a. The product itself may be a substrate for a subsequent reaction, which is then unable to proceed.
   b. The "feedback inhibition" type of control mechanism may be impaired.

Consider the hypothetical sequence of biochemical reactions shown in Figure 5–14 (Rosenberg, 1974). This sequence includes a membrane transport enzyme T_A and three intracellular enzymes E_{AB}, E_{BC} and E_{CD}, each catalyzing a specific reaction in the sequence $A \rightarrow B \rightarrow C \rightarrow D$. A minor pathway $A \rightarrow F \rightarrow G$ exists. The final product of the sequence, D, exerts feedback control over the enzyme E_{AB}.

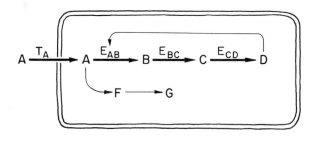

A, B, C, D	–	Substrate and Products of Major Pathway
F, G	–	Products of Minor Pathway
T_A	–	Transport System for A
E_{AB}, E_{BC}, E_{CD}	–	Enzymes Catalyzing Conversion of A to B, B to C, and C to D
‖	–	Cell Membrane

Figure 5–14 A hypothetical sequence of biochemical reactions, discussed in the text. From Bondy and Rosenberg, eds., *Duncan's Diseases of Metabolism.* 7th ed., W. B. Saunders Company, Philadelphia, 1974.

A defect of any of these enzymes will have consequences that can be classified either as precursor accumulation or as product deficiency. These are illustrated below with examples of specific diseases.

1. The membrane transport enzyme T_A is responsible for active transport of precursor A into the cell. In Hartnup disease, there is defective transport of the amino acid tryptophan, which is a precursor of nicotinamide, across the intestinal mucosa. Unless the diet is supplemented with niacin, symptoms of nicotinamide deficiency develop.

2. Deficiency of any of the intracellular enzymes E_{AB}, E_{BC} or E_{CD} could cause accumulation of the immediate precursor. In galactosemia, deficiency of galactose-1-phosphate uridyl transferase causes accumulation of galactose, which is the direct cause of the severe clinical disease. More remote precursors can also accumulate. Thus, in homocystinuria, deficiency of the enzyme cystathionine synthetase causes accumulation both of homocystine, the immediate precursor, and of methionine, a precursor of an earlier reaction.

3. Opening of the minor pathway $A \rightarrow F \rightarrow G$, because of a defect in E_{AB} preventing conversion of A to B, could lead to difficulty if the products of the alternate pathway are toxic. It is believed that in phenylketonuria the severe mental retardation of the disorder results from the toxic effect on the developing brain of abnormal metabolites of alternate pathways of phenylalanine metabolism, which are normally unimportant but become significant when the main pathway is closed.

4. Deficiency of product due to blockage of an earlier synthetic step could have a variety of pathological consequences, depending upon the particular physiological role of the product. There are many examples. One form of albinism results from a failure of melanin formation that is due to deficiency of the enzyme tyrosinase, which normally acts at an earlier stage in the biosynthetic pathway (Fig. 5–15).

Deficiency of product D could also remove feedback inhibition of E_{AB}, causing overproduction of product B. This may be the cause of the great overproduction of uric acid in Lesch-Nyhan syndrome.

The inborn errors of metabolism are clinically very diverse, representing as they do the secondary pathophysiological consequences of a wide assortment of enzyme defects. They range in severity from those that are relatively harmless to those that cause death in early infancy. All are, of course, compatible with life at least to the time of birth; consequently, they must be milder in their effects than the unknown mutations that are lethal in the prenatal period.

The secondary, clinical consequences of metabolic blocks are often unpredictable and hard to interpret in terms of the primary defect. One example, Pompe's disease (glycogen storage disease Type II), is a disorder of early childhood, usually fatal in the first year of life and marked by gross accumulation of glycogen, especially in the heart. The basic defect was a puzzle for many years, but was eventually found to be a specific deficiency of α-1,4 glucosidase, a hydrolytic enzyme of the lysosomes that was not previously thought to be involved in glycolytic degradation at all. In affected infants, glycogen fragments are taken up by the lysosomes but cannot be degraded because the appropriate enzyme is lacking. Pompe's disease is the prototype of the lysosomal storage disorders.

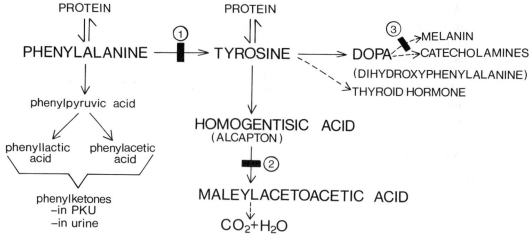

Figure 5–15   Scheme of pathways of phenylalanine and tyrosine metabolism. For discussion, see text.

DISORDERS OF AMINO ACID METABOLISM

Phenylketonuria

Phenylketonuria (PKU) is a disorder of metabolism of the amino acid phenylalanine, which Jervis (1953) found to be caused by a defect in phenylalanine hydroxylase, the enzyme that normally converts phenylalanine to tyrosine as the first step in its degradation, as shown in Figure 5–15. (Note that this statement implies that the structure of phenylalanine hydroxylase is determined by the normal allele at the PKU locus.) The discovery of PKU by Følling in 1934 marked the first demonstration of a genetic defect as a cause of mental retardation. Because phenylalanine is not converted to tyrosine in individuals with PKU, phenylalanine in the diet or formed by normal tissue breakdown is degraded through an alternate pathway, which produces phenylpyruvic acid and other abnormal metabolites. These metabolites are excreted in the urine. The disorder is believed to have an incidence of about one in 10,000 births in North America, so the heterozygote incidence is about one in 50.

The primary effect of the enzyme block is an increased level of phenylalanine in plasma (hyperphenylalaninemia). Severe mental retardation is usually present in untreated patients, who fortunately are now rare. Occasionally, the intelligence is within the normal range. Other neurological features are not marked; there is some accentuation of reflexes, and convulsive seizures may occur. Because melanin synthesis is impaired, affected children may have fairer hair and skin than their normal sibs, but the effect of PKU on pigmentation is variable and depends to some extent on the background genes for pigmentation.

Because children with PKU are normal at birth and become retarded only when they ingest phenylalanine, PKU is a prototype of those inborn errors of metabolism for which mass screening of the newborn population is appropriate. A standard test for PKU has been the ferric chloride test: a few drops of ferric chloride are placed on a wet diaper, and if the urine contains phenylpyruvic acid, a bright green color develops. The color fades to brown within minutes, so the test must be read promptly, and even then false positives or false negatives may occur. A more accurate screening test, the Guthrie bacterial inhibition assay, makes use of the high serum phenylalanine (about 30 times normal) and the principle that the growth of bacteria on agar containing β-2-thienylalanine, a competitive inhibitor of phenylalanine, is a function of the phenylalanine added to the medium in a dried droplet of blood on filter paper. An advantage of the blood test is that it can be used a few days after birth, usually before the infant leaves the hospital, whereas the ferric chloride test is usually not suitable until the baby is a few weeks old. There is also a fluorimetric method for direct determination of phenylalanine in blood after elution from filter paper, and this method is also applicable to mass screening.

The consequence of the metabolic block in PKU may be partly circumvented by reducing the amount of precursor, i.e., by giving the patient a diet low in phenylalanine. Phenylketonuric children are normal at birth because the maternal enzyme protects them during prenatal life. The results of treatment are best when the diagnosis is made soon after birth and treatment is begun promptly. If the child is fed phenylalanine for some time, irreversible mental retardation occurs, apparently because of toxic accumulations of abnormal metabolites of phenylalanine in the brain.

Genetics. PKU is an autosomal recessive disorder, but recently a mild variant has been discovered that is probably caused by a different mutation of the same gene locus responsible for "classic" PKU. There is also a transient type, present in early infancy but not later, and a benign but persistent type. The genetic relationships of these types to classic PKU and the relationship of the degree of retardation to the primary biochemical lesion are not well characterized as yet. Because the gene is not expressed in cell culture, prenatal diagnosis is not feasible at present.

The occurrence of a number of different types of PKU and hyperphenylalaninemia causes difficulty in management. Phenylalanine is an essential amino acid. It is extremely difficult to maintain the concentration in serum at a level high enough to provide for normal growth and development but low enough to prevent brain damage in homozygotes for classic or mild PKU. Fortunately, after the age of about three years the brain has matured sufficiently that rigid adherence to the low-phenylalanine diet is no longer essential.

Maternal Phenylketonuria. PKU can now be treated well enough to allow homozygotes an independent life and almost normal prospects for marriage and parenthood. Because phenylketonurics who have been effectively managed at birth can function adequately in many ways, it has been disconcerting to learn that almost all the children of female phenylketonurics are mentally retarded. Most of these children are heterozygotes (a few are homozygotes, having heterozygous fathers). Their retardation is apparently produced not by their own genetic constitution but by their intrauterine development in mothers who have unduly high plasma phenylalanine levels.

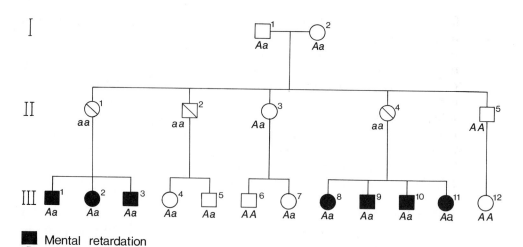

Mental retardation
PKU, controlled

Figure 5–16 Hypothetical pedigree of mental retardation caused by a metabolic disorder, phenylketonuria, in homozygous mothers. Based on Richards, Ann. Hum. Genet. 39:189–191, 1975.

About 25 percent of these children (vs. about 3 percent of controls) also have congenital malformations of various types.

It is possible that the mechanism that leads to mental retardation in the offspring of phenylketonuric mothers is a general mechanism that accounts for certain families in which all the children are retarded. (Fig. 5–16).

Alcaptonuria

In the pathway of phenylalanine metabolism, several other enzymatic blocks are known to be associated with clinical problems. One of these is alcaptonuria. The relationship of these to one another and to albinism is shown schematically in Figure 5–15. Block 1 produces phenylketonuria; 2, alcaptonuria; and 3, albinism. Alcaptonuria is relatively benign. It is characterized by excretion of alcapton (homogentisic acid) in urine, which oxygenates and causes the urine to turn dark on standing. There may also be characteristic depositions of dark pigment in cartilages, which lead to eventual development of mild arthritis (ochronosis).

A DISORDER OF CARBOHYDRATE METABOLISM:

Galactosemia

Galactosemia results from inability to metabolize galactose, a monosaccharide that is a component of lactose (milk sugar). The inheritance is autosomal recessive. Affected infants completely lack the enzyme galactose-1-phosphate uridyl transferase (gal-1-PUT), which normally catalyzes the conversion of galactose-1-phosphate to uridine diphosphogalactose.

Infants with galactosemia are usually normal at birth but begin to develop gastrointestinal problems, cirrhosis of the liver and cataracts as soon as they are given milk. If untreated, galactosemia is usually fatal, though in older children an alternative pathway for galactose metabolism eventually develops. Complete removal of milk from the diet can protect against the harmful consequences of the enzyme deficiency.

In addition to the normal allele Gt^+ and the galactosemia allele gt, there is another well-known allele Gt^D that when homozygous produces the "Duarte variant," with only half the normal enzyme activity but without associated clinical problems. The approximate frequencies, enzyme activities and phenotypes of the six possible genotypes are as follows:

	Frequency (%)	Enzyme Activity (%)	Phenotype
Gt^+Gt^+	91.2	100	Normal
Gt^+Gt^D	7.6	75	Normal
Gt^DGt^D	0.16	50	Normal
Gt^+gt	0.96	50	Normal
Gt^Dgt	0.04	25	Borderline
$gt\ gt$	0.0025	0	Galactosemia

Multiple allelism at a locus with a relatively high frequency of compound genotypes, such as Gt^Dgt in the list above, may be an important source of clinical heterogeneity for many disorders. Unless the affected person is the product of a cousin mating, it is possible that he or she may have a compound genotype rather than a homozygous one.

Heterozygotes and compounds for the various alleles at the galactosemia locus are identified chiefly by assaying for the enzyme in red cells or cultured skin fibroblasts. Because the enzyme is expressed in cultured cells, prenatal diagnosis is feasible. Mass screening for galactosemia in newborns is also feasible and is done in a number of areas though less widely than PKU screening.

LYSOSOMAL STORAGE DISEASES

Tay-Sachs Disease

There are many different disorders in which the characteristic defect is the accumulation of complex substances that in normal cells are degraded to their constituents by specific hydrolytic enzymes segregated within lysosomes. One group of these lysosomal storage disorders is the sphingolipidoses, in which the stored substances are sphingolipids. Gangliosides are a form of sphingolipid in which the basic structure is a ceramide linked to a polysaccharide chain, as shown in Figure 5–17. Tay-Sachs disease (GM_2 gangliosidosis or infantile amaurotic idiocy) is a lysosomal storage disease in which the stored substance is a ganglioside.

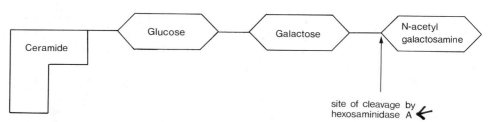

Figure 5-17   Diagram of the structure of GM_2 ganglioside. The ceramide subunit consists of sphingosine linked to a long-chain fatty acid.   ↑accumulates in lysosome

Tay-Sachs disease is one of several different forms of amaurotic familial idiocy, not all of which are gangliosidoses. The term *amaurotic* is derived from *amauros* (dark or obscure) and referred originally to the obscure cause of the progressive blindness that is one of the characteristic features of the disorder. The condition has been mentioned in Chapter 4 as an example of autosomal recessive inheritance.

The enzyme lesion in Tay-Sachs disease is a marked deficiency of the enzyme hexosaminidase A in a wide variety of tissues. The function of this enzyme is to cleave the terminal N-acetyl-β-galactosamine residue from the polysaccharide chain of the ganglioside molecule, and in its absence the ganglioside accumulates, especially in brain tissue. The disorder is much more common in one population (Ashkenazi Jews) than in most other population groups. The frequency is about one affected child in 3600 births among Ashkenazi Jews, and about one in 360,000 in the general population. The frequency of heterozygotes is about one in 30 among the Ashkenazi, one in 300 in other populations. Heterozygotes can readily be detected by screening blood samples for hexosaminidase activity in serum, and the disease can be diagnosed prenatally by biochemical analysis of cultured amniotic fluid cells. Because of these characteristics (high frequency in a particular population group, feasibility of mass screening for heterozygotes and possibility of prenatal detection), Tay-Sachs disease has become the prototype of those inborn errors in which marriages of two heterozygotes can be identified and each pregnancy monitored in order to identify potentially affected fetuses, which in many jurisdictions can then be aborted if the parents so desire. In other words, Tay-Sachs disease has become in theory an almost entirely preventable disease. Many screening programs have already been carried out, both in adults and in high school students, but the long-range effects of these programs have not yet been assessed.

The Mucopolysaccharidoses

Acid mucopolysaccharides are normal constituents of many body tissues, especially connective tissues. The mucopolysaccharidoses (MPS) are a heterogeneous group of lysosomal storage diseases in which the stored substances are partially degraded acid mucopolysaccharides, especially breakdown products of dermatan sulphate and heparan sulphate. They are of particular genetic interest because as a group they include mutations at different loci and more than one mutation at one of the loci (McKusick et al., 1972).

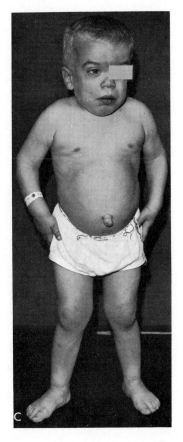

Figure 5–18 A child with Hurler syndrome. From Smith, *Recognizable Patterns of Human Malformation.* 2nd ed., W. B. Saunders Company, Philadelphia, 1976.

The common feature of the structure of mucopolysaccharides is that they are large polymers with a protein core to which are attached extensive polysaccharide branches. In patients with a particular MPS, the protein portion of the molecule has been completely hydrolyzed, but the polysaccharide portion only partly so. Characteristic breakdown products are stored within the lysosomes and appear in large quantities in the urine.

The first two mucopolysaccharidoses to be recognized were the X-linked Hunter syndrome (in three young Winnipeg brothers) in 1917 and the more severe autosomal recessive Hurler syndrome in 1919. Originally each of these conditions was called "gargoylism" because of the coarseness of the facial features (Fig. 5–18). Affected children are mentally retarded and skeletally abnormal with short stature. The corneal clouding that develops in the Hurler but not in the Hunter syndrome is a useful distinguishing sign. Both disorders are progressive, and Hurler syndrome is usually fatal in childhood.

The difference in the pattern of inheritance of these two conditions indicated a biochemical difference, and this was demonstrated in a striking way by Neufeld and Fratantoni (1970), who showed that in cell culture, although fibroblasts from patients of either type accumulated mucopolysaccharides in the culture medium, the accumulation could be corrected by co-cultivation (i.e., by growing cells of both types together in the same culture vessel). This type of experiment is known as a complementation test. The correction factors proved to be lysosomal enzymes, α-iduronidase in the case of Hurler syndrome and sulphoiduronate sulphatase in the case of Hunter syndrome. It has been possible to effect at least temporary clinical improvement by transfusion of normal plasma or leucocytes as a source of the missing enzyme.

Scheie syndrome is clinically much milder than Hurler syndrome, so the finding (by failure to cross-correct in co-cultivation) that the same enzyme deficiency was involved in both was unexpected. The Scheie and Hurler genes appear to be alleles at the same locus but with different effects on α-iduronidase activity.

In contrast, the Sanfilippo syndrome, which was earlier regarded as a single entity, has been split into two nonallelic disorders on the basis of their different enzyme deficiencies and ability to cross-correct. Clinically, they remain indistinguishable. In both, the intelligence is severely impaired, but the physical abnormalities are very mild.

Several other mucopolysaccharidoses have been identified and characterized with varying degrees of completeness. A number of important ones are not listed in Table 5–4. One of these is Morquio syndrome, a form of spondyloepiphyseal dysplasia in which there is excessive excretion of keratan sulphate but not an excess of total mucopolysaccharide excretion. Another is Maroteaux-Lamy dwarfism, which resembles Hurler syndrome in many clinical features but differs in that intelligence is not impaired. Still others are known, and their relationship to previously identified types and to one another is an area of active research.

THE GENETIC HYPERLIPOPROTEINEMIAS

Among the various genetic disorders of lipid metabolism, the genetic hyperlipoproteinemias are of special interest and clinical significance because of their role in myocardial infarction, a major cause of death and disability. Hyperlipoproteinemias are characterized by elevated levels of plasma lipids (cholesterol and triglycerides) and specific plasma lipoproteins. Six genetically distinct forms have been defined, differing in their biochemical and clinical phenotypes, though in some cases the phenotypes are not yet completely defined. These six disorders are believed to result from mutations in single autosomal genes at different loci. At each locus there may well be more than one mutant allele.

To complicate the genetic picture further, it is known that not all hyperlipoproteinemias are caused by single-gene mechanisms; on the contrary, epidemiological studies (Goldstein et al., 1973a and 1973b, Hazzard et al., 1973) have shown that only a minority of persons with significantly elevated plasma lipid and lipoprotein levels have a genetic disorder. In the general population, the background levels are continuously distributed and appear to be determined by a complex interaction of environmental and minor genetic factors. The variable expression of the specific genes responsible for the genetic hyperlipoproteinemias against this multifactorial background makes it difficult to define the phenotypes of the various genetic forms until the basic defect in each is known.

TABLE 5-4 EXAMPLES OF MUCOPOLYSACCHARIDOSES

Syndrome	Enzyme Deficiency	Genetics	Mucopolysaccharides Stored/Excreted	References
Hurler	α-iduronidase	Autosomal recessive	Dermatan and heparan sulphates	Matalon and Dorfman 1972, Bach et al. 1972
Hunter	iduronidate sulphatase	X-linked	Dermatan and heparan sulphates	Bach et al. 1973
Scheie	α-iduronidase	Autosomal recessive	Dermatan and heparan sulphates	Wiesmann and Neufeld 1970, Bach et al. 1972
Sanfilippo A	heparan N-sulphatase	Autosomal recessive	Heparan sulphate	Kresse and Neufeld 1972.
Sanfilippo B	N-acetyl-α glucosaminidase	Autosomal recessive	Heparan sulphate	O'Brien 1972, Figura and Kresse 1972

Only one of these disorders will be discussed. **Familial hypercholes-terolemia** (also known as Type II hyperlipoproteinemia) is an autosomal dominant disorder characterized by elevation of plasma cholesterol carried in the low-density lipoprotein (LDL) fraction. Heterozygotes have premature coronary heart disease, xanthomas (cholesterol-containing tumors) and early development of arcus corneae (deposits of cholesterol around the edge of the cornea). Homozygotes have exceptionally high plasma cholesterol levels and may have clinically significant coronary heart disease in childhood. The homozygous form is rare, but the heterozygous form, with a population frequency of at least one in 500, may be among the most common human single-gene disorders.

Because the disorder is expressed in cultured fibroblasts, the basic defect has been found, and through its analysis much light has been cast on the normal process of cholesterol regulation in the cell. Brown and Goldstein (1974) have shown that fibroblasts from homozygotes exhibit either absence or profound deficiency of the cell surface receptors that normally bind LDL. Heterozygotes have half the normal number of receptors. In normal LDL metabolism, the lipoprotein is taken up from the plasma by LDL receptors, undergoes endocytosis and is hydrolyzed in the lysosomes, with release of free cholesterol from the cholesteryl component of LDL. Cholesterol not required for cellular metabolism may be re-esterified for storage. The free intracellular cholesterol suppresses the activity of the microsomal enzyme 3-hydroxy 3-methylglutaryl coenzyme A reductase (HMG CoA reductase), thus causes reduction of cholesterol synthesis and also activates an enzyme that re-esterifies cholesterol for storage (Fig. 5–19). In familial hypercholesterolemia, since no LDL (or only a small amount) enters the cell, feedback

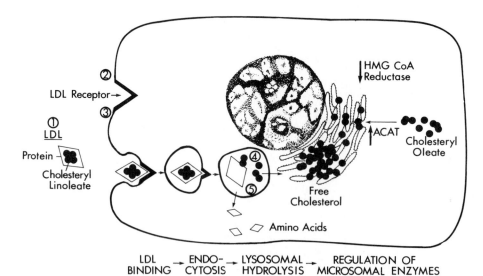

Figure 5–19   The pathways of LDL metabolism in cultured human fibroblasts. The numbers indicate specific known mutations in the pathway. Numbers ② and ③ represents two different forms of familial hypercholesterolemia, one lacking LDL receptors and the other having defective LDL receptors. Numbers ①, ④ and ⑤ represent other single-gene defects in LDL metabolism. ACAT, fatty acyl CoA: cholesterol acyltransferase; other abbreviations in text. From Fredrickson, Goldstein and Brown. *In* Stanbury, Wyngaarden and Fredrickson. *The Metabolic Basis of Inherited Disease*, 4th ed., McGraw-Hill Book Company, New York, 1978, p. 626, by permission.

regulation cannot take place, and cholesterol synthesis is not suppressed, nor are cholesteryl esters formed. Thus there is overproduction of cholesterol, much more severe in homozygotes than in heterozygotes.

Several other mutations are known that have the effect of interfering with cholesterol regulation at different points in the process.

A DISORDER OF PURINE AND PYRIMIDINE METABOLISM:

Lesch-Nyhan Syndrome

Lesch and Nyhan (1964) were the first to describe this rare X-linked disorder, characterized by uric aciduria, cerebral palsy, mental retardation and a compulsive behavior of self-mutilation by gnawing the lips and fingers. The affected boys completely lack the enzyme hypoxanthine-guanine phosphoribosyltransferase (HPRT), which plays a role in the regulation of purine synthesis (Fig. 5–20), converting guanine and hypoxanthine to their respective nucleotides. Deficiency of HPRT leads to overproduction of purines and consequent excessive excretion of uric acid. Not all forms of HPRT deficiency completely lack the enzyme, and a much less severe disorder results if even a very small amount of enzymatic activity remains. HPRT deficiency is of

ROLE OF HYPOXANTHINE-GUANINE PHOSPHORIBOSYLTRANSFERASE
IN PURINE METABOLISM

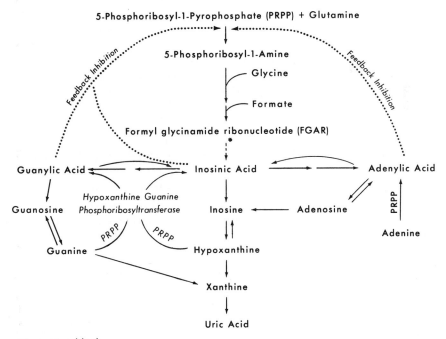

*Azaserine block

Figure 5–20   Feedback control system for regulation of purine synthesis. From Seegmiller et al., Science 155:682, 1967. Copyright © 1967 by the American Association for the Advancement of Science.

particular significance because of the usefulness of the mutation in somatic cell genetics (discussed in Chapter 10).

Many patients who survive for several years with HPRT deficiency develop gouty arthritis. Gout is characterized by recurrent and disabling inflammation, usually involving the first metatarsophalangeal joint. The underlying biochemical finding is hyperuricemia; not all hyperuricemics develop clinical gout, but perhaps 25 percent of them have urate crystals deposited in and around the joints, in the cartilage of the outer ear (tophi) and sometimes in the kidney as kidney stones. The hyperuricemia of gout has a heterogeneous group of causes, but many cases appear to be due to genetic defects of HPRT.

Heterozygotes can be detected, and the defect can also be identified in early prenatal life by assaying for the enzyme in cultured amniotic fluid cells.

A GENETIC DEFECT OF TRANSPORT:

Cystinuria

"Active" transport, as opposed to passive transport, means transport against a gradient and requires energy. Active transport mechanisms are involved in the transport of amino acids and many other substances from the extracellular fluid to the interior of the cell, or across a cell layer such as the gut mucosa or the lining of the proximal tubules of the kidney.

Though membrane transport is by no means fully understood, genetic disorders of transport systems are helping to clarify the process. In amino acid transport, genetically determined proteins in or near the cell surface are thought to bind amino acids, assisting their entry into the cell. These proteins are known as "carriers," "permeases" or simply "reactive sites."

Cystinuria was originally regarded by Garrod as a prototype inborn error of metabolism. Cystinurics form renal calculi composed of cystine and excrete large amounts of cystine in the urine. Although for many years it was believed that the defect in cystinuria was a block in cystine catabolism, with accumulation of cystine prior to the block and its eventual urinary excretion, we now know that the defect is in the reabsorption of cystine by the proximal portion of the renal tubule.

Additional features of cystinuria are that the defect in absorption involves the dibasic amino acids, lysine, arginine and ornithine, as well as cystine, and that the transport defect involves uptake from the gut as well as renal tubular reabsorption.

Genetics.   Though originally cystinuria was considered to be an autosomal recessive disorder, three subtypes are now recognized. Homozygotes for the three types are clinically indistinguishable, but heterozygotes differ. Type I, by far the most common type, is "completely recessive" in that heterozygotes have normal urinary amino acid excretion and cannot be distinguished biochemically from normal individuals. Types II and III are "incompletely recessive," i.e., heterozygotes excrete elevated amounts of cystine

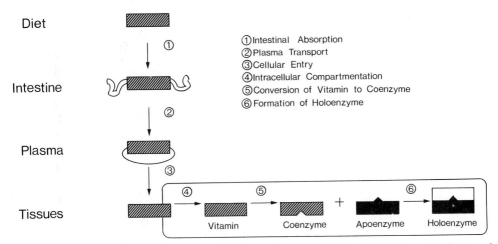

Diet

Intestine

Plasma

Tissues

① Intestinal Absorption
② Plasma Transport
③ Cellular Entry
④ Intracellular Compartmentation
⑤ Conversion of Vitamin to Coenzyme
⑥ Formation of Holoenzyme

Vitamin   Coenzyme   Apoenzyme   Holoenzyme

Figure 5–21 Scheme of sites at which mutations could affect vitamin metabolism and function. From Rosenberg, Vitamin-responsive metabolic disorders. Adv. Hum. Genet. 6:1–74, 1976.

and the dibasic amino acids, and may even form cystine stones. Types II and III are distinguished from one another chiefly by the degree of clinical severity in heterozygotes, which is greater in Type II than in Type III.

The genes for the three types of cystinuria may be allelic, and compound genotypes (i.e., I/II, I/III and II/III) have been described (Rosenberg, 1967). If the different forms are allelic, the same carrier protein must be abnormal in the different types, though the specific alteration varies. However, it is not yet definite that the various mutations are all at the same locus. One alternative possibility is that they affect different subunits of a protein containing two or more different polypeptide chains, a situation analogous to mutations of the α and β hemoglobin genes.

VITAMIN-RESPONSIVE DISORDERS

A category of metabolic disorders is recognized in which the clinical and biochemical abnormalities of the phenotype can be ameliorated by administration of a single vitamin in amounts far above normal physiological requirements. Such disorders are variously described as vitamin-responsive or vitamin-dependent.

Two main groups of vitamins are recognized: fat-soluble (A, E, K and D) and water-soluble. The members of the fat-soluble group vary in their physiological role, but each of the water-soluble ones acts as a coenzyme or, in certain cases, as a coenzyme precursor. Though any vitamin, by definition, cannot be synthesized in the body and must be supplied in the diet, there are still many steps in vitamin transport, metabolism or catalytic function at which a mutation could cause a block. These steps are illustrated schematically in Figure 5–21. Within such a schema, about 25 rare disorders have already been identified, of which some are concerned with vitamin transport and coenzyme synthesis and others with the related apoenzyme. Nearly all are autosomal recessive, though perhaps the first case of a vitamin-responsive disorder to be described was an X-linked dominant, hypophosphatemic rickets (Albright et al., 1937), and there is also an X-linked recessive pyridoxine-responsive anemia. Few of these disorders are as yet thoroughly elucidated at the molecular level. Nevertheless, at least one, cobalamin-responsive methyl-malonicaciduria, has already been diagnosed prenatally (Ampola et al., 1975). In this case, prenatal therapy was instituted by giving the mother large doses of cobalamin, which appeared to cross the placenta and be utilized by the fetus to correct his metabolic abnormality. Whether such a therapeutic approach might be effective or even necessary in other vitamin-responsive disorders remains unknown.

Rosenberg (1976) has reviewed the current state of knowledge of the vitamin-responsive disorders and has pointed out the many genetic and clinical problems posed by them. The heterogeneity within the group is so extensive that at this point few general principles of management can be drawn up. Some, but not all, require neonatal (or earlier) identification and prompt therapy. Some require continuing therapy, but in others it appears that vitamin supplements can eventually be discontinued without ill effects. There is risk of overdosage with possible toxic side effects; in particular, "megavitamin therapy," the indiscriminate use of massive doses of multiple vitamins in the treatment of patients in whom no individual vitamin requirement has been demonstrated, is potentially dangerous.

PHARMACOGENETICS

Pharmacogenetics is the special area of biochemical genetics that deals with drug responses and their genetic modification. Broadly speaking, pharmacogenetics can be said to encompass any genetically determined variation in response to drugs; for instance, the effect of barbiturates in precipitating attacks of porphyria in genetically susceptible persons, and the effect of cortisone in pregnant mice upon the incidence of cleft palate in the progeny. In a narrower sense, pharmacogenetics can be restricted to those genetic variations that are revealed *only* by response to drugs. Classic examples of pharmacogenetic variations include atypical serum cholinesterase, slow isoniazid inactivation, primaquine sensitivity and inability to taste phenylthiocarbamide.

The origin of polymorphisms for drug response and the mechanisms by which they are maintained pose a problem. Obviously they have not developed in response to drugs, since they antedate the drugs concerned. The biotransformation of drugs requires many specific biochemical reactions, and the enzymatic sequences involved may be used in the metabolism of ordinary food substances as well as drugs. For example, an extract of the potato, solanine, is an inhibitor of serum cholinesterase. It is interesting to speculate that possibly, in some ancient age, an atypical serum cholinesterase phenotype might have conferred a selective advantage. Although there is no definite proof that any one genetic polymorphism in drug response has become established in response to a food, it is a reasonable assumption. At least one drug response, primaquine sensitivity, may confer protection against malaria.

Pharmacologists recognize that there is normal variation in response to drugs by defining the "potency" of a drug as that dose that produces a given effect in 50 percent of a population. For genetic traits, continuous variation is usually best explained on the basis of multifactorial inheritance or by a combination of genetic and environmental factors. But response to drugs can also show discontinuous variation, with sharp distinctions between different degrees of response. Discontinuous variation is easier to analyze genetically than continuous variation, and the examples used here are all of the discontinuous type.

SERUM CHOLINESTERASE AND SUCCINYLCHOLINE SENSITIVITY

Serum cholinesterase is an enzyme of human plasma. It has the property of hydrolyzing choline esters, such as acetylcholine, to form free choline and the corresponding organic acid. It used to be known as pseudocholinesterase because its hydrolytic action on acetylcholine is slow, compared to the speed with which "true" cholinesterase, a red cell enzyme, destroys acetylcholine at the neuromuscular junction (endplate).

The function of serum cholinesterase is obscure. It may be a "protective enzyme" because it can hydrolyze some choline esters that in high concentrations inhibit acetylcholinesterase. A low level of serum cholinesterase, or even its complete absence, is fully compatible with normal development and health, so it cannot play a major physiological role.

Succinylcholine (suxamethonium) is a drug widely used as a muscle relaxant in anesthesia and in connection with electroconvulsive therapy. Chemically it is made up of two molecules of acetylcholine, so it is rapidly hydrolyzed by serum cholinesterase. The rapidity of its hydrolysis effectively reduces the amount of succinylcholine that reaches the motor endplates, and this hydrolysis is allowed for in the dosage given to the average patient. Occasionally, however, a patient will respond to the administration of succinylcholine by developing prolonged apnea lasting from one to several hours.

An abnormal response to succinylcholine is not always genetic. Some 50 percent of patients who respond abnormally have an abnormal genotype, but in the remaining 50 percent the underlying problem is a nongenetic, pathological change, or even a technical problem in administration of the anesthetic.

Genetics of Serum Cholinesterase Variants

The activity of cholinesterase in the plasma is determined by two codominant alleles, known as E_1^u and E_1^a. (E_1 signifies the first esterase locus to be described; the superscripts u and a denote genes responsible, respectively, for the usual and atypical forms of the enzyme.) Cholinesterase alteration occurs in persons who are homozygous for the mutant allele E_1^a; the enzyme produced by $E_1^a E_1^a$ homozygotes is qualitatively altered and has lower activity than the usual type. The E_1 locus is on chromosome 1.

Serum cholinesterase phenotypes cannot be determined with certainty on the basis of cholinesterase levels in serum, because the values thus obtained show considerable overlap. Kalow and his collaborators (1957) distinguish between normal and abnormal phenotypes by using an inhibitor of cholinesterase, dibucaine (Nupercaine), which is a well-known local anesthetic. The "dibucaine number" (DN) of a serum sample expresses its percent inhibition by dibucaine. The following relationships exist:

DN	Phenotype	Genotype	Approximate Frequency (Canadian Data)
About 80	Usual	$E_1{}^u E_1{}^u$	0.9625
About 60	Intermediate	$E_1{}^u E_1{}^a$	0.0370
About 20	Atypical	$E_1{}^a E_1{}^a$	0.0005

The Silent Allele. In about 20 percent of atypical individuals, the cholinesterase phenotypes observed cannot be explained by a two-allele hypothesis. Very rarely, individuals have been observed whose serum completely lacks cholinesterase activity. To explain these rare instances, a third allele $E_1{}^s$ (silent) has been postulated. Either $E_1{}^a E_1{}^a$ or $E_1{}^a E_1{}^s$ can produce the atypical phenotype, and $E_1{}^s E_1{}^s$ produces no activity at all.

The "Fluoride-Resistant" Allele. A fourth allele at the E_1 locus, $E_1{}^f$, determines the structure of a type of cholinesterase in which the enzyme is unusually resistant to inhibition by sodium fluoride. Most serum samples are classified identically regardless of whether dibucaine or fluoride is used as the inhibitor, but there are a few rare sera that are resistant to inhibition by fluoride but not by dibucaine. The enzyme determined by the $E_1{}^f$ allele has lower activity than the usual type, and some degree of suxamethonium sensitivity is present in individuals who are $E_1{}^a E_1{}^f$, $E_1{}^s E_1{}^f$ or $E_1{}^f E_1{}^f$.

The E_2 Locus. A locus on chromosome 16 determining cholinesterase activity, E_2, accounts for a serum cholinesterase isozyme that is detectable only by electrophoresis. About 10 percent of Caucasians have a variant allele $E_2{}^+$, which produces an extra isozyme band, C_5 (the four usual isozymes are C_{1-4}), and 25 percent higher activity than the more common $E_2{}^-$ type.

Other Genetic Problems in Anesthesia. Though perhaps the most common and best known, succinylcholine sensitivity is by no means the only example of an abnormal response to anesthesia with a genetic basis. In particular, various genetic forms of malignant hyperthermia have been identified (Kalow and Britt, 1973). In affected persons general anesthesia, especially with halothane, produces hyperpyrexia, which is often associated with hypertonicity of voluntary muscles. In most but not all of the families described, inheritance is autosomal dominant.

TABLE 5-5 RACIAL DIFFERENCES IN SPEED OF INACTIVATION
OF ISONIAZID*

Racial Origin	Percent Rapid Inactivators	Percent Slow Inactivators
American and Canadian white	45.0	55.0
American black	47.5	52.5
Eskimo	95.0	5.0
Latin American	67.0	33.0

*After Kalow, 1962.

ISONIAZID METABOLISM

Isoniazid is a drug used in the treatment of tuberculosis. The rate of inactivation of isoniazid after a test dose shows polymorphism, either in the quantity of free isoniazid excreted in the urine during the following 24 hours or in the plasma concentration after six hours. "Slow inactivators" are homozygous for a recessive gene that is believed to lead to lack of the hepatic enzyme acetyltransferase, the enzyme that normally acetylates isoniazid as one step in the metabolism of this drug. "Rapid inactivators" are normal homozygotes or heterozygotes. The same enzyme acetylates the drugs sulfamethazine, hydralazine and sulfisoxazole, but not all sulfa drugs; sulfanilamide, for example, has been shown to be acetylated in the red blood cells, not in the liver.

The gene for slow inactivation shows marked differences in geographic distribution (Table 5–5). It is very low in Eskimos, for example; but up to 80 percent of African blacks and of some European populations are either heterozygous or homozygous for the slow-inactivation gene.

Surprisingly, the isoniazid inactivation phenotype appears to have no influence upon the response of patients with tuberculosis to isoniazid treatment. However, it does seem to have a bearing on the development of side effects; slow inactivators are more likely than rapid inactivators to develop polyneuritis as a complication of isoniazid treatment. Slow inactivators develop toxic reactions at the dosage necessary to maintain adequate blood levels in fast inactivators, and this should be considered when isoniazid is used in the treatment of tuberculosis. On the other hand, there is a danger that rapid inactivators (a group that includes most of the Canadian native populations) may not always be able to maintain adequate blood levels of the drug unless they receive their medication on a carefully regulated schedule (Jeanes et al., 1972).

GLUCOSE-6-PHOSPHATE DEHYDROGENASE VARIANTS

Primaquine is a drug used in the treatment of malaria. From the time it was first introduced, it was known to be capable of inducing hemolytic anemia in some patients, especially in black males. Further investigation of this phenomenon showed that the red cells of primaquine-sensitive subjects are deficient in glucose-6-phosphate dehydrogenase (G6PD), an ubiquitous en-

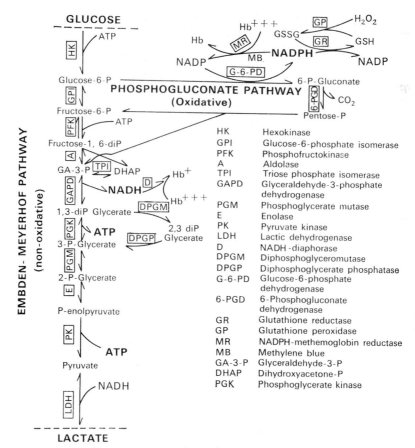

Figure 5-22 Glycolysis in the red cell. G6PD catalyzes the oxidation of G6P to 6PG in the hexosemonophosphate shunt pathway. From Giblett, *Genetic Markers in Human Blood.* Blackwell Scientific Publications, Oxford, 1969, p. 445, by permission.

zyme involved in glucose metabolism. The trait is inherited as an X-linked recessive.

Favism, a severe hemolytic anemia in response to ingesting the broad bean *Vicia faba*, has been known since ancient times in parts of Italy, and its hereditary nature has been recognized. The basis of favism, like that of primaquine sensitivity, is G6PD deficiency.

Primaquine sensitivity and favism are only two of well over 100 genetic variants of G6PD, many of which are associated with some degree of deficiency of activity of the enzyme. About 100 million people throughout the world have G6PD deficiency. For a deleterious gene to reach such a high population frequency, a selective advantage must be postulated, and it is believed that many G6PD variants, like sickle cell hemoglobin, may confer some protection against malaria.

The specific biochemical role of G6PD is in the hexosemonophosphate shunt pathway, which is the minor pathway for red cell glycolysis (Fig. 5–22). Ninety percent of red cell glycolysis takes place through the anaerobic pathway, in which no enzyme polymorphisms are known. It may be that as a general rule, polymorphisms are biologically more acceptable in minor pathways than in major ones.

TABLE 5–6   COMMON G6PD PHENOTYPES

Gd Type	Gene Symbol	Electropho-retic Mobility	Enzyme Activity (% normal, approx.)	Approximate Population Distribution
B	Gd^B	Normal	100	Normal
A	Gd^A	Fast	90	20% of American black males
A–	Gd^{A-}	Fast	15	10% of American black males
B–	Gd^{B-}	Normal	4	Common in Mediterranean areas

The normal and common abnormal variants of G6PD are listed in Table 5–6. The normal is known as type B, which is the conventional label given to normal electrophoretic variants of a number of genetic markers. About 20 percent of American black males have the faster, nearly normal A variant. The two deficient types migrate electrophoretically at the same rate as A and B but have much lower activity, and so are called A– and B– respectively. Because the Gd locus is on the X chromosome, males have only a single *Gd* gene, but heterozygous females (e.g., AB heterozygotes) show two bands. The electrophoretic patterns are shown in Figure 5–23.

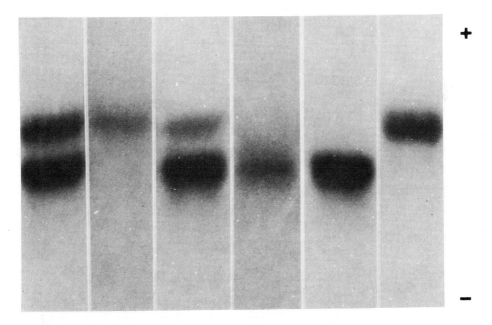

BA   A-   BA-   B-   B   A

Figure 5–23   Starch gel electrophoretic patterns of six different G6PD phenotypes in red blood cell hemolysates. From Giblett, *Genetic Markers in Human Blood.* Blackwell Scientific Publications, Oxford, 1969, facing p. 452, by permission.

X Inactivation

The Lyon hypothesis of X inactivation is discussed elsewhere, but must be briefly noted here. In females, an X chromosome is randomly inactivated early in embryonic life, and this event leads to mosaicism in heterozygous females and also provides a mechanism for "dosage compensation," i.e., for equalizing the amount of product of X-linked genes in males and females. Because G6PD is X-linked and its variants are so numerous and relatively common, it has been very useful in testing the Lyon hypothesis.

Linkage Relations

The Gd locus is very close to the locus for hemophilia A, and to the protan and deutan loci for red-green color blindness, but is distant from the Xg blood group locus. Because variants at the locus are so frequent, it is one of the most useful markers on the X chromosome.

Clinical Aspects

Many of the G6PD variants are associated with a high risk of hemolytic crises on exposure to certain drugs that stress the shunt pathway. These drugs include sulfanilamide, trinitrotoluene, quinidine, primaquine, naphthalene (moth balls) and many others. When an infection or some other precipitating factor is present, an even longer list of drugs can cause hemolysis — for example, acetylsalicylic acid. It is noteworthy that the A– type of deficiency is much less susceptible to hemolysis that the B– type; for example, favism is common in the Mediterranean region but almost unknown in blacks. Deficient infants of the Mediterranean type can even develop hemolytic crises if dressed in clothes that have been in moth balls and not thoroughly cleaned.

Congenital nonspherocytic hemolytic anemia associated with very low enzyme activity is a consequence of some of the rare variants, especially in whites.

Though G6PD deficiency is far more common in males, it is not impossible for females to inherit two abnormal alleles and consequently to be affected. About 2 percent of American black females are genetically $Gd^{A-}Gd^{A-}$ and clinically susceptible to drug-induced hemolysis.

TASTE SENSITIVITY TO PTC

The human polymorphism for taste sensitivity in response to the chemical phenylthiourea (also known as phenylthiocarbamide and, commonly, as PTC) has already been described as a classic example of a difference determined by a single pair of alleles, designated as T and t, t being recessive to T. Tasters are TT or Tt, and nontasters are tt. About 70 percent of whites are tasters and 30 percent are nontasters.

When taste sensitivity to PTC was analyzed more precisely, it became apparent that discrimination between the two phenotypes is not absolute. Harris and Kalmus (1949) have therefore designed a more elaborate test, in which the ability to taste PTC at different dilutions is measured. By this test it is possible to show that the threshold of taste sensitivity is distributed over a

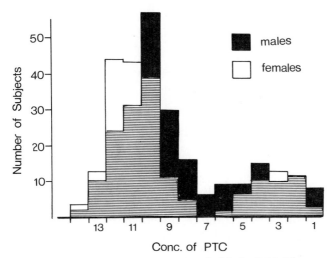

Figure 5–24 Population distribution of taste thresholds for PTC. The concentration of PTC was 0.13 percent in solution 1, and each successive solution was diluted to half the strength of the preceding one. Redrawn from Kalow, *Pharmacogenetics: Heredity and the Response to Drugs.* W. B. Saunders Company, Philadelphia, 1962, p. 121. Data from Leguebe, 1960.

rather wide range of variation, but is bimodal (Fig. 5–24). About two-thirds of subjects fall in the group able to taste very dilute concentrations of PTC, and the remainder can taste it only in very high concentrations. Some overlap exists between the two groups, so a few people cannot be classified accurately. The distribution of taste levels within each genotype probably depends on modifying genes. There is also a slight sex difference in taste sensitivity, with a slightly higher frequency of tasters among females than among males, and a slight loss in sensitivity with increasing age.

A number of other drugs (e.g., thiouracil and its derivatives) evoke a similar bimodal taste response. The chemical grouping $NC{=}S$ is found in all these drugs, including PTC, and so is believed to be responsible for the taste response.

The distribution of PTC sensitivity varies strikingly in different populations. Nontasters are almost unknown among American Indians and Eskimos and are rare among blacks. Some typical figures are given in Table 5–7. An intrepid group of investigators went so far as to test PTC sensitivity in anthropoid apes and reported that of 27 chimpanzees, 20 were tasters. By chance or otherwise, this is close to the relative frequency of the two genotypes in whites.

TABLE 5–7 RACIAL VARIATION IN THE ABILITY TO TASTE PTC

Race	Percent Tasters
American Indian	90–98
American black	91
American white	65–75
African black	91–97
Chinese	89–94

TABLE 5-8   ASSOCIATION OF ABILITY TO TASTE PTC
AND THYROID DISEASE

Condition	Percent Tasters
Normal	70
Nodular goiter	60
Athyrotic cretinism	15

PTC occurs in nature in some plants and may be found in the milk of cattle pastured on these plants. PTC is therefore a normal constituent of food, but its concentration is below the taste threshold even for tasters.

The substances most similar to PTC are goitrogenic (i.e., antithyroid) drugs, and PTC itself has been shown to have a goitrogenic effect in mice. This observation has led to research into the PTC sensitivity of patients with various types of thyroid disease, and several curious relationships have been revealed. Nontasters are more frequent than expected among patients with hypothyroid types of disorder (athyrotic cretinism and nodular goiter) (Table 5-8). It may be that the PTC-tasting polymorphism has some relation to protection against thyroid disease, but the mechanism, if it exists, is unknown.

GENERAL REFERENCES

Bondy, P. K., and Rosenberg, L. E., eds., 1974, *Duncan's Diseases of Metabolism*. 8th ed. W. B. Saunders Company, Philadelphia.

Gardner, L. I., ed., 1975. *Endocrine and Genetic Diseases of Childhood and Adolescence*. 2nd ed. W. B. Saunders Company, Philadelphia.

Giblett, E. R. 1969. *Genetic Markers in Human Blood*. Blackwell Scientific Publications Ltd., Oxford.

Harris, H. 1975. *The Principles of Human Biochemical Genetics*. 2nd ed. North-Holland Publishing Company, Amsterdam and London; American Elsevier Publishing Co., Inc., New York.

Kalow, W. 1962. *Pharmacogenetics: Heredity and the Response to Drugs*. W. B. Saunders Company, Philadelphia.

McKusick, V. A. 1978. *Mendelian Inheritance in Man: Catalogs of Autosomal Dominant, Autosomal Recessive and X-Linked Phenotypes*. 5th ed. Johns Hopkins University Press, Baltimore.

Stanbury, J. B., Wyngaarden, J. B., and Fredrickson, D. S., eds. 1978. *The Metabolic Basis of Inherited Disease*. 4th ed. McGraw-Hill Book Company, New York.

Weatherall, D. J., ed., 1976. Haemoglobin: Structure, function and synthesis. Br. Med. Bull. 32(3).

a good reference for Biochemical disease

PROBLEMS

1. A man is heterozygous for Hb M Saskatoon. His wife is heterozygous for Hb M Boston. Heterozygosity for either of these alleles produces methemoglobinemia. Referring to Table 5-2 for details, outline the possible genotypes and phenotypes of their offspring.

2. A woman whose father had G6PD deficiency of the B— type marries a man from a population in which the frequency of the A— type of G6PD deficiency is 5 percent.
   a) What is the probability that their son would be G6PD-deficient?
   b) What type of G6PD deficiency would an affected son have?
   c) What is the probability that a child of these parents would have a compound genotype?

3. In view of the lethal effect of sickle cell anemia, what are the phenotypes of the parents of affected children?

4. If the brother of a child who dies of sickle cell anemia marries the sister of another such child, what is the probability that their firstborn child will be affected?

5. A man has sickle cell trait and his wife is heterozygous for HbC. What is the probability that their child has no abnormal hemoglobin?

6. A woman is the sister of a boy who died in childhood of a disorder described as "gargoylism," but the exact nature of his disorder was unknown. The woman's husband's brother has Scheie syndrome. Referring to Table 5–3, answer the following questions:
   a) If the first child of this couple is a male with Hunter syndrome, state the most likely genotypes of the child, his parents and his maternal uncle.
   b) If the first child of this couple has a disorder somewhat resembling Hurler syndrome but less severe with demonstrable α-iduronidase deficiency, state the most likely genotypes of the child, his parents and his maternal uncle.

7. If all fetuses homozygous for Tay-Sachs disease were aborted, what would be the expected effect on the population frequency of the Tay-Sachs gene?

6

CHROMOSOMAL ABERRATIONS

A new era in medical genetics opened in 1959 with the demonstration by Lejeune and his colleagues that "enfants mongoliens" (children with mongolism or, as it is now commonly called, Down syndrome) have 47 chromosomes instead of the usual 46 in their body cells. Cytogenetic abnormalities are much more frequent and varied than was originally anticipated. Chromosomal aberrations are a significant cause of birth defects and fetal loss, being present in an estimated 0.7 percent of live births, half of all spontaneous abortions, and about 7.5 percent of all conceptuses. There are many excellent reviews of medical cytogenetics, some of which are listed in the General References of this chapter.

Patients with chromosomal aberrations usually have characteristic phenotypes and often look more like other patients with the same karyotype than like their own brothers and sisters. The phenotypic anomalies result from genetic imbalance, which perturbs the normal course of development, but little is known about how any specific imbalance produces its phenotypic consequences. Developmental retardation and multiple dysmorphic features are common to all the autosomal aberrations, regardless of which chromosome is involved and whether the chromosomal material is deleted or present in excess. Balanced structural rearrangements, in which the genetic material is all present but abnormally arranged, are usually associated with normal phenotypes, though there is an excess of *de novo* balanced rearrangements in mentally retarded populations.

Although in any two people the majority of the genes on any chromosome are probably identical, each of us carries an assortment of variant genes, either for common polymorphisms or for rare mutations. Therefore, the general phenotypic similarity of subjects with a specific chromosome abnormality is modified by individual differences determined by the individual genotype. If the abnormality is the loss of part or all of a chromosome, the subject is then hemizygous for all the gene loci that have been deleted, and any recessive mutants on the chromosome or chromosome segment homologous to the deletion are then expressed. This may explain why, in general, monosomy is more damaging than trisomy.

142

CLASSIFICATION OF CHROMOSOMAL ABERRATIONS

Abnormalities of the chromosomes may be either numerical or structural and may affect either autosomes or sex chromosomes or, rarely, both simultaneously. A given abnormality may be present in all body cells, or there may be two or more cell lines, one or more of which are abnormal; the latter situation is termed mosaicism. In this section the more common types of chromosomal aberration are defined, and the terms used to describe them are introduced.

NUMERICAL ABERRATIONS

Numerical changes arise chiefly through the process of **nondisjunction** (failure of paired chromosomes or sister chromatids to disjoin at anaphase, either in a mitotic division or in the first or second meiotic division) (Fig. 6–1). **Anaphase lag,** when the members of a chromosome pair fail to synapse and therefore do not move apart correctly on the spindle, is a type of nondisjunction that can result in one member of the pair failing to be included in either daughter cell.

Any species has a characteristic chromosome number (in man, $2n = 46$ and $n = 23$). Any number that is an exact multiple of the haploid number is **euploid.** Euploid numbers need not be normal; $3n$ (triploid) and $4n$ (tetraploid) chromosome numbers are known in man, though only a few triploids have been born alive and tetraploids have been seen only in early abortuses. Chromosome numbers such as $3n$ and $4n$, which are exact multiples of n but greater than $2n$, are said to be **polyploid.** Polyploidy can arise by a variety of mechanisms; triploidy probably results from failure of one of the maturation divisions, either in ovum or sperm, and tetraploidy results from failure of completion of the first cleavage division of the zygote.

Any number that is not an exact multiple of n is **aneuploid.** Some types of aneuploids are **trisomics,** with $2n+1$ chromosomes and three members of one

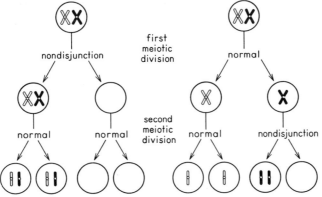

Figure 6–1   Nondisjunction occurring at the first and second meiotic divisions. Nondisjunction at meiosis I results in gametes with both members of a chromosome pair or with neither member. Nondisjunction at meiosis II results in gametes containing (or lacking) two identical chromosomes both derived from the same member of the homologous pair.

particular chromosome, as in Down syndrome; **monosomics,** with $2n-1$ chromosomes and only one member of some chromosome; double trisomics $(2n+1+1)$, with an extra member for each of two chromosomes and so forth.

Any chromosome number that deviates from the characteristic n and $2n$ is **heteroploid,** whether it is euploid or aneuploid.

Aneuploidy

The chief cause of aneuploidy is nondisjunction in a meiotic division, leading to unequal distribution of one pair of homologous chromosomes to the daughter cells so that one daughter cell has both and the other has neither chromosome of a pair. Failure of pairing of two homologous chromosomes, followed by their random assortment rather than segregation, can have the same result.

If nondisjunction of chromosome 21 occurs at meiosis, the gametes formed have either an extra chromosome 21 ($n+1=24$ chromosomes in all) or one too few ($n-1=22$). Fertilization of the 24-chromosome gamete by a normal gamete produces a 47-chromosome zygote, trisomic for chromosome 21.

Nondisjunction can take place at either the first or the second meiotic division. The consequences are rather different (Fig. 6–1). If nondisjunction occurs at meiosis I, the gamete with $n+1$ chromosomes will contain both the paternal and the maternal representatives of that chromosome; if it involves the two chromatids of one chromosomes at meiosis II, the gamete with $n+1$ chromosomes will contain a double complement of *either* the paternal *or* the maternal chromosome. (This simplified explanation omits the effect of meiotic crossing over on the gene content of the chromosomes.) Nondisjunction can occasionally occur at successive meiotic divisions, or in both male and female gametes, so that zygotes with bizarre chromosome numbers may be formed, although these "multisomics" have been described only with respect to the X chromosome. Mitotic nondisjunction also produces multisomy in some malignant cell lines and some cell cultures. Double aneuploidy (trisomy for two different chromosomes at once) has been observed.

Nondisjunction can also occur at *mitosis* after formation of the zygote, in which case the nondisjoining objects are the chromatids of a single chromosome, as in meiosis II. If this happens at an early cleavage division, a trisomic and a monosomic cell line are established; the trisomic one might persist, but the monosomic one ususally does not. Again the X chromosomes are an exception, because lines with a single X are viable.

ABERRATIONS OF CHROMOSOME STRUCTURE

Much of our knowledge of human structural aberrations is based on work with other organisms, especially the fruit fly and certain plants. Structural rearrangements result from chromosome breakage, followed by reconstitution in an abnormal combination. Chromosome breaks occur normally at a low frequency, but may also be induced by a wide variety of breaking agents (clastogens) such as ionizing radiation, some virus infections and many chemicals.

The changes in chromosome structure resulting from breakage may be either stable (i.e., capable of passing through cell division unaltered) or unstable. The stable types of aberration are **deletions, duplications, inversions, translocations, insertions** and **isochromosomes.** The unstable types, which fail to undergo regular cell division, are **dicentrics, acentrics** and **rings.** These aberrations are shown diagrammatically in Figure 6–2 and elsewhere in this chapter.

Deletion

Deletion is loss of a portion of a chromosome, either terminally following a single chromosome break, or perhaps more often interstitially between two breaks. The deleted portion, if it lacks a centromere, is an acentric fragment, which because it has no centromere fails to move on the spindle, and is eventually lost at a subsequent cell division. The structurally abnormal chromosome lacks whatever genetic information was present in the lost fragment. The common example of deletion in humans is the *cri du chat* syn-

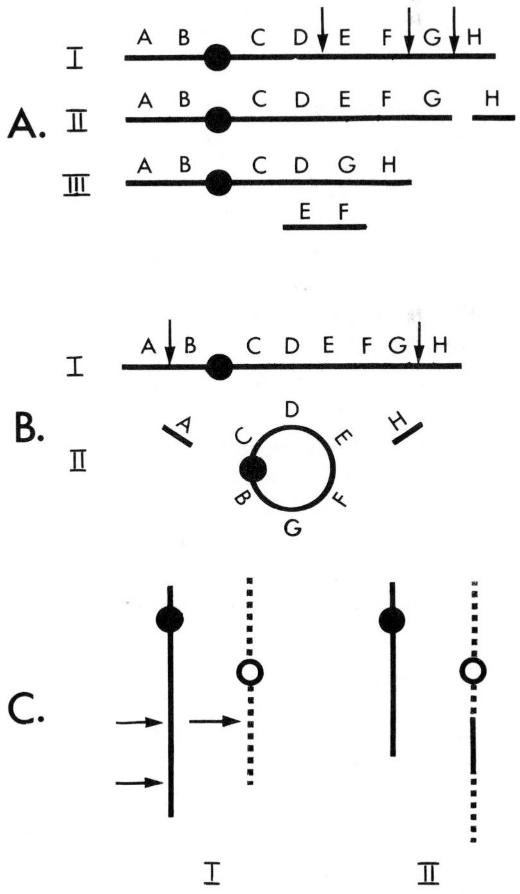

Figure 6–2 Structural rearrangements of chromosomes.
A. Deletion. I: Arrows indicate break points. II: Single break produces acentric fragment, usually lost in a subsequent division because it lacks a centromere. III: Interstitial deletion of fragment EF, with reconstitution of the deleted chromosome.
B. Ring chromosome. A type of deletion in which terminal fragments are deleted and the broken ends of the chromosome reunite to form a ring.
C. Insertion. I: Arrows show break points. II: Deleted segment is inserted into a nonhomologous chromosome.

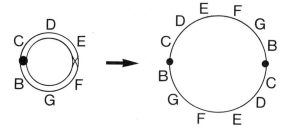

Figure 6-3 Origin of a double-sized dicentric ring after sister chromatid exchange (SCE) in mitosis at X between E and F. Ring opens up to form dicentric ring, which may be broken at anaphase.

drome, in which part of the short arm of chromosome 5 is deleted (see later).

Figure 6–2A shows a chromosome in which the order of the genes is given as ABCDEFGH (I). A terminal deletion might produce a chromosome lacking H (II). A deletion produced by two breaks, between D and E and between F and G with loss of the fragment between them, might produce a chromosome ABCDGH (III). If the deleted portion does not involve the centromere, the chromosome replicates and divides in the normal way at subsequent divisions, but the acentric fragment will probably be lost.

A ring chromosome is a type of deletion chromosome in which both ends have been lost and the two broken ends have reunited to form a ring (Fig. 6–2B). If it has a centromere, a ring chromosome can pass through cell division, but it may undergo alteration in structure (Fig. 6–3).

Duplication

Duplication is the presence of an extra piece of chromosome, which usually has originated by unequal crossing over. The reciprocal product is a deletion. Duplications are more common and much less harmful than deletions. In fact, small duplications ("repeats") may be an evolutionary mechanism for the acquisition of new genes, which may then evolve into genes with quite different functions from the genes from which they originated. Duplications of whole genes or, much less frequently, of parts of genes are considered to be important factors in evolution (see hemoglobin loci, Lepore hemoglobin, haptoglobin).

Duplication of parts of chromosomes may occur as a consequence of various structural rearrangements. For example, if a patient is a translocation heterozygote, unbalanced gametes may be formed which have, in effect, a duplication of one segment and deletion of another (Fig. 6–4). Partial duplications also result from crossing over in inversion heterozygotes, or from isochromosome formation (see later).

Inversion

Inversion involves fragmentation of a chromosome by two breaks, followed by reconstitution with inversion of the section of the chromosome between the breaks (ABCDEFGH might become ABCFEDGH). If the inversion is in a single chromosome arm it is **paracentric** (beside the centromere), but if it involves the centromere region it is **pericentric** (around the centromere).

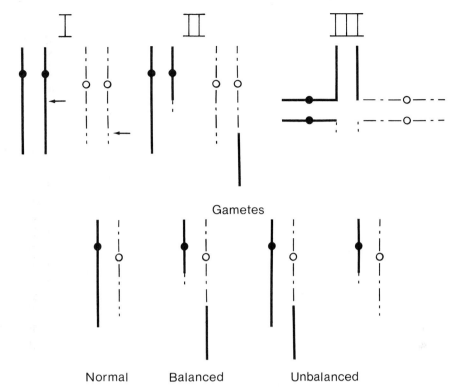

Gametes

Normal   Balanced   Unbalanced

Figure 6–4 Reciprocal translocation and its consequences. Above, origin by breakage of two nonhomologous chromosomes (I), and reconstitution with the broken ends interchanged (II). At meiosis, a cross-shaped figure is formed (III). Below, gametes formed by this translocation heterozygote: normal, balanced and unbalanced. It is also possible, though less common, for segregation to occur so that both homologous centromeres pass to one gamete, again producing unbalanced gametes. Normal or balanced gametes result in phenotypically normal progeny, but unbalanced gametes result in zygotes that are partly trisomic and partly monosomic and consequently develop abnormally.

Because inversions interfere with pairing between homologous chromosomes in inversion heterozygotes, crossing over may be suppressed within them. This can lead to the retention by a species of groups of genes that can then evolve as units. Inversions are therefore of evolutionary significance.

Usually a change in gene order produced by an inversion does not lead to an abnormal phenotype. The medical significance of inversions is for the subsequent generation. It arises from the consequences of crossing over between a normal chromosome and one with an inversion (Fig. 6–5).

For the homologous chromosomes to pair, one of them must form a loop in the region of the inversion. If the inversion is paracentric, the centromere lies outside the loop. When a crossover occurs within the loop, a dicentric chromatid and an acentric fragment are formed, as well as a normal and an inverted one. Both aberrations are unstable. If the inversion is pericentric, the centromere lies within the loop. If a crossover now takes place, each of the two chromatids involved in the crossover has both a duplication and a dele-

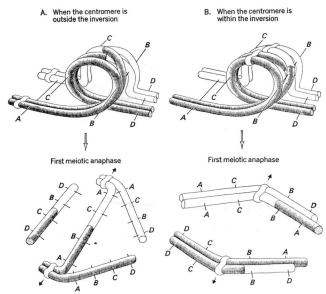

A. When the centromere is
outside the inversion

B. When the centromere is
within the inversion

First meiotic anaphase

First meiotic anaphase

Figure 6–5 Crossing over within the loop of an inversion heterozygote results in gametes with duplications or deficiencies as well as normal and balanced gametes. From Srb, Owen and Edgar, *General Genetics*. 2nd ed., San Francisco, W. H. Freeman & Company. Copyright © 1965.

tion. If gametes are formed with these abnormal chromosomes, the resulting progeny will be monosomic for one part of the chromosome and trisomic for another part.

Translocation

Translocation is the transfer of part of one chromosome to a nonhomologous chromosome. The process requires breakage of both chromosomes, with repair in an abnormal arrangement. Translocations are often, but not always, reciprocal. A balanced translocation does not necessarily lead to an abnormal phenotype, but, like inversions, translocations can lead to the formation of unbalanced gametes and therefore carry a high risk of abnormal offspring.

Robertsonian translocations are a special type in which the breaks occur at the centromeres and whole chromosome arms are exchanged. This process is also called centric fusion. When it occurs in man it involves two acrocentric chromosomes, e.g., t (21q22q) or t (14q21q). The relation of translocation to Down syndrome is described later.

An **insertion** is a type of translocation in which a broken part of a chromosome is inserted into a nonhomologous chromosome (Fig. 6–2C). This process requires three breaks, and is relatively unusual but has been identified in man.

Gametogenesis in Heterozygotes for Reciprocal Translocations. Because translocations interfere with normal chromosome pairing and segregation at meiosis I, they can lead to unbalanced gametes and unbalanced offspring. The consequences of gametogenesis in an individual carrying a reciprocal translocation are shown in Figure 6–4. The two normal and the two translocated chromosomes synapse as a cross-shaped figure, which may open up into a ring or chain unless the arms of the chromosomes are held together by chiasmata.

The most frequent types of gametes formed include a normal combination, an abnormal but balanced combination and two abnormal unbalanced combinations. The first two types can lead to normal progeny, but the last two can produce unbalanced progeny with a duplication and a deletion of parts of chromosomes. There is also an increased risk of nondisjunction in translocation heterozygotes.

Gametogenesis in Carriers of Robertsonian Translocations. A model of synapsis and disjunction in carriers of Robertsonian translocations is shown in Figure 6–6. The clinical significance of this phenomenon is that carriers of translocations between a D and a G chromosome, [symbolized as t(DqGq)], or between two G's [t(GqGq)] have a high risk of producing offspring with Down syndrome. In these carriers, one of the chromosomes involved in the translocation is always a 21; the D chromosome is usually 14, sometimes 15 and only rarely 13; and in t(GqGq) the second chromosome is nearly always 22. The theoretical risk is that one-third of the offspring of such translocation carriers have Down syndrome, but as noted later the observed risk is much lower. The 21q21q translocation is a special case, because all the offspring inevitably have either Down syndrome or monosomy 21, which is usually lethal in early development.

Consider gametogenesis in a carrier of a translocation involving 14q and 21q. The carrier has 45 chromosomes including t(14q21q) and is phenotypically normal, though the short satellited arms of both chromosomes are missing. Theoretically, he or she forms six types of gametes in equal proportions:

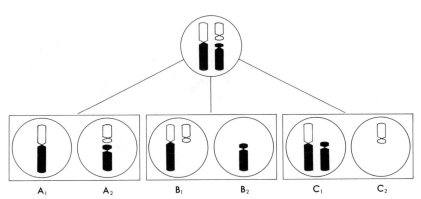

Figure 6–6 Gametogenesis in a carrier of a Robertsonian translocation involving the long arms of chromosome 14 (black) and chromosome 21 (outline). For details, see text.

A$_1$ Balanced, 22 chromosomes including t(14q21q)
A$_2$ Normal, 23 chromosomes

B$_1$ Abnormal, 23 chromosomes including t(14q21q) and 21
B$_2$ Abnormal, 22 chromosomes with no 21

C$_1$ Abnormal, 23 chromosomes including t(14q21q) and 14
C$_2$ Abnormal, 22 chromosomes with no 14

Two of these types (B$_2$ and C$_2$) lack rather large amounts of chromosomal material. Assuming union with a normal gamete, they would produce zygotes monosomic for chromosome 21 and chromosome 14 respectively. These are inviable (except for rare instances of monosomy 21 in liveborn infants). C$_1$ would lead to a zygote with trisomy 14, which also seems to be inviable but is responsible for many spontaneous abortions.

The three remaining types of gametes can produce viable offspring. Type A$_2$ is entirely normal. Type A$_1$ leads to a phenotypically normal balanced translocation carrier, like the parent. Type B$_1$ produces translocation Down syndrome, that is, a child with 46 chromosomes one of which is a 14q21q translocation. The karyotype is fully described as 46, 14−, t(14q21q)+, indicating that there are 46 chromosomes in all, a chromosome 14 is missing, and a Robertsonian translocation of the long arms of a 14 and a 21 is present.

Isochromosomes

During cell division the centromere of a chromosome sometimes mistakenly divides so that it separates the two arms rather than the two chromatids (Fig. 6–7). The chromosomes so formed are isochromosomes. The most common kind of isochromosome is for the long arm of the X and is designated Xqi. A woman with a normal X and an Xqi is monosomic for the genes on the short arm of the X and trisomic for the genes on the long arm. About 15 to 20 percent of women with Turner syndrome have this karyotype (see Chapter 7).

MOSAICS

If nondisjunction happens at an early cleavage division of the zygote rather than during gametogenesis, an individual with two or more cell lines with different chromosome numbers is produced. Such individuals are termed **mosaics.**

A chromosomal mosaic has at least two cell lines, with different karyotypes, derived from a single zygote. The alterations in the karyotype may be either numerical or structural. Many different mosaics have been described, most of which have cell lines with different sex chromosome constitutions. About 1 percent of patients with Down syndrome have a mixture of 46-chromosome and 47-chromosome tissues.

The proportion of normal and abnormal cells may vary from tissue to tissue within the same patient, as well as from patient to patient. It may also

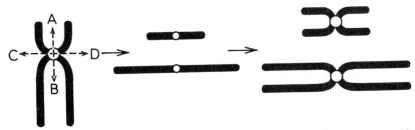

Figure 6–7 Possible means of formation of isochromosomes. The centromere divides in the plane C-D instead of the normal plane A-B. Both products are metacentric chromosomes, each with duplication of one arm and deficiency of the other arm of the original chromosome.

change during development; patients are known in whom mosaicism was apparent in lymphocyte cultures at birth but had disappeared a few months later. On the average mosaics are less severely affected than their nonmosaic counterparts.

There are many practical difficulties in the investigation of mosaicism. The only tissues that are studied frequently are peripheral blood, marrow and skin, and only blood can be repeatedly sampled, so even if mosaicism cannot be demonstrated, it cannot definitely be ruled out. Because normal and abnormal cells may survive and multiply at different rates in culture, the relative proportions in a chromosome culture may not mirror closely the proportions in the patient. In fact, mosaicism may arise independently *in vitro*, and this can cause a serious problem in interpretation if it happens in an amniotic fluid cell culture (Rudd et al., 1977). In any case, what is seen at the time of study may be quite different from what was present during the critical period of development, the first trimester of prenatal life.

CAUSES OF CHROMOSOMAL ABERRATIONS

Although by now the mechanisms that produce aberrations of chromosome number and structure are understood in a general way, little is known about the predisposing genetic and environmental factors. Figure 6–8 demonstrates the nonrandom nature of chromosome abnormalities and strongly suggests a common underlying mechanism or mechanisms.

Late maternal age is a major factor in the etiology of Down syndrome and, to a lesser extent, of the other trisomies, but the reason for the correlation of late maternal age with the nondisjunctional event that underlies Down syndrome is unknown. Because paternal age probably has no effect upon nondisjunction, attempts have been made to explain it in terms of the differences between the processes of oogenesis and spermatogenesis, described in Chapter 2. However, the basis of the nondisjunction is still obscure. The risk of Down syndrome in live births and in fetuses, at different maternal ages, is shown in Table 6–1.

Genes predisposing to nondisjunction probably exist in man, since such genes are known in other organisms. There are case reports of patients trisomic for two different chromosomes at once (double aneuploids), and pedigrees with clusters of aneuploids of the same or different kinds. Such reports suggest a predisposing genetic mechanism.

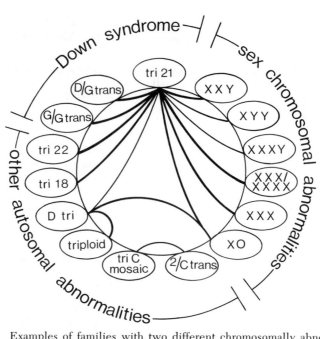

Figure 6-8 Examples of families with two different chromosomally abnormal members. Heavy lines indicate sibs, light lines indicate relatives other than sibs. Redrawn from Hecht et al., N. Engl. J. Med. 271:1081–1086, 1964, by permission.

Autoimmune disease seems to have some role in the pathogenesis of nondisjunction, in view of an observed correlation between high thyroid autoantibody levels and chromosomal anomalies in families.

Radiation has been postulated as a cause of nondisjunction. Uchida (1977) has reported experimental data and has reviewed a number of epidemiological studies of the association between radiation, late maternal age and nondis-

TABLE 6-1 RISK OF DOWN SYNDROME IN FETUSES AT AMNIOCENTESIS AND IN LIVE BIRTHS*

Maternal Age*	Frequency of Down Syndrome	
	Fetuses	Live Births
–19	–	1/1550
20–24	–	1/1550
25–29	–	1/1050
30–34	–	1/700
35	1/350	1/350
36	1/260	1/300
37	1/200	1/225
38	1/160	1/175
39	1/125	1/150
40	1/70	1/100
41	1/35	1/85
42	1/30	1/65
43	1/20	1/50
44	1/13	1/40
45–	1/25	1/25

*Approximate (rounded) estimates from data of Hook and Chambers, 1977. Birth Defects: Orig. Art. Ser. 13(3A): 123–141; Trimble and Baird, 1978. Am. J. Med. Gen. 2: 1–5; Hook, 1978. Lancet 1: 1053–1054; and Spielman et al., 1978. Lancet 1: 1306–1307.

junction. In her experiments, when oocytes in metaphase of meiosis II from irradiated mice were compared with those of unirradiated age-matched controls, the frequency of nondisjunctional products was found to be higher in the irradiated group, especially in the older mice. When human lymphocytes were either irradiated or grown in irradiated serum, the frequency of trisomy was four times as high in the irradiated series as in the controls. Finally, in 11 epidemiological studies of the radiation histories of the mothers of Down patients as compared with mothers of controls, nine of the studies showed an increase in radiation exposure prior to conception in the mothers of the Down patients, though the difference was significant in only four. Uchida concludes that though it may be premature to state with conviction that radiation, as a cause of nondisjunction, increases the frequency of trisomy 21, it seems logical to avoid unnecessary radiation exposure.

Viruses have been shown to cause breakage of chromosomes. For example, Nichols (1963) has demonstrated that the measles virus causes visible fragmentation. The possible effect of this process on the genetic material requires further investigation.

Chromosome abnormalities themselves lead to abnormal segregation of chromosomes. The most obvious example is the transmission of a translocation, as described above, but there are other aspects of the phenomenon. The chromosomes most commonly involved in rearrangements are members of the acrocentric groups, which are satellited. In mitotic metaphase spreads these chromosomes can frequently be seen in satellite association. Satellite association could interfere with normal chromosomal segregation.

CLINICAL ASPECTS OF AUTOSOMAL DISORDERS

There are several well-defined chromosome disorders in which an abnormality of either an autosome or a sex chromosome is present. There is also a growing list of very rare chromosomal syndromes, usually involving loss or gain of only a segment of a chromosome arm instead of a whole chromosome. In this section the best known autosomal disorders are described, and some of the less common syndromes involving parts of the same chromosomes are briefly noted. The sex chromosome disorders are discussed in Chapter 7.

The phenotypes of the various chromosomal disorders all show considerable variation between patients, and the characteristic features listed are not all present in every patient. Nevertheless, there is a striking similarity among patients with the same chromosomal defect, and photographs show better than words the individuality of the phenotypes.

Cri du Chat Syndrome (5p—)

Deletion of part of the short arm of chromosome 5 results in a syndrome, originally described by Lejeune et al. (1963), which has been named because of the resemblance of the cry of an affected child to the mewing of a cat. The facial appearance (Fig. 6–9) is distinctive, with microcephaly, hypertelorism,

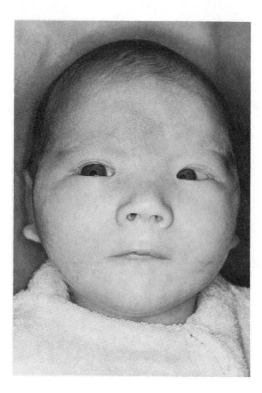

Figure 6–9 Infant with cri du chat syndrome, resulting from deletion of part of the short arm of chromosome 5 (5p–). Note characteristic facies with hypertelorism, epicanthus and retrognathia.

antimongoloid slant of the palpebral fissures, epicanthus, low-set ears sometimes with preauricular tags, and micrognathia. The dermal patterns of the palms, fingers and soles are also characteristic, with simian creases, a high total ridge count and a high frequency of thenar patterns. (Dermatoglyphic terminology is given in Chapter 17.)

Severe mental retardation, failure to thrive, hypotonia and low birth weight despite normal gestation time are characteristic of cri du chat infants. The growth retardation and hypotonia persist through adult life. The disorder accounts for about 1 percent of institutionalized retardates with intelligence quotients below 35.

Most cases are sporadic, but 10 to 15 percent are the offspring of translocation carriers. Mosaicism, with a normal cell line and a 5p– line, has been observed.

TRISOMY 13 (D TRISOMY)

The striking phenotype of trisomy 13 is shown in Figure 6–10. Though the pattern of malformations characteristic of this disorder was probably recognized at least 300 years ago, the chromosome anomaly was first identified in 1960 by Patau et al. Because originally the three pairs of D group chromosomes were indistinguishable, trisomy 13 is often referred to in the older literature as D or D_1 trisomy.

Trisomy 13 is a severe disorder, fatal in about half the liveborn infants within the first month. It is rare, the estimates of its frequency varying from about one in 7000 to one in 20,000. It is very rare or unknown in first-trimester abortions and is not often seen in prenatal diagnosis even though the mean maternal age is advanced.

About 20 percent of the cases are caused by translocation, far above the frequency of translocation in Down syndrome. Even when one parent is a translocation carrier, the empiric risk of the same defect in a subsequent child seems to be below 2 percent. Even so, the risk is sufficiently high to justify prenatal diagnosis in subsequent pregnancies.

The phenotype of trisomy 13 includes severe central nervous system malformations such as arhinencephaly and holoprosencephaly. Growth retardation and severe mental retardation are present. The forehead is sloping, there is ocular hypertelorism, and there may be microphthalmia, iris coloboma or even absence of the eyes. The ears are malformed. Cleft lip and cleft palate are often present. The hands and feet may show postaxial polydactyly and an unusual clenching pattern, with the second and fifth fingers overlapping the third and fourth. The feet are "rocker-bottom," with prominent calcanei. The dermal patterns are unusual, with simian creases on the palms, distal axial triradii and unusual hallucal patterns. Internally there are usually congenital heart defects of specific types and urogenital defects including cryptorchidism in males, bicornuate uterus and hypoplastic ovaries in females and polycystic kidneys. Of this constellation of defects, the most distinctive

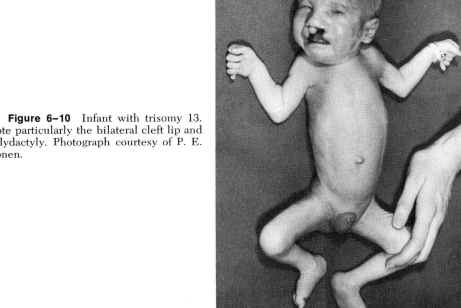

Figure 6–10 Infant with trisomy 13. Note particularly the bilateral cleft lip and polydactyly. Photograph courtesy of P. E. Conen.

are the general facial appearance with cleft lip and palate and ocular abnormalities, polydactyly, and the clenched fists and rocker-bottom feet.

Other Abnormalities of Chromosome 13.   Both partial trisomies and partial deletions of chromosome 13 have been seen. The affected children show a variety of phenotypes, but careful analysis of the phenotypic features in relation to the additional or missing segments is leading to the beginning of a "phenotype map" of chromosome 13 (Fig. 6–11). Similar maps are being constructed for other chromosomes as well.

TRISOMY 18 (E TRISOMY)

The syndrome caused by trisomy 18 was first described by Edwards et al. in 1960. Its incidence is about one in 8000 newborns. Probably 95 percent of trisomy 18 fetuses abort spontaneously. Postnatal survival is also poor. Among those who are liveborn, the mean survival time is only two months, though a few survive for 15 years or more. About 80 percent of the patients are female, perhaps because of preferential prenatal survival. As in other trisomies, maternal age is advanced. Usually the cause is primary nondisjunction, though rarely a sporadic or familial translocation is present. About 10 percent of cases are mosaics, and these display milder manifestations, survive longer and are born to mothers of normal age distribution.

The features of trisomy 18 are shown in Figure 6–12. Mental retardation and failure to thrive are always present. Hypertonia is a typical finding. The head has a prominent occiput, and the jaw recedes. The ears are low-set and malformed. The sternum is short. Though the general appearance is quite

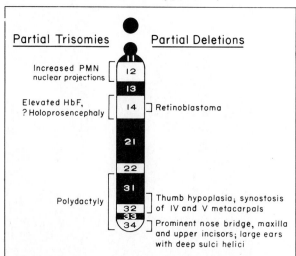

CHROMOSOME 13
Provisional Phenotypic Map

Partial Trisomies   Partial Deletions

Increased PMN nuclear projections — 12

13

Elevated HbF, ? Holoprosencephaly — 14   Retinoblastoma

21

22

31

Polydactyly — 32   Thumb hypoplasia; synostosis of IV and V metacarpals

33 34   Prominent nose bridge, maxilla and upper incisors; large ears with deep sulci helici

Figure 6–11   A provisional phenotypic map of chromosome 13. From Yunis and Chandler, Am. J. Pathol. 88:466–496, 1977. PMN, polymorphonuclear leucocyte; HbF, fetal hemoglobin.

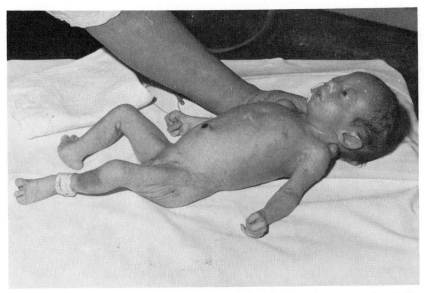

Figure 6–12 Infant with trisomy 18. Note the clenched fist with the second and fifth digits overlapping the third and fourth, the rocker-bottom feet with prominent calcanei, the large simple helix and the dorsiflexion of the big toe. Photograph courtesy of D. H. Carr.

unlike trisomy 13, the fists are similarly clenched and the feet are rocker-bottom. The dermal patterns are very distinctive, with simian creases on the palms and simple arch patterns on most or all digits. The nails are usually hypoplastic. Severe congenital malformations of the heart are present in almost all cases.

Other Abnormalities of Chromosome 18. Other anomalies of chromosome 18 have been identified: partial deletion of the short arm and both partial deletions and partial trisomies of the long arm. There are also ring 18's lacking part of each arm. It is difficult to generalize briefly about the variety of phenotypes associated with these different karyotypes. All share mental retardation and growth retardation.

TRISOMY 21 (DOWN SYNDROME)

Because the autosomal disorders are discussed here in the numerical order of the chromosome involved, the one which is the most important clinically has been left to the last. Down syndrome is by far the most common and best known of the chromosome disorders. It was first described by Langdon Down in 1866, but its cause remained a deep mystery for nearly a century. Two noteworthy features of its population distribution drew attention: the late maternal age and the peculiar pattern within families — concordance in all monozygotic twins but almost complete discordance in dizygotic twins and other relatives. It was suggested by Waardenburg in 1932 that a chromosomal anomaly could explain these observations; in 1959 this was

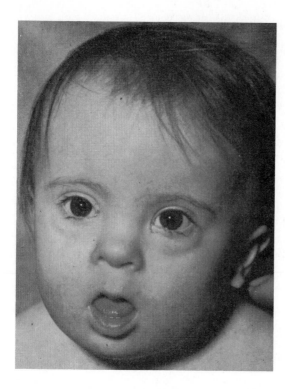

Figure 6–13 A child with Down syndrome. From Smith, *Recognizable Patterns of Human Malformation.* 2nd ed., W. B. Saunders Company, Philadelphia, 1976.

verified when it was found that children with Down syndrome have 47 chromosomes, the extra member being a small acrocentric that has since been designated chromosome 21.

The older name of mongolism, now falling into disuse, refers to the somewhat Oriental cast of countenance produced by the characteristic epicanthal folds, which give the eyes a slanting appearance.

Down syndrome can usually be diagnosed at birth or shortly thereafter by its phenotypic features (Fig. 6–13). Hypotonia is often the first abnormality noticed. Mental retardation is present, the intelligence quotient usually being in the 25 to 50 range when the child is old enough to be tested. The head is brachycephalic, with a flat occiput. The eyes have epicanthal folds, and the iris shows speckles (Brushfield spots) around the margin. The nose has a low bridge. The tongue usually protrudes and is furrowed, lacking a central fissure. The hands are short and broad, usually with a simian crease and clinodactyly (incurving) of the fifth finger. There may be a single crease on the fifth finger. The dermal patterns are characteristic and are described in Chapter 17. On the feet, there is often a wide gap between the first and second toe and a furrow extending proximally along the plantar surface. In about half the patients the hallucal dermal pattern is an arch tibial, which is rare in normal persons. About one-third of the patients have congenital malformations of the heart. Radiologically, the acetabular and iliac angles are decreased. The stature is below average. Often the diagnosis presents no particular difficulty, but karyotyping is nevertheless indicated for confirmation and for determining whether the child has the typical trisomy 21 karyotype (95 percent of cases), a translocation (4 percent) or mosaicism (1 percent).

The Chromosomes in Down Syndrome

Though Down syndrome always involves trisomy for chromosome 21, in about 4 percent of the cases the extra chromosomal material is present not as a separate chromosome but as a translocation of 21q to either a D or another G chromosome, as described earlier. Karyotypes are shown in Figures 6–14 and 6–15.

A child with a translocation, t(14q21q) for example, has 46 chromosomes in all, but the karyotype is effectively trisomic for chromosome 21, and the phenotypic consequences are indistinguishable from those of standard trisomy 21.

About 1 percent of Down syndrome patients are mosaics, usually 46/47 mosaics (i.e., with a mixture of 46- and 47-chromosome cells). Such patients have relatively mild stigmata and are less retarded than the typical trisomies. However, they themselves may be at high risk of having Down children if the mosaicism extends to the cells of the germ line.

Risk of a Down Child

A frequent problem in genetic counseling is the assessment of the risk that a woman will have a Down child. This is especially important now that in many centers guidelines are used to select for prenatal diagnosis only those women in whom the risk of a Down child outweighs the risk of amniocentesis itself. The risk varies with the woman's age, her karyotype and that of her

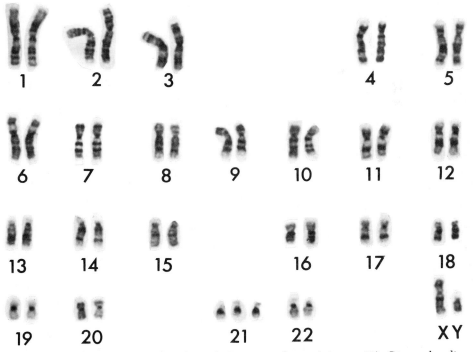

Figure 6–14 Karyotype of a male with Down syndrome (trisomy 21). Giemsa banding. Photomicrograph courtesy of R. G. Worton.

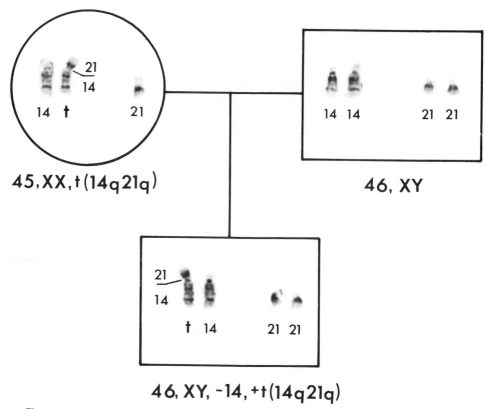

Figure 6–15 Translocation Down syndrome transmitted from a carrier mother to a male child. The father's chromosomes are normal. Only chromosomes 14, 21 and t(14q21q) are shown. Giemsa banding. Courtesy of R. G. Worton.

husband and her reproductive history with respect to Down syndrome and other trisomies.

Estimates of the population incidence of Down syndrome in newborns currently average about 1 in 800. The incidence seems to have remained about the same in recent years despite the decline in mean maternal age (Lowry et al., 1976). Since affected children now survive much longer than before, to an average age of about 40 years, the prevalence of living cases has risen sharply.

The average maternal age at the birth of a Down child is about 34 years, compared with the current mean maternal age in North America of about 26 years. Though the risk is lower at early maternal ages, there are so many more births at the younger maternal ages that the absolute number of Down babies born to young mothers is quite high. Mothers of translocation Down patients have the same age distribution as control mothers. Of all the Down children born to women under 30, about 9 percent have a translocation, whereas for mothers over 30 the proportion is only 1.5 percent.

The increasing risk of a Down child, or a Down fetus, with increasing maternal age is shown in Table 6–1 and has already been referred to repeatedly. As shown, the incidence of the syndrome in fetuses karyotyped at the sixteenth week of gestation or near that time is much higher than one would expect from the estimates for newborns. The difference in incidence does not seem explicable by fetal loss later in pregnancy, so it appears that the only reasonable explanation is that in the past the diagnosis was often missed in newborns, and perhaps especially often in stillbirths. However, the incidence in newborn surveys, in which only infants dying within the first 24 hours after birth were missed, is still much lower than the figures obtained by prenatal diagnosis, so the discrepancy is still puzzling.

At present, in the United States and Canada, 5 to 10 percent of women over the age of 40 and many younger women receive amniocentesis for fetal chromosome analysis.

Risk to Offspring of Translocation Carriers. Most of the translocations seen in Down syndrome are of the Robertsonian type, either Dq21q or 21q22q. As discussed earlier, the theoretical risk for the progeny of carriers of this type of translocation is 1/3 normal, 1/3 carrier and 1/3 affected. The observed proportions are much lower, especially for the offspring of male carriers, as shown in Table 6–2. Preliminary results of a collaborative survey of prenatal diagnosis data, in which the specific D chromosome involved in each case has been identified, agree well with the risks shown in this table.

Previous Trisomic Child. On average, the risk of Down syndrome (or some other trisomy) after the birth of one such child appears to be about 1 percent regardless of maternal age. This represents a large increase in risk for mothers under the age of 30, but not for older mothers. It is possible that some of the younger mothers are unrecognized Down mosaics, or are predisposed for some other cause that is unknown.

Other Abnormalities of Chromosome 21

Monosomy 21 ("antimongolism") is rare and has been reported only a few times. The infants are severely abnormal and retarded in growth and mental development. Partial monosomy and partial trisomy have both been described, and the trisomies in particular have indicated that the part of the chromosome chiefly responsible for the phenotypic features of Down syn-

TABLE 6–2 PROGENY OF TRANSLOCATION CARRIERS

	Normal	*Carrier*	*Affected*
Theoretical (if parent is a translocation carrier)	0.33	0.33	0.33
Observed			
Mother Dq Gq carrier	0.49	0.40	0.11
Father Dq Gq carrier	0.39	0.59	0.02
Mother Gq Gq carrier	0.46	0.52	0.01
Father Gq Gq carrier	0.34	0.66	—

drome, when present in triplicate, is the light-staining band at the distal end of the long arm.

Source of the Abnormal Gamete in Down Syndrome

To understand the causes of Down syndrome, it would be helpful to know what proportion of Down children receive their extra chromosome from the father and what proportion from the mother, as well as the relative frequency of nondisjunction at the first and second meiotic divisions. Analysis of polymorphisms of chromosome 21 in Down children and in their parents is beginning to provide some answers. Preliminary observations indicate that perhaps about one-third of all nondisjunctional events happen in spermatogenesis rather than oogenesis, and that most such accidents occur at meiosis I.

POPULATION CYTOGENETICS

Though population cytogenetics encompasses a wide range of studies of cytogenetic abnormalities and variants — their phenotypic consequences, causation and segregation in families as well as their frequency — this discussion is restricted mainly to incidence as found at different ages, especially in spontaneous abortions, 16-week fetuses and newborns.

SPONTANEOUS ABORTIONS

Chromosome abnormalities are an important cause of spontaneous abortion. About 15 percent of all recognized pregnancies end in spontaneous abor-

TABLE 6–3 RELATIVE FREQUENCIES OF TYPES OF ABERRATION IN CHROMOSOMALLY ABNORMAL ABORTUSES*

Type	Frequency (%)
Trisomy	52
Trisomy 5	—
14	3.7
15	4.2
16	16.4
18	3.0
21	4.7
22	5.7
Other	14.3
45,X	18
Triploid	17
Tetraploid	6
Unbalanced translocations	3
Other	4
Total	100

Data summarized from Carr, D. H., and Gedeon, M. Population cytogenetics of human abortuses. In: Hook, E. B. and Porter, I. H., eds. 1977. *Population Cytogenetics: Studies in Humans.* Academic Press, Inc., New York, San Francisco and London.

TABLE 6–4   INCIDENCE OF CHROMOSOMAL ABERRATIONS SEEN
IN NEWBORN SURVEYS[1]

Sex Chromosome Abnormalities in *37,779 Males*	*Number Seen*	*Approx. Incidence*
XYY	35	1/1000
XXY	35	1/1000
Other	28	1/1300
Sex Chromosome Abnormalities in *19,173 Females*		
45,X	2	1/10,000
XXX	20	1/1000
Other	7	1/3000
Autosomal Aberrations in 56,952 Babies		
+D (trisomy 13)	3	1/20,000
+E (trisomy 18)	7	1/8000
+G (nearly all trisomy 21)	71	1/800
Other trisomies	1	1/50,000
Rearrangements		
Balanced	110	1/500
Unbalanced	34	1/2000
Total Chromosomal Aberrations	353	1/160

[1]Summarized from Hook, E. B. and Hamerton, J. L. The frequency of chromosome abnormalities detected in consecutive newborn studies—differences between studies—results by sex and by severity of phenotypic involvement. In: *Population Cytogenetics: Studies in Humans.* (Hook, E. B., and Porter, I. H., eds.) Academic Press, Inc., New York, San Francisco and London, 1977.

tion, 80 percent of these during the first trimester. The incidence of chromosome abnormalities in the early abortions is about 61.5 percent (Boué et al., 1975) and in the later ones about 5 percent. Combining these figures, we see that the overall frequency of chromosome abnormalities in spontaneous abortions is very close to 50 percent and that one in 13 conceptuses has a chromosome abnormality of some kind.

The kinds of abnormalities seen in these abortuses (Table 6–3) differ in a number of ways from those seen in liveborn infants (Table 6–4). The single most common abnormality is 45,X (18 percent of chromosomally abnormal spontaneous abortions as compared with about one in 20,000 live births). It appears that more than 99 percent of all 45,X conceptuses abort spontaneously. Sex chromosome abnormalities other than 45,X are unusual in abortions. Trisomies have been seen for all autosomes except 1 and 5, but the distribution of types is quite different in abortions and live births. Trisomy 16, the most common trisomy in abortions, is not seen at all in live births. Trisomy 21 and trisomy 22 are about equally common in abortions, but in live births trisomy 21 is common and trisomy 22 very rare. If 9 percent of trisomic abortions (4.5 percent of all chromosomally abnormal abortions, 2.25 percent of all spontaneous abortions) and only one live birth in 800 are 21 trisomic, then about 75 percent of all Down syndrome conceptuses must abort spontaneously. Of the three D trisomies, only trisomy 13 is seen in liveborn children, but both trisomy 14 and trisomy 15 rank high in frequency in spontaneous abortions. Trisomy 18 occurs in 5 or 6 percent of trisomic abortions and one in 8000 newborns, so it appears that over 95 percent of all trisomy 18's abort. Triploids and tetraploids are common in abortions, but triploids are rarely born alive and only a few who are mosaic with a diploid line survive any length of time, and tetraploids have even worse survival, being seen most frequently in the very earliest abortions.

FETUSES AT 16 WEEKS OF DEVELOPMENT

Since most spontaneous abortions in which the fetus is chromosomally abnormal happen before the sixteenth week of pregnancy, the frequency of the chromosome abnormalities seen in prenatal diagnosis should be much the same as the frequency at birth. On the whole this is true, but there seems to be an exception for Down syndrome at maternal ages 40 to 44. For this maternal age range the incidence of Down syndrome is substantially higher in the 16-week fetus than at birth.

NEWBORN SURVEYS

The incidence of chromosome abnormalities in newborns has been estimated in at least six large studies, which included 56,952 children of whom 37,779 were male and 19,173 female. (In some series, only males were examined.) The findings are summarized in Table 6–4.

CONCLUSION

In summary, consider the fate of 10,000 conceptuses. About 800 are chromosomally abnormal; these include about 140 with 45,X, 112 with trisomy 16, 20 with trisomy 18 and 45 with trisomy 21.

Of these, 750 abort spontaneously as do 750 chromosomally normal fetuses. The abortions include about 139 with 45,X, all 112 with trisomy 16, 19 with trisomy 18, 35 with trisomy 21 and a variety of rare types, chiefly trisomies.

The remaining 50 abnormals represent about 0.6 percent of the 8500 liveborns. The liveborn group includes about one each with 45,X and trisomy 18; 10 with trisomy 21; 15 with sex chromosome trisomy (XYY, XXY or XXX); and 20 with rearrangements, of which 16 are balanced and 4 unbalanced.

Other surveys in special populations have revealed a higher than normal incidence of chromosome abnormality in stillbirths (about 5 percent, almost 10 times the incidence in live births), in institutionalized retarded children and in women with a history of repeated abortion or in their spouses.

GENERAL REFERENCES

Carr, D. H. 1971. Chromosomes and abortion. *Adv. Hum. Genet.* 2:201–258.

Hamerton, J. L. 1971. Human Cytogenetics. Vol. I. *General Cytogenetics.* Vol. II. *Clinical Cytogenetics.* Academic Press, New York.

Hook, E. B., and Porter, I. H., eds. 1977. *Population Cytogenetics: Studies in Humans.* Academic Press, New York.

Smith, D. W., and Wilson, A. C. 1973. *The Child with Down's Syndrome (Mongolism).* W. B. Saunders Company, Philadelphia.

Smith, G. F., and Berg, J. M. 1976. *Down's Anomaly.* 2nd ed. Churchill Livingstone, Edinburgh.

Yunis, J. J., ed. 1977. *New Chromosomal Syndromes.* Academic Press, New York.

PROBLEMS

1. a) A phenotypically normal woman has the karyotype 45,XX, t(21q21q). Assuming fertilization by a normal gamete, state the karyotypes and phenotypes possible in her offspring.

   b) A woman's karyotype, examined after solid staining, is interpreted as 45,XX, t(GqGq). She has a normal child. How would this affect your interpretation of her karyotype?

2. A pregnant woman of age 37 has a normal karyotype. The karyotype of her fetus is found to be 45,XY, t(14q21q). Her husband's karyotype is the same as that of the fetus.

   a) On this evidence, is termination of this pregnancy indicated?

   b) Is this strong evidence that the husband is actually the father of the fetus?

   c) What would your advice to this woman be concerning any subsequent pregnancies?

3. About 3 percent of trisomic abortuses have trisomy 13. Using information given in this chapter, calculate the proportion of all trisomy 13 conceptuses aborted spontaneously.

4. What proportion of all conceptuses are chromosomally abnormal?

7

THE SEX CHROMOSOMES AND THEIR DISORDERS

Man and the other mammals have a sex determination mechanism in which the crucial factor is the presence or absence of the Y chromosome. In this respect we differ from the fruit fly *Drosophila,* in which both X and Y chromosomes are present, but the sexual phenotype is determined by the ratio between the number of X's and the set of autosomes, a single X conferring a male phenotype even in the absence of a Y. The difference between man and fruit fly in this fundamental characteristic was not understood until chromosome analysis in man became possible in 1959, even though the early cytologist Painter had stressed the role of the human Y (Painter, 1921, 1923). Because the sex chromosome constitution underlies normal sexual development and many of its disorders, it is appropriate to begin this chapter with a brief review of the embryology of the human reproductive system.

EMBRYOLOGY OF THE REPRODUCTIVE SYSTEM

The embryonic germ cells are wandering cells derived from the gut endoderm. By about the eighth week of intrauterine life, they settle down in the cortex and medulla of the primitive, bipotential gonad. Normally, if a Y chromosome is present the gonad differentiates in the male direction, the medullary tissue forming typical testes with seminiferous tubules and Leydig cells which become capable of androgen secretion. If two X chromosomes are present, or even if there is one X and no Y, the gonad differentiates in the female direction, the cortex developing and the medulla regressing. If two X's are present, oogonia develop in the cortex, and before the time of birth the first meiotic division has begun in the sex cells.

By about the fourth week of intrauterine life thickenings in the genital ridges indicate the positions of the wolffian (male) and müllerian (female) duct systems. In the presence of androgen-secreting Leydig cells the wolffian duct normally differentiates to form the typical male duct system. At the same time the Leydig cells produce a second male organizer which causes regression of the müllerian duct system. If an ovary is present, or if the gonad remains undifferentiated, the müllerian system normally develops toward a typical female duct system and the wolffian system regresses (Fig. 7–1).

In the early embryo the external genitalia consist of a genital tubercle, two labioscrotal swellings and two urethral folds. The genital tubercle can form either penis or clitoris. In the male, under the influence of androgens, the labioscrotal folds fuse to form the scrotum and the urethral folds fuse to form the urethra including its penile portion, while the genital tubercle takes on a male (penile) configuration. In the absence of a differentiated gonad, or in the presence of an ovary, the labioscrotal folds remain separate to form the labia majora, and the urethral folds remain unfused to form the labia minora. The genital tubercle assumes the female (clitoral) configuration.

Abnormalities of genes, chromosomes or environment can affect any of these stages. In either 46,XX or 46,XY individuals, if no differentiated gonad forms, the phenotype remains female. If the androgenic male organizer is deficient, the wolffian system does not develop, but if the second male organizer is present the müllerian system also fails to develop and the internal genitalia are rudimentary, typical of neither sex. Similarly, refractoriness of the tissues to androgen may produce female external genitalia in an XY individual. In an XX fetus, androgen production (as from an abnormal suprarenal gland) or exogenous hormones crossing the placenta from the mother may lead to varying degrees of masculinization of the external genitalia.

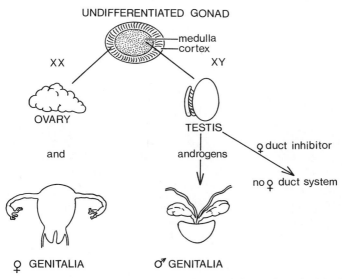

Figure 7–1 Development of the gonads and genital ducts, described in the text.

THE SEX CHROMOSOMES

THE Y CHROMOSOME

The distal part of the long arm of the Y chromosome fluoresces so vividly by quinacrine staining that it can be seen even in interphase cells. Polymorphism of the Y is very frequent; about 10 percent of males have a Y that is longer or shorter than usual, the difference being entirely in the length of the fluorescent portion, and this trait is hereditary. Apparently the length of the Y has little significance in fertility, though in one study a long Y seemed to be associated with an increased risk of abortion.

SEX CHROMATIN

Since 1921 it has been known that male and female cells differ in their sex chromosome complement, but it was not until 1949 that a sex difference in interphase cells was detected. In that year Barr and Bertram noted that a previously recognized mass of chromatin in the nuclei of some nerve cells was frequently present in females but not in males (Fig 7–2). This mass is now known as the sex chromatin or Barr body. Cells (or people) are said to be chromatin-positive if sex chromatin is present and chromatin-negative if it is absent.

Barr and his students later found sex chromatin in the cells of most of the tissues of females of many species of mammals including humans. The Barr body can be seen in many cell types, but the epithelium of the buccal mucosa is the most convenient. A buccal smear is made simply by scraping a few cells from the inside of the cheek, smearing them on a slide and staining. The slide is then examined microscopically to determine the percentage of cells that show Barr bodies (Fig. 7–3).

The association of the sex chromatin with the number of X chromosomes and its lack of association with the Y chromosome is shown in the following

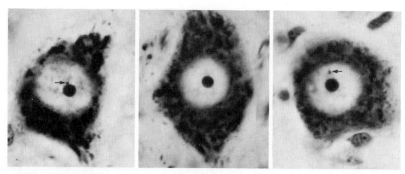

Figure 7–2 Sex chromatin (arrows) in nerve cells of a female cat. The center cell, from a male, shows no sex chromatin. This illustration appeared in the original description of sex chromatin by Barr and Bertram, Nature 163:676–677, 1949.

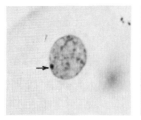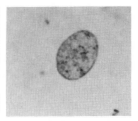

Figure 7–3   Sex chromatin in epithelial cells of human buccal mucosa. Arrows indicate the sex chromatin at the nuclear membrane. A male cell (right) has no sex chromatin. This illustration appeared in the original description of the buccal smear technique by Moore and Barr, Lancet 2:57–58, 1955.

table. The phenotypes typical of some anomalous chromosome complements are described later in this chapter.

Chromosome Complement	Number of Barr Bodies
45,X 46,XY 47,XYY	0
46,XX 47,XXY 48,XXYY	1
47,XXX 48,XXXY 49,XXXYY	2
48,XXXX 49,XXXXY	3
49,XXXXX	4

Note that sex chromatin is present if there are two or more X chromosomes and that the number of sex chromatin bodies is one less than the number of X's. Not all cells show the characteristic number. Normal females exhibit recognizable Barr bodies in only about 50 percent of the epithelial cells at any one time. Whether or not a Barr body can be seen in any one cell depends on the stage of the cell cycle and the orientation of the nucleus on the slide.

The Barr body is now known to represent one of the two X chromosomes of female cells, which completes its replication later than its homologue and remains condensed and genetically inactive throughout interphase. This is true only of somatic cells; in the germ line of the female both X's remain active (Ohno et al., 1962; Epstein, 1969), and in the germ line of the male the single X is inactivated (Lifschytz and Lindsley, 1972).

X INACTIVATION

For many years the action of X-linked genes was a puzzle to geneticists. How could it be that females with two representatives of every gene on the X chromosome formed no more of the product of these genes than did hemizygous males with only a single X? And why were females homozygous for an

X-linked mutant gene no more severely affected than were hemizygous males? Some mechanism of "dosage compensation" was indicated.

A hypothesis to explain dosage compensation in terms of a single active X chromosome was arrived at independently by several groups of workers in 1961; it is usually known as the Lyon hypothesis, named after Mary Lyon, who was the first to state it explicitly and in detail. Lyon based her hypothesis of X inactivation in part on genetic observations of X-linked coat color genes in the mouse and in part upon cytological data. She noted that in a female mouse heterozygous for X-linked coat color genes, the coat was neither like that of either homozygous phenotype nor intermediate between the two; instead, it was mottled, i.e., made up of patches of the two colors, the patches being random in arrangement and rarely crossing the midline, as shown in Figure 7–4. (Female tortoiseshell cats show the same effect.) On the other hand, males never had this patchy phenotype but instead had coats of a uniform color. Mice with one X and no Y chromosome also expressed X-linked coat color genes as a uniform, not mottled, coat color.

The chief cytological observation on which Lyon based her hypothesis has already been mentioned: the number of sex chromatin bodies in interphase cells is always one less than the number of X chromosomes seen at metaphase. Additional cytological evidence is found in the observations that at prophase one X chromosome is heteropyknotic and late-replicating. Thus, it is clear that the sex chromatin is a heteropyknotic (condensed) X chromosome; only one X is active in cellular metabolism, and the second X (or in abnormal cells with extra X chromosomes, any extra X) appears as a sex chroma-

Figure 7–4 Female mouse heterozygous for the X-linked coat color gene *Tortoiseshell*. The mosaic phenotype of such females provided one of the first lines of evidence for the Lyon hypothesis (see text). From Thompson, Can. J. Genet. Cytol. 7:202–213, 1965.

tin body. In the male, the single X is uncoiled and active at all times, and consequently there is no sex chromatin.

The Lyon hypothesis states that:

1. In the somatic cells of female mammals, only one X chromosome is active. The second X chromosome is condensed and inactive and appears in interphase cells as the sex chromatin.

2. Inactivation occurs early in embryonic life.

3. The inactive X can be either the paternal or maternal X (X^P or X^M) in different cells of the same individual; but after the "decision" as to which X will be inactivated has been made in a particular cell, all the clonal descendants of that cell will "abide by the decision"; i.e., they will have the same inactive X. In other words, inactivation is *random* but *fixed*.

The time of onset of X inactivation in the embryo is still unknown, despite much experimental work. To summarize the findings briefly, it seems that the X inactivation does not occur at the first cleavage division (Hoppe and Whitten, 1972), and that it may take place as late as at the time of determination of tissue primordia (Deol and Whitten, 1972). On the whole, the evidence favors the blastocyst stage, not before the 4.5-day-old stage, as demonstrated by Gardner and Lyon (1971) when they transplanted single cells from embryos of different early stages into host blastocysts, allowed development to proceed and eventually found that most of the resultant products showed the type of mosaicism characteristic of the mice from which the single cells had been derived.

The randomness of inactivation is evident from the mosaic coat color itself. Exceptions occur in some specialized circumstances. In females with loss of material from one X chromosome (a deletion, ring or isochromosome), the structurally abnormal X always forms the Barr body, which may then appear appropriately larger or smaller than usual. In contrast, if a female has a translocation between an X and an autosome, the rule is that the intact X becomes inactivated, though other inactivation patterns may occur. These patterns are readily explained. In a deletion, if the intact X were to become inactivated, the functional X, lacking a chromosome segment, would render the cells containing it inviable. In X-autosome translocations, inactivation of a translocated part of one X might bring about inactivation of the autosome to which it is translocated, thus rendering the cell concerned inviable because it has only a single functional copy of that autosome. In either case, nonrandom inactivation has the effect of making the female with the X chromosome abnormality, in effect, hemizygous for any allele on her active X. By this mechanism a female can express an X-linked disorder, and rare cases are known of women who are color blind or have a severe disorder such as Duchenne muscular dystrophy as a consequence of having the deleterious gene on the X that is functional. Clinically, this means that a woman who has an X-linked trait should have chromosome analysis as part of the investigation of her case.

In the female mule, which is a natural hybrid between horse and donkey, the two X chromosomes differ in the position of the centromere, so that it is possible to determine which one is late-replicating in autoradiographs. The G6PD enzyme types in blood and tissues can be distinguished electrophoretically. The proportion of horse G6PD formed in any tissue correlates closely with the percentage of late-replicating donkey X chromosomes (Fig. 7–5). In most mules, the horse (maternal) enzyme preponderates, and the donkey X is inactivated more than half the time, but it is not certain whether this is due to preferential rather than random inactivation or whether it represents a process of selection after inactivation (Cohen and Rattazzi, 1971).

An extreme example of nonrandom expression was shown by Nyhan et al. (1970) in women heterozygous both for the common A and B types of G6PD and for HPRT deficiency. Only one type of G6PD is present in their tissues, and this is explained on the basis of random inactivation followed by selection against the HPRT-deficient cells.

GENETIC CONSEQUENCES OF X INACTIVATION

The Lyon hypothesis has three principal genetic consequences: dosage compensation, variability of expression in heterozygous females and mosaicism.

Dosage Compensation

The amount of product of X-linked genes, such as G6PD or antihemophilic globulin (the substance deficient in classic hemophilia), is equivalent in the two sexes. For many years there was no satisfactory explanation as to how compensation for the dosage effect of the two X chromosomes of the female as compared with the male's single X was accomplished. X inactivation adequately explains this phenomenon. However, a number of problems remain. One of these is the abnormal phenotype shown by individuals with abnormal sex chromosome complements. If one and only one X is active regardless of

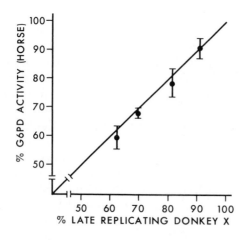

Figure 7–5 X inactivation and G6PD synthesis in the mule, described in the text. After Cohen and Rattazzi, Proc. Natl. Acad. Sci. USA 68: 544–548, 1971.

Figure 7–6 Distribution of G6PD phenotypes in 42 women heterozygous for the G6PD alleles A and B. From Nance, *Genetic Studies of Human Serum and Erythrocyte Polymorphisms: Glucose-6-Phosphate Dehydrogenase, Haptoglobin, Hemoglobin, Transferrin, Lactic Dehydrogenase and Catalase.* Ph.D. Thesis, University of Wisconsin, 1968.

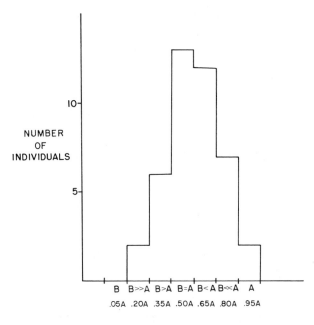

the number present, it is difficult to understand why a 45,X or XXY individual should show any phenotypic abnormality at all, and why more than one extra X produces a progressively more abnormal phenotype.

Variability of Expression in Heterozygous Females

Since inactivation is random, females heterozygous for X-linked genes should have varying proportions of cells in which a particular allele is active. In other words, among such females there should be considerable phenotypic variability.

A demonstration that this really does happen in human females has been given by Nance (1964) for G6PD. In 30 women heterozygous for the A and B variants of G6PD, he estimated the relative proportions of the two enzyme variants and found a wide range and approximately normal distribution, as shown in Figure 7–6. For many, if not all, of the X-linked eye disorders, "manifesting heterozygotes" have been observed, and these patients may be regarded as extreme examples of "unfavorable Lyonization," in whom the deleterious allele is functional in a majority of the cells. Color blindness, classic hemophilia, Christmas disease and Duchenne muscular dystrophy are other conditions in which heterozygote variability has been noted.

Mosaicism

It should be possible to demonstrate that women heterozygous for X-linked genes actually do have two populations of cells, one population with one X active, the other with the alternative X active. The first demonstration of mosaicism at the cellular level was provided by cloning of cultured fibroblasts from a woman heterozygous for two different G6PD alleles. Two different clonal populations were demonstrated, differing as to which G6PD allele was functioning (Davidson et al., 1963). This is direct evidence for mosaicism. More recently, mosaicism has been shown at the cellular level in women heterozygous for the X-linked type of mucopolysaccharidosis (Hunter syndrome) and for HPRT deficiency. The demonstration of mosaicism in cultured fibroblasts by cloning is an important potential tool for genetic counseling in X-linked disorders. It provides evidence that the disorder under investigation is X-linked rather than autosomal, and shows whether the mother is heterozygous or has transmitted a new mutation to her affected son.

An exceptional property of the Xg blood group locus is that it apparently does not undergo inactivation. Evidence of mosaicism in Xg^aXg women has been searched for vigorously without success. Failure of the Xg locus to inactivate and abnormality of phenotype in some X-chromosome aneuploids, described in the following section, both suggest that part of the X does not undergo inactivation.

SEX CHROMOSOME ABNORMALITIES

INCIDENCE

The incidence of the common sex chromosome disorders in newborn surveys and in spontaneous abortions is discussed in Chapter 6. In general, the types found in males are quite common, with a total incidence of about one in 400 births. In females the frequency is a little lower, about one in 650 births, of which the great majority are XXX. The only sex chromosome abnormality that appears in abortions with appreciable frequency is 45,X; this is the most common cytogenetic abnormality seen in abortuses and, as noted earlier, it accounts for 18 percent of all chromosomally abnormal abortions and is estimated to be present in 9 percent of all spontaneous abortions and in about 1.4 percent of all conceptions.

CLINICAL ASPECTS

Klinefelter Syndrome

Figure 7–7 illustrates this syndrome, which was described by Klinefelter et al. in 1942. Because the phenotype is male, but Barr bodies are present, the condition was a prime candidate for chromosome analysis, and its XXY chromosome complement was found shortly after human chromosome studies became possible (Jacobs and Strong, 1959).

The syndrome is not recognizable before puberty unless the child's chromosomes are analyzed for some unrelated reason, such as participation in a newborn survey. After puberty, the chief characteristics are small testes and hyalinization of the seminiferous tubules. Usually the secondary sexual char-

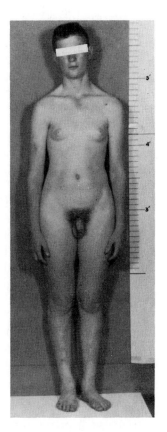

Figure 7-7 Phenotype of 47,XXY Klinefelter syndrome. Gynecomastia, present in this patient, occurs in only about 20 percent of patients. Photograph courtesy of M. L. Barr.

acters are poorly developed, gynecomastia may appear, and many patients are tall and eunuchoid. Subnormal mentality is not unusual in older Klinefelter patients, but prepubertal XXY males identified in population surveys appear to have normal intelligence.

There are several variants of Klinefelter syndrome: XXYY, XXXY, XXXXY and others. As a rule, the additional X's cause a correspondingly more abnormal phenotype, with a greater degree of dysmorphism, a more severe defect in sexual development and more severe mental retardation.

About 15 percent of Klinefelter patients are mosaics, with two or more distinct cell lines. The most common mosaic form is XY/XXY, and in these patients it is noteworthy that though the buccal smear may be chromatin positive, testicular development and mental status may not be abnormal. A few patients with 46,XX karyotypes have been found, and these are probably best explained as mosaics in whom the line with the Y chromosome has been lost in early development, or at least has not been identified.

Maternal age is advanced in XXY patients, and about 60 percent of the group owe their origin to either meiotic or postzygotic maternal nondisjunction, i.e., they are X^MX^MY. The remaining 40 percent are X^MX^PY, indicating nondisjunction of the X and Y in the first meiotic division of spermatogenesis.

XYY Syndrome

Males with a second Y chromosome have aroused great interest since they were found to be frequent among males in a maximum security prison (Jacobs et al., 1968). About 3 percent of males in prisons and mental hospitals are XYY, and among the group over six feet tall the proportion is much higher, over 20 percent. In newborn surveys the incidence is about one per 1000 births, so most XYY males must be indistinguishable from XY males on the basis of their behavior or phenotype.

The relationship between XYY and aggressive, psychopathic or criminal behavior has aroused great public interest. XYY males are perhaps six times as likely to be imprisoned as XY males. When an XYY infant is identified in a newborn survey, or even prenatally, it is difficult to know how to handle the problem. Should an XYY fetus be aborted? Should the parents of an XYY infant be told that the child has an increased risk of behavioral difficulties? If so, how should the child's upbringing be modified? The answers to these questions await better understanding of the range of variation of XYY's, since the majority of those who have been studied so far have been ascertained because of their antisocial behavior.

The origin of the XYY karyotype is paternal nondisjunction at the second meiotic division, which produces YY sperm. The less common XXYY and XXXYY variants, which share the features of the XYY and Klinefelter syndromes, probably also originate in the father, by a sequence of nondisjunctional events.

Turner Syndrome (Gonadal Dysgenesis)

The syndrome of sexual infantilism, short stature, webbing of the neck and cubitus valgus (reduced carrying angle at the elbow) described by Turner (1938) is shown in Figure 7–8. The phenotype is female though the patients are (usually) chromatin negative. The discrepancy suggested a chromosome abnormality, and this was confirmed by Ford et al. (1959), who demonstrated the 45,X karyotype. In medical literature the standard Turner karyotype is sometimes written as XO; either 45,X or XO is acceptable.

Other features of the phenotype include the low hairline at the nape of the neck, characteristic facial appearance, unusual dermatoglyphics with high total ridge count, wide chest with broadly spaced nipples, coarctation of the aorta and, especially in newborns, edema of the feet, which together with neck webbing should alert the physician to the need for chromosome studies. The external genitalia are juvenile, and the internal sexual organs are female although the ovary is usually only a streak of connective tissue; however, the streak may be arranged in the manner of ovarian stroma and ovarian follicles may be present in fetal life, though usually not postnatally. Axillary and pubic hair are usually present but sparse. Primary amenorrhea is usual, though not invariable.

The high incidence of the 45,X karyotype in spontaneous first-trimester abortions was noted in Chapter 6. It is surprising that 45,X is so severe a defect prenatally, yet relatively benign after birth.

The increase in twinning in the sibships of 45,X females to five to 10

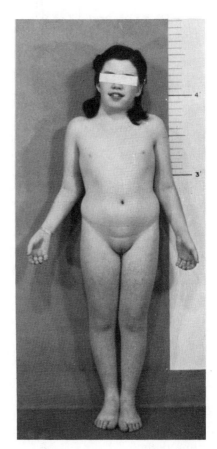

Figure 7–8   Phenotype of 45,X Turner syndrome. For details, see text. From Barr, Am. J. Hum. Genet. 12:118–127, 1960.

times the normal rate has been noted by Nance and Uchida (1964). They have speculated that 45,X individuals may originate in postzygotic loss of a chromosome. Several sets of MZ twins in each of which only one member was 45,X have been reported (summarized by Benirschke, 1972), and all of these must of course have arisen by postzygotic accidents. The error in gametogenesis, when it can be traced, is usually in the paternal gamete and maternal age is not advanced.

Only about 60 percent of patients with the Turner syndrome have monosomy X. The remainder have a variety of karyotypes with either a structural variant of the X or mosaicism. The most frequent of these is XXqi, that is, an isochromosome for the long arm of the X, referred to in Chapter 6. XXp–, XXq– and XXr are other structural variants, in which on the whole, the greater the loss of material from the short arm, the closer the resemblance to the 45,X phenotype. Deletion of the long arm is less likely to be associated with short stature.

Mosaicism, especially of the X/XX or X/XXX type, is very common in Turner syndrome, accounting for about 15 percent of all cases. As usual with mosaicism, the phenotype in such patients varies depending on the time of the postzygotic accident and the proportion of abnormal cell lines in different tissues. X/XY mosaicism also leads to variable phenotypic changes.

XXX, XXXX and XXXXX

Trisomy X and the other karyotypes with additional X's are the counterparts in the female of Klinefelter syndrome in the male. Usually XXX females are not phenotypically abnormal, though some are sterile and some are mentally retarded. Some are first identified in infertility clinics, others in institutions for the mentally retarded, but probably many remain undiagnosed. Some have borne children, virtually all of whom have normal karyotypes, indicating that XX ova are probably selected against. In some of the buccal smear cells of XXX women there are two sex chromatin bodies.

The presence of four X chromosomes often leads to retardation in both physical and mental development, and XXXXX usually causes severe developmental retardation with multiple physical defects.

DISORDERS OF SEXUAL DEVELOPMENT WITH NORMAL CHROMOSOMES

In some newborn infants, assignment of sex is difficult or impossible because the genitalia are ambiguous, with anomalies that tend to make them resemble those of the opposite sex. The anomalies vary through a spectrum from a mild form of hypospadias in the male to enlarged clitoris in the female, with many intermediate states and many degrees of severity. Such problems do not usually involve abnormalities of the sex chromosomes, but determination of the child's karyotype is an essential part of their investigation.

Homosexuality, transvestitism and other sexual psychological variations do not as a rule have their basis in either chromosomal aberrations or single-gene defects.

CLINICAL EXAMPLES

True Hermaphroditism

True hermaphroditism is very rare, fewer than 100 cases having been reported (Hamerton, 1971). A true hermaphrodite has both testicular and ovarian tissue, either as two separate organs or as a single ovotestis, not necessarily functional but histologically identifiable. The internal and external sexual organs are very variable and not in any way diagnostic. Sex hormone studies are not helpful.

The great majority of hermaphrodites are XX (contrary to the rule that a Y is needed for testicular development), some are XY and others are not chromosomally normal but are XX/XY chimeras or mosaics.

Pseudohermaphroditism

Pseudohermaphrodites, unlike true hermaphrodites, have gonadal tissue of only one sex. Male pseudohermaphrodites are XY and have testicular tissue; female pseudohermaphrodites are XX and have ovarian tissue. Both types have ambiguous genitalia, in which the anomalies range from hypospadias or enlarged clitoris to variants in which assignment of the individual to either sex is virtually impossible.

Male Pseudohermaphroditism

Testicular Feminization.   This condition was mentioned in Chapter 4 as an example of the difficulty in distinguishing X linkage from autosomal inheritance with sex-limited expression when affected males do not reproduce. The external genitalia are female, but the patient is XY and has testes that may be in the abdominal cavity, inguinal canal or labia. Many patients are well-developed females who seek medical advice because of amenorrhea, sterility or inguinal hernia (in which at operation a testis may be found).

Female Pseudohermaphroditism

Congenital Adrenal Hyperplasia.   This is the most common cause of female pseudohermaphroditism. The incidence is about one in 25,000 births. Several distinct genetic and clinical forms are known, all inherited as autosomal recessives and each characterized by a block in a specific step in cortisol biosynthesis, resulting in increased secretion of ACTH (adrenocorticotropic hormone) and hyperplasia of the adrenal glands. This in turn leads to masculinization of female fetuses. Affected baby girls frequently have major anomalies of the external genitalia, often to the point that sex assignment may be impossible. The clitoris may be enlarged and the labia majora rugose and even fused. In males the same genotype produces premature virilization, but there is no difficulty in identifying the sex. In one form of congenital adrenal hyperplasia, salt-losing crises may be very severe.

Other Forms of Female Pseudohermaphroditism.   The external genitalia of a female fetus may be masculinized if the fetal circulation contains excessive amounts of either male or female sex hormones. These hormones reach the fetal circulation from the maternal circulation, and may originate either endogenously or exogenously. Androgenic hormones may be present in excessive amounts if the mother's adrenal cortex is overactive or if she has received hormone therapy, e.g., some progestins used to prevent spontaneous abortion. [When therapy itself produces a disorder, the disorder is said to be iatrogenic (*iatros*, physician).]

MANAGEMENT OF DISORDERS OF SEXUAL DEVELOPMENT

There are certain general principles that can serve as useful guidelines in the management of these disorders.

The chromosomal sex is *not* the final indicator of the sex to which the patient should be assigned. Much more important factors in the decision are the presenting phenotype and the sex of rearing. If by medical or surgical intervention the patient can be allowed to lead a relatively normal though infertile life as a member of one sex, this is usually the best option. Altering the sex of rearing is usually unwise, except in very early infancy.

The diagnosis should be made as early as possible, both to prevent clinical problems that might arise immediately or later, and to avoid the severe psychological trauma of possible sex reversal at a later age. These disorders cause great concern and embarrassment to the family, and the concern is intensified if the disorder is one that might recur in a subsequent child. Thus these cases must be handled with tact and good judgment.

GENERAL REFERENCES

Most of the General References cited for Chapter Six are also useful for Chapter Seven. The following references apply particularly to the biology of the X chromosome.

Gartler, S. M., and Andina, R. J. 1976. Mammalian X-chromosome inactivation. Adv. Hum. Genet. 7:99–140.

McKusick, V. A. 1964. *On the X Chromosome of Man*. American Institute for the Biological Sciences, Washington, D.C.

Ohno, S. 1967. *Sex Chromosomes and Sex-Linked Genes*. Springer-Verlag, New York Inc., New York.

Simpson, J. L. 1976. *Disorders of Sexual Differentiation: Etiology and Clinical Delineation*. Academic Press, New York.

PROBLEMS

1. A color-blind woman with Turner syndrome and a 45,X karyotype has a color-blind father. From which parent did she receive a chromosomally abnormal gamete?

2. If a man is color blind and his 45,X daughter has normal color vision, can the source of the abnormal gamete be determined? Explain.

3. In a color-blind male with 47,XXY Klinefelter syndrome:
   a. What is the probable genotype?
   b. If his parents have normal color vision, what are their genotypes? In which parent and at which meiotic division did nondisjunction occur?

4. a. In a triple-X female, what types of gametes should one expect to find and in what proportions?
   b. If each type of gamete is equally viable and equally likely to result in a living child, what are the theoretical karyotypes and phenotypes of her progeny? (Actually, women who are XXX almost always have normal offspring only.)

5. The inheritance of the Xg blood groups has been described in Chapter 4.
   a. An XXY, Xg(a+) male has an Xg(a−) mother. In which parent and at which meiotic division was the abnormal gamete formed?
   b. An Xg(a+) man married to an Xg(a−) woman has an Xg(a−) son with an XXY karyotype. Which parent produced the abnormal gamete? At which meiotic division?

6. The birth incidence of XXY and XYY anomalies is roughly equal. Is this to be expected on the basis of theoretical considerations? Explain.

8

IMMUNOGENETICS

Immunogenetics is concerned with the genetics of **antigens, antibodies** and their interactions. Four areas of immunogenetics are of medical significance: (1) blood groups and clinical problems related to blood group incompatibilities, (2) transplantation, (3) immune deficiency diseases and (4) autoimmune diseases. Because of the importance of the blood groups as genetic markers as well as in clinical medicine, they are discussed separately in Chapter 9.

An antigen is a molecule that can stimulate antibody production and react specifically with the antibody so produced. To be antigenic it must have a surface conformation that is recognized by the host as different from any of its own molecules. Most proteins and some other macromolecules (some polysaccharides and some nucleic acids) are antigenic.

Antibodies are serum proteins of the **gamma globulin class,** referred to as **immunoglobulins.** The antigen-antibody reaction depends upon mutually interlocking sites specific for each antigen-antibody pair. Formation of the antigen-antibody complex initiates events that serve to remove the foreign antigen from the body. The whole sequence of events constitutes the **immune response.**

The immune response is first elicited in the **lymphoid tissue** (lymph node or tonsil). Lymphoid cells may respond to an antigenic stimulus either by production of humoral antibody (circulating immunoglobulin) or by the production of **T effector** ("killer") cells with antibodies on the surface. In general, humoral antibodies effect rejection of bacteria and viruses, whereas T effector cells cause rejection of malignant cells or transplants.

IMMUNOGLOBULINS

A gamma globulin molecule is composed of two identical **light** and two identical **heavy chains** held together by disulfide bonds (Fig. 8–1). The light (**L**) chains occur in two types, **kappa** (κ) and **lambda** (λ), differing in amino acid sequence. They are alike in all classes of immunoglobulins, but each type has its own characteristic heavy (**H**) chain. The five classes, their chain compositions and their molecular weights are as follows:

Ig Class	Heavy Chain	Light Chain	MW
IgG	γ	κ or λ	150,000
IgM	μ	"	900,000
IgA	α	"	160,000, 320,000, 480,000
IgD	δ	"	185,000
IgE	ϵ	"	200,000

The differences in molecular weight shown above depend partly on the carbohydrate content of the molecules, but chiefly upon their unit structure; IgG is a monomer ($\gamma_2\kappa_2$ or $\gamma_2\lambda_2$), but IgM is a pentamer, composed of five basic units. IgA may have one, two or three basic units.

The various classes of immunoglobulins have different functions. IgG is by far the most abundant of the immunoglobulins and is largely responsible for combating infections. It is easily transported across the placenta to give passive immunity to the fetus. It makes its appearance late in the primary response. IgM precedes IgG in the immune response, is effective in activating complement and forms the rheumatoid factor in autoimmune disease. IgA's main role is in the defense of body surfaces and protection against microorganisms in the gut. So far the function of IgD is unknown. IgE is responsible for allergic reactions and the release of histamine and may also combat intestinal parasites.

Each chain of the immunoglobulin molecule has two regions: a **variable (V)** region at the N-terminal end, which is part of the antibody combining site, and a **constant (C)** region at the C-terminal end. Each combining site involves the V regions of one H and one L chain; thus there are two combining sites

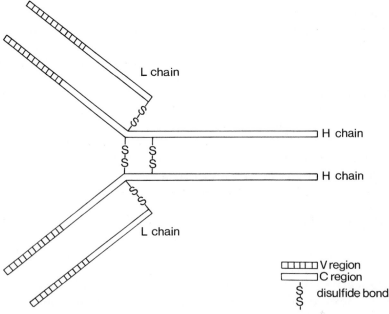

Figure 8-1 Diagrammatic representation of a gamma globulin molecule. Note that both light and heavy chains have a variable (V) region and a constant (C) region.

per immunoglobulin molecule. The variable region has three or four hyper-variable regions of about 10 amino acids each which form the active site.

Genetic Aspects of Immunoglobulin Synthesis

The complicated genetic coding that determines immunoglobulin struc-ture and allows for antibody specificity and diversity is not yet fully under-stood. There appear to be three systems of genes that are autosomal and not closely linked to one another coding for kappa chains, lambda chains and heavy chains, respectively. Within each system, the genes for the constant (C) and variable (V) regions are separately encoded. Each class of gene exists in several subgroups, which are closely linked to one another and are believed to have originated through extensive gene duplication; for example, there are 10 loci for the 10 known types of heavy-chain C regions (four γ, two α, two μ, one δ and one ϵ). Each type of C region is encoded only once. In contrast, the genes for the V regions, which hold the key to antibody diversity, must allow for the synthesis of at least a million different amino acid sequences, and there are three current hypotheses as to how this might be achieved:

1. Separate coding of all the possible V region genes in the genome.

2. Coding of a smaller number of V region genes that undergo somatic mutation, especially within the "hypervariable" regions that form their antibody-combining sites.

3. Somatic recombination within segments of V region genes.

All three of these processes appear to be involved in the generation of anti-body diversity, but their relative importance remains a matter of debate.

New evidence suggests that in addition to the genes for the V and C re-gions, there exists an independent gene or gene cluster that codes for a short region, the J region, that joins the V and C regions of the molecule.

An important additional feature of the genetics of immunoglobulin structure is that although the genes for the V and C regions of a particular immunoglobulin chain are not contiguous in the chromosome, they are tran-scribed into a single molecule of messenger RNA, which in turn is translated into a single immunoglobulin chain. The synthesis of a single chain from separate genes coding for its V, C and J regions constitutes an important exception to the principle of "one gene, one polypeptide chain." Just how the V and C regions are brought into juxtaposition, with loss of an intervening sequence, is not clear, but the process appears to be important in the commit-ment of a particular cell to the synthesis of a particular antibody. At an early stage of development there are already lymphocytes committed to the produc-tion of numerous different antibodies, though any one lymphocyte may make antibody of a single specificity. When such a cell encounters an antigen that binds to it, it locks in on production of its antibody and undergoes repeated division to generate a clone of cells similarly committed.

COMPLEMENT

After formation of an antigen-antibody complex, cell destruction depends upon a sequential series of reactions involving a complicated "cascade" of proteins known collectively as **complement.** Alper and Rosen (1976) have reviewed the genetics of the complement system in man and certain animals. Complement is ineffective without the antigen-antibody complex, but once the complex has been formed complement is responsible for the destruction of the antigen-bearing cells, usually by facilitating phagocytosis. Complement is active in the destruction of microorganisms, the production of inflammation and the rejection of implanted tissue by second-set response.

Hereditary deficiencies of almost all the components of complement are known, each usually associated with autoimmune disease or increased susceptibility to infection.

Perhaps the most common genetic disorder related to complement is deficiency of the **inhibitor** of the activated C1 component. This is the cause of the autosomal dominant disorder, hereditary angioneurotic edema.

Several components of complement have polymorphic variants, so their linkage relations have been analyzed. So far, all the components that have been mapped have proved to be on chromosome 6, closely linked to the HLA complex.

TRANSPLANTATION

In the earliest attempts at transplantation, the grafted tissue survived for only a few days; the only exception was for grafts between monozygotic (MZ) twins, which were usually accepted. This exception reflects the genetic basis of antigenic specificity. Today, because of improved knowledge of the immune reaction and methods of circumventing it, transplantation of tissues and organs has become an important area of medicine. Damaged organs such as the kidney, heart or liver can now be replaced with reasonable survival. However, much remains to be learned.

The mechanism of rejection is the immune response. The implanted tissue is antigenic to the host, which reacts by the production of T effector lymphocytes. These invade the tissue surrounding the graft and, in a manner as yet obscure, cause graft rejection.

LYMPHOID TISSUE AND THE IMMUNE REACTION

The regional lymph nodes respond to a graft of foreign tissue by cellular proliferation. The important cells in this response are small lymphocytes of two types, **T cells** (thymus-dependent cells) and **B cells** (bursa-equivalent cells). Their names come from their similarity to two cell types in birds, which have, in addition to the thymus, a separate lymphoid organ called the bursa of Fabricius at the distal end of the alimentary canal. T cells are important in cell-mediated reactions such as tissue rejection, and B cells produce humoral antibodies. There is no bursa in man or other mammals, but bone marrow appears to contain precursors of both T and B cells.

Comparison of T and B Cells

T cells are sensitive to antigens on the surface of implanted cells or of malignant cells that may arise *de novo*. They respond by proliferation and differentiation to form **blast** cells (lymphoblasts). Blast cells produce **T effector** cells that travel to the graft site and bring about rejection; **T helper** cells, which are abundant among the circulating lymphocytes and assist the B cells in response to various antigens; **primed** antigen-sensitive cells (memory cells), responsible for the second set response (see later) and probably also **T suppressor** cells, which inhibit the response of the B cells to some antigens.

B cells, which are sensitive to bacterial and viral antigens, are transformed by antigenic stimulation into **plasma** cells, which mature by proliferation and differentiation. The plasma cells of one clone synthesize only one antibody.

The relationship between these cell types is shown in Figure 8–2.

The Mixed Lymphocyte Culture Test

Foreign antigens have a **mitogenic** (mitosis-stimulating) effect on lymphocytes. This property has been exploited in the mixed lymphocyte culture (MLC) test (Bain and Lowenstein, 1964). Lymphocytes of a prospective donor are mixed in culture with those of the prospective host, and the mitogenic response is measured by the incorporation of tritiated thymidine. The greater the antigenic disparity between prospective donor and prospective host, the greater the mitogenic activity. The prospective donor whose cells elicit the least reaction is the one whose tissue is most likely to be accepted.

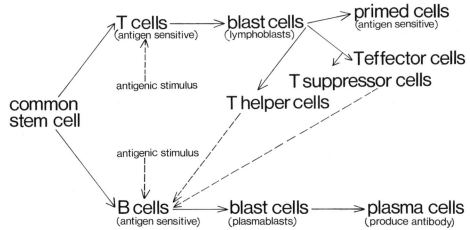

Figure 8–2   The relationships between T and B cells and their offspring. Note that T helper and T suppressor cells both act on B cells, either stimulating (helper) or inhibiting (suppressor) their responses to antigens.

NOMENCLATURE OF TYPES OF GRAFT

There is a standard nomenclature for different types of grafts. An **autograft** is a graft of the host's own tissue. An **isograft** (isogeneic graft) is a graft between genetically identical individuals, i.e., MZ twins, members of an inbred strain, or the F_1 of a cross between two inbred strains. Both autografts and isografts are normally accepted. An **allograft** (allogeneic graft) is a graft between two genetically different members of a species, and is rejected unless special precautions are taken (see p. 190). A **xenograft** is a graft between members of different species, e.g., from ape to man; this type of graft is rejected more rapidly than an allograft. The old term *homograft* is now rarely used because it is imprecise, referring to either isograft or allograft. Human grafts, except for those between MZ twins, are allografts.

THE ALLOGRAFT REACTION

The **allograft reaction,** by which a normal host rejects an allograft, is also termed the **primary response.** The graft at first may appear to be accepted and a blood supply may be established, but within a few days, the duration depending on the antigenic differences between host and donor, the graft dies and is sloughed. Different tissues show different strengths of antigenic reaction, connective tissue being the most easily accepted.

After rejection of one implant the host rejects a second graft from the same donor even more rapidly, with a **second set** (hyperacute or secondary) response brought about by the primed antigen-sensitive cells (memory cells) of the T series. The second set response is more rapid and stronger than the primary response.

TRANSPLANTATION GENETICS

The principles of transplantation genetics as they apply in man and other mammals have been uncovered largely through experimental work on inbred mice. Members of an inbred strain, some of which have been developed by brother-sister mating for well over 100 generations, are virtually identical (**isogenic**) and consequently do not reject tissue from animals of the same strain.

Tissue from a mouse of an inbred strain can, with rare exceptions, be successfully implanted into the F_1 of a cross between a mouse of the donor strain and one of another inbred strain (Figs. 8–3 and 8–4). An inbred mouse is homozygous at all loci (except for rare mutations), so its progeny receive one copy of each of its genes. Thus an F_1 hybrid does not regard tissue from either parent as foreign and therefore does not mount an immune attack against it. However, the tissue of the AB hybrid has antigens from the other parent and will not be accepted by either parent strain.

Inbred strain A × inbred strain B ⟶ Hybrid AB

Donor	Host		
	A	B	AB
A	+	−	+
B	−	+	+
AB	−	−	+

+, acceptance
−, rejection

GRAFT-VERSUS-HOST REACTION

Not only does the host react against antigenically different donor cells, but the donor cells, if they are immune-competent, also react against the antigens of the host. This is the **graft-versus-host** reaction. For many years the graft-versus-host reaction prevented marrow transplantation because the donor tissue, containing immune-competent cells, reacted to the antigens of the host with serious and often fatal consequences. Modern techniques of donor selection (see later) now allow marrow grafts to be used therapeutically in certain immune disorders, aplastic anemias or even in leukemias after destruction of the marrow cells of the host by radiation.

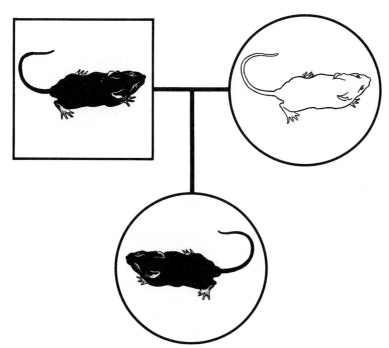

Figure 8–3 Cross between mice of two different inbred strains, C57BL/6J (black) and A/HeJ (white) producing black F₁ progeny having all the histocompatibility antigens of both parents. See also Figure 8–4.

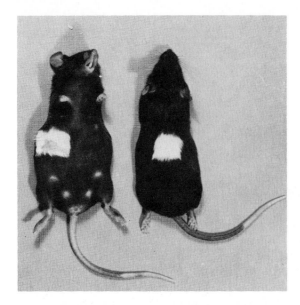

Figure 8–4 Two F_1 progeny of cross shown in Figure 8–3. The animal on the left has received a mammary gland graft (complete with skin and nipple) and the animal on the right has received a dorsal skin graft, both from donors of the white parental strain. Note that several months after transplantation both grafts are still flourishing.

If a graft-versus-host reaction occurs in a young animal, it may not be fatal but will inhibit growth and render the host weak and sickly; this is runt disease (Fig. 8–5).

DEVELOPMENT OF IMMUNE COMPETENCE

Tissue antigens form throughout fetal development, but the immune response does not develop until shortly before birth. By this time the antigens that might cause an immune response are normally present, so do not elicit an immune response. Tolerance to one's own antigens is **immunological homeostasis.** Indeed, a fetus or in some species a newborn will accept cells from a genetically different donor and allow them to persist and proliferate, making him a **chimera** with lifelong tolerance to tissue from that donor. This phenomenon, which is known as **acquired tolerance,** might eventually be exploited as a means of overcoming problems of graft rejection.

A few cases are known of human DZ twins who have exchanged hematopoietic stem cells in utero, and thus have a population of blood cells derived from the co-twin as well as their own. Such a person, known as a twin chimera, will accept tissue from the co-twin.

An exception to immunological homeostasis is autoimmunity, in which antibodies are formed against antigens of self.

HISTOCOMPATIBILITY GENES IN MOUSE AND MAN

The principles of histocompatibility revealed by studies in the mouse are of general significance in mammals, including man. In the mouse, there are 25 to 45 **histocompatibility loci** (H loci), each of which has several alleles. The

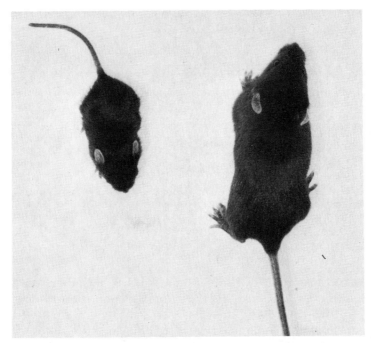

Figure 8–5   Runt disease in F_1 hybrid of a cross between C57BL/6J ♀ and A/HeJ ♂. Within 24 hours of birth, the animal on the left received an intraperitoneal injection of 10^7 spleen cells from a C57BL/6J adult. These immune-competent cells formed antibody against the A antigens of the host, producing runting. The animal on the right is a normal littermate control. Photograph courtesy of R. Escoffery.

loci are known as **H-1**, **H-2** and so forth, the alleles at a locus are designated *H-2*^{a}, *H-2*^{b} and so forth, and the corresponding tissue antigens (H antigens) are H-2^a and H-2^b. The **major histocompatibility complex** (MHC) is the highly complex H-2 locus, which determines powerful antigens that produce strong T cell responses.

Though animals of a given strain normally accept tissue of the same strain, there is one exception: in some mouse strains, females reject tissue from males of the same strain. The antigen concerned, determined by a gene on the Y chromosome, is known as the H-Y antigen. H-Y is weak in comparison to the H-2 antigens. There appears to be a comparable H-Y locus in man and other mammals.

The HLA Complex

The **HLA** (human lymphocyte A) complex in man is homologous to the H-2 locus of the mouse and is the major histocompatibility complex of man, though not the only one. The HLA complex (Fig. 8–6) is a region on chromosome 6 that contains at least four HLA loci as well as several loci related to complement synthesis, the Chido and Rodgers blood group loci and some enzyme loci (see also chromosome map, p. 247).

The four HLA loci are designated **HLA-A**, **HLA-B**, **HLA-C** and **HLA-D**.

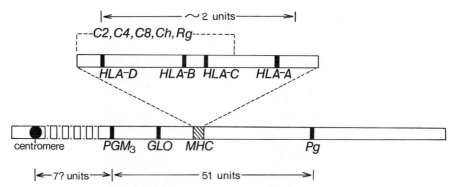

Figure 8–6 The HLA major histocompatibility complex of chromosome 6. The complex itself is just over two units long. The exact locations of complement loci *C2*, *C4* and *C8* and of blood group loci Chido (*Ch*) and Rodgers (*Rg*) are unknown. *PGM₃*, phosphoglucomutase 3 locus; *GLO*, glyoxalase locus; *Pg*, urinary pepsinogen locus.

Each locus has a number of alleles, some known but others only tentatively identified (Table 8–1). The set of HLA genes on one chromosome constitutes a haplotype; thus each individual has two haplotypes, each with four HLA determinants, for a total of eight HLA antigens. Since the genes of a haplotype move as a unit (except for rare recombinants arising by crossing over), there is one chance in four that two sibs will have identical HLA antigens (Fig. 8–7).

PREVENTION OF GRAFT REJECTION

Donor tissue is more likely to be accepted by a host if it is antigenically identical (or as closely matched as possible) or if the host's immune mechanism can be bypassed or suppressed.

If the graft is an autograft, e.g., a graft of tissue such as skin or bone from one part of the body to another, obviously no immune response occurs. There are certain types of graft in which the immune response is not important; corneal transplants survive because of certain anatomical and physiological

TABLE 8–1   DEFINED ALLELES AT THE HLA LOCI*

HLA-A	HLA-B	HLA-C	HLA-D
A1	B5	None firmly assigned at time of writing	
A2	B7		
A3	B8		
A9	B12		
A10	B13		
A11	B14		
A28	B18		
A29	B27		

*There are also 5 or more alleles at each locus tentatively identified and designated W (for Workshop) e.g., BW16, DW3.

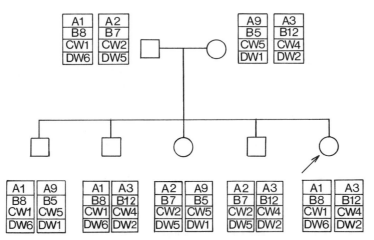

Figure 8–7 The inheritance of HLA haplotypes. A haplotype consists of four loci (HLA-A, HLA-B, HLA-C and HLA-D), each with different antigenic specificities. Since haplotypes are transmitted as units, a child has only one haplotype in common with each of his parents, but there is a one-in-four probability that two sibs will match exactly. A theoretical distribution of haplotypes in a family of five sibs is shown. The child indicated by the arrow has one matching sib. W = "Workshop," indicating that the identification of the particular antigen is still tentative.

peculiarities, and a bone graft merely acts as a framework on which new host growth can occur. Plastic prostheses, such as those used in blood vessel surgery, are not antigenic. However, if a graft must survive and function and there are no special circumstances that would permit the immune reaction to be avoided, the best approach is to select a host whose tissue will elicit the mildest possible reaction and then to suppress even this reaction.

Donor Selection

Donor selection now becomes a matter of major importance, since the less antigenicity donor tissues have for the host, the less the immune response need be suppressed. Current methods usually indicate with reasonable accuracy the likelihood that a particular host will accept tissue from a particular donor. To ease the problem of finding suitable donors there is an international register of patients needing transplants, with their HLA and blood types; when a cadaver donor becomes available, the most suitable host can be promptly determined and the required organ made available to his medical team.

Matching Tests

Matching tests are done *in vitro* to remove the possibility of causing a hyperacute reaction in the recipient. There are two general types of tests:

1. **Mixed lymphocyte culture (MLC) test.** The principle of this test has already been described. Since not all histocompatibility differences can yet be identifed by tissue typing, the MLC test is used to determine directly the presence and degree of antigenic differences between donor and host.

2. **Leucocyte typing by the lymphocytotoxicity test.** This test depends on the ability of living lymphocytes to remain impervious to certain dyes such as trypan blue that stain dead lymphocytes. An antiserum to an antigen borne by a sample of lymphocytes will, in the presence of complement, kill the cells. Antisera are prepared from women who have been pregnant or from recipients of blood transfusions. Lymphocytes of the host and each of the prospective donors are incubated in a battery of antisera, after addition of complement. Stain is then added and the number of dead lymphocytes determined. If host and prospective donor react differently to any antiserum, they differ in one or more antigens.

Relative Versus Cadaver As Donor

Since the HLA system is highly polymorphic, with rare exceptions everyone has two different haplotypes and parents are unlikely to have a haplotype in common. Thus a parent usually has only one haplotype in common with his child. Sibs, however, as noted earlier, have a one in four chance of having identical HLA haplotypes. Thus sibs are preferable to parents as donors.

Cadaver donors matched at all four HLA loci are very rarely obtained. Matches at the A and B loci had been thought to be adequate, but it is now apparent that a DR ("D-related," status at present indefinite) locus is also important and that DR typing in cadaver tissue is at present unsatisfactory (Ting and Morris, 1978). This reduces still further the prospect of finding a perfect match in a cadaver donor.

The ABO Blood Group System in Transplantation

Donor and host must be compatible, not only with respect to HLA, but also with respect to the ABO blood group system. Recall that a host who does not have antigen A or B on his red cells has the corresponding anti-A or anti-B in his serum, so is already geared to attack implanted A or B cells. The Lewis blood group system also appears to be a factor in graft rejection (Oriol et al., 1978).

Immunosuppression

To reduce the risk of graft rejection, a number of immunosuppressant drugs have been developed. The two most commonly used are azathioprine and prednisone. When azathioprine is used, the dosage must be kept as low as possible to avoid damage to cells of the host's own immune system and thus reduce his resistance to infections.

Antilymphocytic globulin, which contains antibodies to human lymphocytes, has also been used as an immunosuppressant. Success has varied, probably because of inconsistencies in the potency of the serum. X-irradiation has also been used, especially in cases of acute rejection, with some degree of success.

Why Are Some Grafts Rejected?

The major cause of graft rejection is unsuitability of the donor. The use of sib donors matched for HLA and ABO enhances the possibility of acceptance. If an unrelated donor must be used, the results are poorer. This difference in graft success is partly caused by imprecision of the testing methods and incomplete knowledge of human transplantation antigens, but often results are poor simply because no suitable donor was available.

If the host has had a previous blood transfusion, he may have had an opportunity to form primed, antigen-sensitive "memory" cells against some antigen of the donor tissue. If so, he will react by second set response to antigens whose effect might otherwise have been manageable by immunosuppressive therapy. In women who have borne children, fetal HLA antigens might similarly have sensitized the mother.

The practical value of knowledge of human transplantation genetics is illustrated by the results of some 25,000 kidney transplants, as recorded by the Human Renal Transplant Registry. Over 90 percent of grafts from HLA-compatible sibs have survived for five years or more, grafts from other living related donors have a 70 percent five-year survival rate, but grafts from cadavers have only a 50 percent chance of surviving for five years.

THE FETUS AS ALLOGRAFT

A human mother rejects tissue from her child, and an inbred female mouse rejects tissue from her offspring unless the father is of the same strain, though in both cases rejection may take longer than would otherwise be expected. However, the trophoblast (fetal tissue), which certainly acts as a graft on the mother, is not rejected. The mechanism that prevents rejection of the fetus is still not known.

Fetal blood cells enter the maternal circulation, allowing the mother an opportunity to make antibodies or T effector cells, or both, directed against fetal tissue. Some IgG antibodies cross the placental barrier from mother to fetus, but they do not appear to harm the fetus except in cases of blood group incompatibility (described in Chapter 9), and some give the newborn passive immunity to infections against which the mother has circulating antibodies.

The depression of the maternal immune response to fetal tissue may be caused by blocking factors, particularly chorionic gonadotropin, which has been shown to inhibit the mixed lymphocyte reaction. Alpha fetoprotein, a glycoprotein present in large amounts in the fetus and placenta, may also inhibit the immune response. Moreover, there is an IgG in maternal serum that may remove fetal HLA antigens and prevent them from alerting the maternal immune response.

It may also be that trophoblast cells have relatively weak tissue antigens. Fetal antigens appear to develop very slowly as the products of conception grow, and their slow development may allow gradual reduction of the function of the maternal immune system for these particular antigens.

IMMUNE DEFICIENCY DISEASES

Immune deficiency diseases, most of which are genetically determined, result from abnormality in some component of the immune response. Since the immune response requires the normal development and functioning of many cell types, it is not surprising that there are many different immune deficiency diseases. Not all the diseases in this group have been clearly defined. The genetic diseases listed here include one in which the primary defect is in the B cell, one in which it is in the T cell, several in which both cell types are deficient, and others in which the site of the defect, if known, does not appear to be primarily related to either cell type. All these disorders are associated with increased frequency and severity of infection, and in most there is an increased incidence of autoimmune disease and of malignancy, especially lymphomas.

X-linked agammaglobulinemia (Bruton disease) is the classic example of B cell deficiency. Onset is typically at six to 12 months of age. Immunoglobulins are virtually absent, and the affected children are highly susceptible to pneumococcal and streptococcal infections, though not to viral infections. The B cells and plasma cells are absent, and the tonsils are very small. There are nearly normal numbers of circulating lymphocytes, but these are T cells; the thymus is normal, and grafts are rejected.

DiGeorge syndrome (congenital thymic aplasia), which apparently has no genetic basis, is the most common type of T cell deficiency. The thymus and parathyroid glands are absent, indicating an embryological defect in the development of the third and fourth pharyngeal pouches. The patient has no T cells and no cell-mediated immune response. There is marked susceptibility to some viral and fungal infections. Effective treatment by thymic transplant has been reported.

There are several disorders in which both B and T cells are absent or greatly reduced, causing **severe combined immunodeficiency** (SCID). **Swiss-type agammaglobulinemia** is an autosomal recessive form of SCID in which the humoral immunoglobulins are markedly decreased or absent and there are very few circulating lymphocytes, and hence no cell-mediated immune response. The thymus is vestigial. Plasma cells are absent. The patient is prone to many kinds of infection and does not reject allografts. Untreated, the disease is almost invariably fatal within the first year, but recent reports indicate some success in management by transplantation of HLA-compatible bone marrow from a sib, or in a few cases by transplantation of thymus or fetal liver.

Heterogeneity in SCID was first clearly demonstrated in 1972, when Giblett et al. found total absence of the enzyme adenosine deaminase (ADA) in two affected children. It is now known that ADA deficiency accounts for about 20 percent of all cases of SCID, and that deficiency of nucleoside phosphorylase or inosine phosphorylase, other enzymes in the pathway of purine metabolism, can also produce SCID. These variants of SCID are autosomal recessive, and there are several other autosomal recessive forms as well as an X-linked form.

In several other genetic disorders of immune function, the specific defect does not appear to be in the pathway of differentiation of T and B cells. **Ataxia-telangiectasia** (AT) is an autosomal recessive marked by the development of cerebellar ataxia and telangiectasia, which are most readily seen in the sclerae. Other features of AT include severe immunodeficiency, chromosomal instability and a notable increase in the risk of malignancy which applies to otherwise unaffected heterozygotes as well as to homozygotes. **Wiskott-Aldrich syndrome** is X-linked, characterized by thrombocytopenia, eczema and susceptibility to many infections, and associated with complex immune deficiencies, both cellular and humoral. **Chronic granulomatous disease** is a disorder of the phagocytes (more correctly, a group of disorders, since there are autosomal recessive forms as well as the typical X-linked form) in which phagocytosis is normal, but phagocytosed microorganisms are not killed.

AUTOIMMUNE DISEASE

Autoimmunity, as the name implies, is the immune response to specific antigens of the self. Normally a state of immunological homeostasis exists, but this can be disturbed on either side of the antigen-antibody relationship; an antigen that has not previously been present may elicit antibody formation, or a new antibody may attack a pre-existing antigen.

Initiation of the autoimmune process can happen as a result of the appearance of a new antigen through viral infection, somatic mutation or other mechanisms. Some proteins, such as those of spermatozoa, are not synthesized until years after birth; normally spermatozoa do not cause an immune reaction because they do not reach the lymphoid tissue, but experimental injection of its own sperm into a guinea pig incites an immune response. Frequently haptens (small molecules, nonantigenic alone but capable of being antigenic when combined with other proteins) enter the body, combine chemically with larger molecules and then act as antigens.

Pre-existing antigens may be released from target tissues by infection or injury, and may then incite antibody synthesis. **Sympathetic ophthalamia** is a state in which severe injury to an eye may result weeks later in blindness of the other eye; the explanation is that sequestered antigens of the injured eye released by the injury provoked the synthesis of a new antibody that then attacked the normal eye. Similarly, injury to the heart by surgery, cardiac infarct or infection can result in autoimmune disease directed against the heart.

Homeostatic balance can also be affected on the antibody side by somatic mutation of a gene involved in antibody production, leading to the development of a "forbidden clone" of cells capable of formation of an altered antibody, which might then react with some pre-existing self antigen. This is believed to be a mechanism of acquired hemolytic anemia. It is possible that viral transformation could lead to the same effect.

Exposure to a foreign antigen resembling a self antigen may result in the formation of a new antibody capable of combining with the self antigen. For example, some streptococci produce an antigen so similar to some of the histocompatibility antigens of the human heart muscle that immunoglobulin formed in response to a streptococcal infection may adhere to human heart cells. This appears to be the cause of cardiac damage in rheumatic fever. Similar phenomena may account for rheumatoid arthritis (see below). Autoimmune thyroiditis appears to be caused by an altered thyroglobulin that incites synthesis of an antibody, which then attacks both the altered and the normal thyroglobulin. The incidence of autoantibodies increases with age; 20 percent of people over age 70 have thyroid autoantibodies.

To establish the autoimmune nature of a disease, patients must be shown to have autoantibodies that react specifically with self tissues. A number of diseases that fit this definition are shown in Table 8–2. Note that some are systemic, but others are organ-specific.

In rheumatoid arthritis there is an IgM autoantibody that combines with IgG of plasma; in other words, this autoantibody, known as rheumatoid factor, is an "antiantibody." Rheumatoid factor can be demonstrated in patients with rheumatoid arthritis, particularly in plasma cells around involved joints and in germinal centers of lymph nodes.

The tendency to form forbidden clones has a genetic basis. Rheumatoid factor may be found in relatives of patients with rheumatoid arthritis or in relatives of patients with an entirely different autoimmune disease, systemic lupus erythematosus. The familial incidence suggests a genetic cause. Patients suffering from one autoimmune disease often show a high incidence of another (e.g., pernicious anemia in patients with Hashimoto thyroiditis, or other autoimmune diseases in 70 percent of patients with autoimmune hemolytic anemia). Moreover, patients with one autoimmune disease may show a very high incidence of autoantibodies to antigens of organs not involved in the original disease, without actually being clinically affected with a second autoimmune disease. Autoimmune phenomena are

TABLE 8–2   EXAMPLES OF AUTOIMMUNE DISEASES

Systemic Diseases

Dermatomyositis, polymyositis	Scleroderma
Polyarteritis nodosa	Sjögren syndrome
Rheumatic fever	Systemic lupus erythematosus
Rheumatoid arthritis	

Organ-Specific Diseases

Disease	Organ Involved
Addison disease, idiopathic	Adrenal gland
Diabetes mellitus, juvenile*	Pancreas
Encephalomyelitis, acute disseminated	Central nervous system
Goodpasture glomerulonephritis	Kidney
Hashimoto thyroiditis	Thyroid gland
Multiple sclerosis**	Central nervous system
Myasthenia gravis	Muscle
Pernicious anemia	Gut
Sympathetic ophthalmia	Eye
Ulcerative colitis	Gut

*Diabetes mellitus is heterogeneous, and the extent of the role of autoimmunity in its etiology is still undetermined.
**Not proven

unusually frequent in immune deficiency diseases, as if a specific defect in the immune response may be an important factor in autoimmunity. The significance of all these observations is not clear, but all in all they suggest a genetic tendency to autoimmune disease.

Autoimmune diseases exhibit a predilection for connective tissue. Many of them constitute the so-called collagen diseases, in which there are quite similar pathological changes. The great variety of tissues and organs affected and the diversity of the resultant problems indicate the importance of autoimmune disorders in modern medicine.

HLA AND DISEASE

Ankylosing spondylitis is a joint disease, chiefly affecting the sacroiliac joints, in which inflammation causes ossification of the ligaments, eventually leading to fusion (ankylosis). In 90 percent of persons with this disease the HLA antigen B27 is present, though in the general population the frequency of B27 is only 7 percent. This is by far the most striking disease association known.

In recent years there have been hundreds of reports of an increased frequency of one or more of the HLA antigens in association with a specific disease. Not all the reports have been reliable, but an appreciable number have been confirmed. Though no other association is as strong as that of ankylosing spondylitis with B27, many are positive. The relative risk (defined as the ratio of the risk of developing the disease in those *with* the antigen to the risk in those *without* it) is about a hundredfold in ankylosing spondylitis, but is about tenfold in most other conditions showing positive associations.

The associations vary in specificity as well as in strength, some diseases showing association with more than one antigen. In general, the kinds of disorders for which positive associations exist show familial clustering, but do not have a clearly Mendelian pattern of inheritance. Many are regarded as immune-related. Some of the best-known associations are shown in Table 8–3.

Before discussing the reasons for these associations, it is necessary to define briefly an unusual feature of the HLA complex, **linkage disequilibrium.**

LINKAGE DISEQUILIBRIUM IN THE HLA SYSTEM

When a new mutation occurs at a locus, it joins a group of neighboring alleles at the nearby loci on the chromosome. For the next few generations the probability is that it will keep the same neighbors and will be transmitted as a unit with them. However, as time and generations pass, the new mutation will inevitably become separated from its original neighbors by crossing over and will form recombinants with other alleles of those originally neighboring loci. Eventually a state of equilibrium should be established in which the new mutation is associated with any allele at a nearby

TABLE 8–3   EXAMPLES OF ASSOCIATIONS BETWEEN HLA
ANTIGENS AND DISEASE

Disease	Antigen	Frequency of Antigen (%)	
		Patients	Controls
Addison disease, idiopathic	B8	68	24
Ankylosing spondylitis	B27	90	7
Celiac disease	A1	41	21
	B8	45–78	24
Diabetes mellitus, juvenile	DW3	–	–
Graves disease	B8	47–53	21–24
Hemochromatosis	A3	69	31
Multiple sclerosis	A3	41	24
	B7	35	26
	DW2	75	18
Myasthenia gravis	B8	52–58	24
Psoriasis	B13	16–18	4
	BW16	15	5
	BW17	29	8
Reiter syndrome	B27	78	8

From Alexander and Good, Fundamentals of Clinical Immunology, W. B. Saunders Company,
Philadelphia, 1977, and other sources.

locus in the same proportion as the frequency of that allele in the general
population.

As a simple example, consider the loci for the ABO blood group system
and the nail-patella syndrome (NP), which are known to be about 10 units
apart on chromosome 9. We do not know which ABO allele was present on
the chromosome on which a particular NP mutation originated, but by now the
alleles are in equilibrium. By this we mean that though in an individual
family NP is linked to a specific ABO allele, in the general population it is
found linked to A, B and O in proportion to the frequencies of the ABO alleles
in the population; in other words, a state of linkage equilibrium exists between the two loci.

The length of time necessary for linkage equilibrium to be achieved
depends on the distance between the two loci concerned; the closer together they are, the fewer recombinations, consequently the greater the number
of generations required to reach equilibrium.

Within the HLA complex, with its many closely linked loci, most of the
separate loci are in equilibrium with one another, but there are some outstanding exceptions. The best known exception is the A1, B8 haplotype in
North Europeans. A1 has a frequency of 0.17 and B8 a frequency of 0.11.
The expected frequency of the A1, B8 haplotype is then $0.17 \times 0.11 =
0.019$, but the observed frequency is 0.09. Thus the two loci are said to
show linkage disequilibrium.

The basis of linkage disequilibrium is unknown, but in general terms it
appears that natural selection favors the A1, B8 combination in preference
to others. Perhaps as more is learned about the nature and function of the
genes in the HLA complex, the reason will become more clear.

REASONS FOR DISEASE ASSOCIATIONS

There are three possible explanations for an association between a genetic marker and a specific disease:

1. Ethnic stratification. The disease could be common in an ethnic subgroup in which, by chance, certain HLA antigens are also common.

2. Direct causation. Susceptibility to the disease could be a direct consequence of the presence of the particular allele.

3. Linkage disequilibrium. Susceptibility to the disease could be determined by an allele at a nearby locus, which shows linkage disequilibrium with the disease-associated HLA locus.

Most workers favor the third explanation and expect that as more is learned about the genes within the HLA complex and as family studies improve our knowledge (since they are not affected by either ethnic stratification or linkage disequilibrium), the true susceptibility genes and their genetic patterns will be clarified.

GENERAL REFERENCES

Alexander, J. W., and Good, R. A. 1977. *Fundamentals of Clinical Immunology.* W. B. Saunders Company, Philadelphia.

Harris, H. 1975. *The Principles of Human Biochemical Genetics.* 2nd ed. North-Holland Publishing Company, Amsterdam and London; Elsevier North-Holland Inc., New York.

Marx, J. L. 1978. Antibodies I: New information about gene structure. *Science* 202:298–299. Antibodies II: Another look at the diversity problem. *Science* 202:412–415.

Snell, G. D., Dausset, J., and Nathenson, S. 1976. *Histocompatibility.* Academic Press, New York.

Watson, J. D. 1976. *Molecular Biology of the Gene.* 3rd ed. W. A. Benjamin Inc., Menlo Park, Calif.

PROBLEMS

1. An F_1 hybrid mouse produced by crossing a female of the C3H strain with a male of the C57BL/6 strain is called a C3B6F₁.

   a. From which of the following will a $C3B6F_1$ normally tolerate skin grafts?
      1) C3H
      2) C57BL/6
      3) $C3B6F_1$
      4) Wild mouse
      5) Mouse of another inbred strain

   b. Which of the following will normally accept a skin graft from a $C3B6F_1$?
      1) C3H
      2) C57BL/6
      3) $C3B6F_1$
      4) Wild mouse
      5) Mouse of another inbred strain

2. Arrange the following relatives of a recipient in order of their suitability as donors of a kidney transplant: Sib, father, MZ twin, uncle, cousin, mother, DZ twin.

9

BLOOD
GROUPS
AND OTHER
GENETIC
MARKERS

Numerous polymorphisms are known in components of human blood, especially in the antigens of the blood group systems. Because of their ready classification into different phenotypes, their simple mode of inheritance and their different frequencies in different populations, these polymorphisms are useful genetic markers in family and population studies and in linkage analysis. Many of them, especially the ABO and Rh blood group systems, have important clinical applications. Though all the genetic polymorphisms of blood, including the HLA system (discussed in Chapter 8), might appropriately be called blood groups, the term is restricted in usage to the red cell antigens.

The usefulness of a genetic trait as a genetic marker depends upon the following characteristics:

1. A simple and unequivocal pattern of inheritance. Most of the blood groups, enzyme variants and serum protein variants are determined by systems of codominant genes, i.e., both alleles of heterozygotes are phenotypically expressed, permitting the genotype to be inferred directly from the phenotype. (Exceptions include the O allele of the ABO blood group system, which has no detectable product.)

2. Accurate classification of the different phenotypes by reliable techniques.

3. A relatively high frequency of each of the common alleles at the locus. If the rarer of a pair of alleles is encountered in only a small percentage of a population, it is not likely to be very useful for family studies and linkage analysis. Equal frequency of the two alleles is the ideal, since this situation provides the best opportunity for segregation to be observed at the locus.

4. Absence of effect of environmental factors, age, interaction with other genes or other variables on the expression of the trait.

The study of the blood group systems has permitted many important contributions to human genetics and to the elucidation of genetic principles. Multiple allelism at a locus was first demonstrated in man by the ABO blood group genes. In linkage studies, the first four autosomal linkages to be found all involved blood groups. The X-linked blood group system Xg is a standard point of reference on the X chromosome. Xg has also been useful in investigation of X inactivation and sex chromosome aneuploidy. Population genetics has made extensive use of the blood groups. Interaction of nonallelic genes is demonstrated by the ABO-H-Secretor-Lewis relationship and by other biosynthetic pathways less thoroughly explored so far. The long debate over whether the C, D and E antigens of the Rh system are determined by different closely linked alleles or are different expressions of a single locus has illustrated the difficulties of deciding between these possibilities in a human system and has provided insight into the nature of genes.

BLOOD GROUPS

The human blood groups occupy a special place in medical genetics both because of their many contributions to the development of genetic principles and because of their clinical importance in blood transfusion and obstetrics.

The first successful transfusion of human blood was performed in 1818, but transfusion for therapeutic purposes did not become reasonably safe until the discovery of the ABO blood group system by Landsteiner in 1900. Even when the donor's blood group is the same, with respect to the ABO system, as that of the patient, careful matching must be carried out to detect antibodies reacting with other red cell antigens.

In a single year, about 1 percent of North Americans receive one or more blood transfusions and thus have an opportunity to form antibodies against antigens on the donor's red cells.

The clinical significance of the blood groups in maternal-fetal incompatibility results from the fact that sensitization of a pregnant woman to antigens of her fetus (usually, though not always, antigens of the Rh system) can produce hemolytic disease in her newborn child. It should also be remembered that the survival of transfused cells may be curtailed by the "immune" antibodies formed by multiparous women. Until recently, hemolytic disease of the newborn ranked among the first 10 causes of perinatal death, and led to physical or mental handicap in surviving children who were inadequately treated. The development of a means of prevention of sensitization of the mother has been one of the most significant advances in obstetrics and pediatrics in recent years.

BLOOD GROUP NOTATION

The notation for blood group genes has been developed in a piecemeal fashion and is riddled with inconsistencies. Three chief kinds of notation are used:

1. Alternative alleles may be designated by letter sequences (A and B, M and N).

TABLE 9–1 THE BLOOD GROUP SYSTEMS MENTIONED IN THE TEXT

System	Main Antigens°	Discovered by°°	Comments
ABO	A_1, A_2, B, H	Landsteiner, 1900	Major clinical importance
MNSs	M, N, S, s	Landsteiner and Levine, 1927	Most useful marker, clinically insignificant
P	P_1, P_2, p^k	Landsteiner and Levine, 1927	–
Rh	C, C^w, c, D, D^u, E, e and many others	Landsteiner and Wiener, 1940	Major clinical importance
Lutheran	Lu^a, Lu^b	Callender et al., 1945	Lu-secretor linkage was the first known.
Kell	K, k, Kp^a, Kp^b, Js^a, Js^b	Coombs et al., 1946	Some clinical significance
Lewis	Le^a, Le^b	Mourant, 1946	Related to ABO; see text.
Duffy	Fy^a, Fy^b	Cutbush et al., 1950	First to be assigned (to chromosome 1)
Kidd	Jk^a, Jk^b	Allen et al., 1951	Occasional clinical importance
Diego	Di^a, Di^b	Layrisse et al., 1955	Oriental marker
Cartwright	Yt^a, Yt^b	Eaton et al., 1956	–
Auberger	Au^a	Salmon et al., 1961	–
Dombrock	Do^a, Do^b	Swanson et al., 1965	Useful marker
Colton	Co^a, Co^b	Heist et al., 1967	–
Sid	Sd^a	Macvie et al., 1967; Renton et al., 1967, and previous workers	–
Scianna	Sc1, Sc2	Lewis et al., 1974, and previous workers	–
Secretor	–	Friedenreich and Hartmann, 1938, and previous workers	An "honorary" system; see text.
I	I, i	Wiener et al., 1956	Chiefly immunological importance
Chido	Ch^a	Harris et al., 1967	Linked to HLA
Rodgers	Rg^a	Longster and Giles, 1976	Linked to HLA
Xg	Xg^a	Mann et al., 1962	X-linked

°Not a complete list.
°°References prior to 1975 are from Race and Sanger, Blood Groups in Man, 6th ed., Blackwell Scientific Publications, Oxford, 1975.

2. Alternative alleles may be designated by large and small letters (S and s, K and k, C and c). Note that use of a lower-case letter does not here imply that the trait is recessive.

3. Alternative alleles may be designated by a symbol with a superscript (Lu^a and Lu^b, Fy^a and Fy^b).

A blood group system is now usually named for a person in whom the antibody was first recognized (e.g., Duffy and Kidd), and some part of this name is used to designate the gene locus concerned (Fy for Duffy, Jk for baby J. Kidd). The Lutheran system, on the other hand, was named for the donor of the provoking antigen. For the more recently discovered blood groups, there is a standard notation that can be described with reference to the Kidd groups. The symbol Jk is used for the locus, and the known genes are Jk^a and Jk^b. A third allele has been postulated and is symbolized by merely Jk without a superscript, to show that the antibody defining the corresponding antigen has not yet been discovered. The two known antigens are called Jk^a and Jk^b, and they are detected by antibodies anti-Jk^a and anti-Jk^b. The reactions of red blood cells are described as Jk(a+b+), Jk(a+b−), Jk(a−b+) or Jk(a−b−), depending upon the results of testing the cells with anti-Jk^a and anti-Jk^b.

The following description of the best-known blood group systems makes extensive use of Race and Sanger's *Blood Groups in Man* (6th edition, 1975), to which readers are referred for additional information. Though most of the known blood group systems receive at least a brief mention in the following pages, not all the known antigens have been included and complications of some systems have been omitted, especially when they do not elucidate new principles.

Twenty blood group systems and the "honorary" blood group system, Secretor, are listed in Table 9–1. The first example of the antibody defining each system was found as follows: naturally occurring in healthy subjects (ABO, Lewis); in immunized animals (MN, P, Rh); in the mothers of infants with hemolytic disease of the newborn (Rh, Kell, Kidd, Diego); or during cross-matching tests.

In addition to the blood group systems shown in Table 9–1, there are several others whose rank as independent systems is still uncertain. An example of the discovery of a "new" system that eventually was shown to be part of an old system is provided by "Gonzales." The antigen Go^a was originally defined in 1962 by an antibody that caused mild hemolytic disease of the newborn, and was later found to be restricted to blacks among whom its incidence was about 2 percent. It was eventually recognized that Go^a belongs to the Rh system, in which it is one of several types of unusual D antigen.

Public and Private Antigens.   A number of other antigens are known that are either very common (public) or very rare (private). Public and private antigens probably differ from ordinary blood group systems only in their incidence. It is hard to know whether an ultrarare or ultracommon antigen belongs to some known blood group system or is actually part of a new system, but in either case its frequency detracts from its utility as a genetic marker, because there are so few families in which segregation is demonstrable at the locus concerned.

THE ABO BLOOD GROUP SYSTEM

The first notable event in blood group history is Landsteiner's discovery of the ABO blood groups. He and his colleagues found that the blood of any individual belongs to one of four different types. These types are distinguished by an agglutination reaction; if the serum of the recipient is mixed with cells from an incompatible donor, the cells are agglutinated. The clumped red cells, or "agglutinates," can usually be observed with the naked eye.

There are four major ABO phenotypes, known as O, A, B and AB. The reaction of each phenotype with specific anti-A (−A) and anti-B (−B) antibodies is as follows:

Red Cell Phenotype	Reaction With	
	−A	−B
O	−*	−
A	+	−
B	−	+
AB	+	+

*−denotes no agglutination; +denotes agglutination.

TABLE 9–2　THE ABO BLOOD GROUPS

Blood Group (Phenotype)	Genotype	Antigens on Red Cells	Antibodies in Serum
O	OO	neither	anti-A, anti-B
A	AA AO	A	anti-B
B	BB BO	B	anti-A
AB	AB	A, B	neither

Group A individuals possess antigen A on their red cells, group B individuals possess antigen B, group AB individuals possess both A and B, and group O individuals possess neither. The genetics of the ABO blood groups has been outlined in Chapter 4 as an example of multiple allelism.

A remarkable feature of the ABO groups is the reciprocal relation between the antigens present on the red cells and the antibodies in the serum of the same individual, summarized in Table 9–2. (Serum and plasma are interchangeable terms in this connection. Serum is more satisfactory than plasma for laboratory tests.) Anti-A is found in the serum of persons whose cells do not contain A, and anti-B is found in persons whose cells do not contain B. The reason for this reciprocal relation is uncertain, although it is believed that formation of anti-A and anti-B is probably a response to A and B substances occurring naturally in the environment. However they originate, the regular presence of anti-A and anti-B explains the failure of many of the early attempts to transfuse blood.

An antibody (immunoglobulin) present in the plasma of the recipient will cause agglutination of donor red cells if the donor cells carry the corresponding antigen (Table 9–3). Hence, there are "compatible" and "incompatible" combinations, a compatible combination being one in which the red cells being tested (or the red cells of a donor) do not carry an antigen corresponding

TABLE 9–3　AGGLUTINATION REACTIONS WITHIN THE ABO BLOOD GROUP SYSTEM

Blood Group	Recipient Antibodies (Serum or Plasma)	O	A	B	AB
O	anti-A, anti-B	—*	+	+	+
A	anti-B	—	—	+	+
B	anti-A	—	+	—	+
AB	neither	—	—	—	—

*— indicates no agglutination; + indicates agglutination of donor cells by antibodies in recipient's serum (or plasma).

to the antibodies in the test (or recipient) serum. Antibodies in the donor's plasma are not usually taken into account in transfusion, presumably because they are greatly diluted in the recipient's circulation. Although there are theoretically "universal donors" (group O) and "universal recipients" (group AB), a patient is always given blood of his own ABO group, except in real emergencies.

Subtypes of the ABO groups can be recognized. The most important is the separation of group A into A_1 and A_2, with a corresponding separation of AB into A_1B and A_2B. A_2 is recessive to A_1, i.e., the A_1A_2 genotype has the same phenotype as A_1A_1 or A_1O. About 85 percent of group A bloods are A_1. Other variants of both A and B are known, but they are rare.

Frequency of the ABO Groups and Genes

Because the relative proportions of the ABO groups vary widely in different populations, frequency figures are valid only for the specific population from which they are derived. In white North Americans, groups O and A usually have frequencies in the 40 to 50 percent range, group B about 10 percent and group AB 3 to 4 percent. The following figures will serve as an illustration of a population in Hardy-Weinberg equilibrium (see Chapter 15).

Blood Group	Frequency
O	0.46
A	0.42
B	0.09
AB	0.03
	1.00

For the example given above, the corresponding frequencies of the alleles O, A and B are 0.68, 0.26 and 0.06, respectively. In other populations, the frequencies might be quite different; for example, group B is much more common in Eastern Europe and Asia than in Western Europe or areas of Western European immigration. Though American Indians are of Oriental origin, almost all lack the B allele and most also lack A, unless it has been introduced recently from some other source.

Linkage Relations of the ABO Locus

One of the first human linkages to be described was that of the ABO locus with the locus for nail-patella syndrome, which has already been mentioned. The recombination frequency between the two loci is about 10 percent. Later, the locus for adenylate kinase, a polymorphic red cell enzyme, was found to be very close to the nail-patella locus. The ABO:nail-patella syndrome:adenylate kinase group was the first triplet linkage group found in man. It has now been assigned to chromosome 9, which like most of the other human chromosomes is rapidly filling up with markers.

TABLE 9-4   EXAMPLES OF SIGNIFICANT ASSOCIATIONS BETWEEN ABO BLOOD GROUPS AND DISEASE*

Disease	Comparison	Relative Risk
Gastric cancer	A:O	1.2
Tumors of salivary glands:		
malignant	A:O	1.6
nonmalignant	A:O	2.0
Cancer of cervix	A:O	1.1
Duodenal ulcer	O:A	1.3
Gastric ulcer	O:A	1.2
Rheumatic diseases	Non-O:O	1.2
Diabetes mellitus	Non-O:O	1.1
Ischemic heart disease	Non-O:O	1.2

*Summarized from Vogel, 1970. Am. J. Hum. Genet. 22:464–475.

ABO and Disease Associations

In Chapter 8 some of the disease associations of alleles at the HLA locus were listed, and the theoretical reasons for the associations were briefly discussed. Similar but weaker associations have been found for the ABO blood groups. Theoretically, if a gene is to be maintained in a population at a frequency higher than can be explained by recurrent mutation, it must confer some selective advantage, yet for the ABO groups three alleles are maintained at quite high frequencies even though there is no obvious advantage conferred by any genotype. Research has so far failed to demonstrate that any particular blood group is advantageous (Reed, 1969), so any differences that exist must be small and statistically undemonstrable except in very large studies. The relation of the ABO groups to disease is well known, but it may not be of great genetic import because the diseases concerned usually affect people in middle or later life, after the peak reproductive period.

Disease Susceptibility.   The idea that certain blood groups might be associated with susceptibility or resistance to certain categories of disease was ridiculed for many years, but by now numerous studies all over the world, with highly consistent results, make it impossible to doubt that certain associations are real. The first strong evidence was produced by Aird et al. (1953), who described an excess of group A among patients with gastric cancer, an observation that has been repeatedly confirmed. An even closer association exists between duodenal ulcer, group O and nonsecretion of ABH substances, and several other associations have been suggested. Table 9–4 summarizes the best-known associations.

It is possible that the ABO blood groups play a part in blood coagulation. In fact, the reported association of group O with duodenal ulcer may be a consequence of the greater risk of hemorrhage in duodenal ulcer patients of group O, who because of bleeding are hospitalized and therefore more likely to be ascertained. In contrast, group A patients are thought to be at increased risk of thrombosis because of an increased clotting tendency. These observations, if confirmed, could have a bearing on the risk of thrombosis in women on oral contraceptives. Because oral contraceptives reduce the level of an-

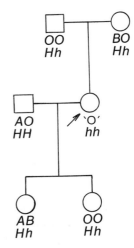

Figure 9-1 A pedigree of the Bombay phenotype (genotype *hh*), described in the text. After Levine et al., Blood 10:1100–1108, 1964, by permission.

tithrombin III, they increase the probability of thrombosis, and this may be serious in group A women.

H Substance

H substance (H antigen) is the substrate from which the A and B substances are made by the action of the *A* and *B* genes. The *O* gene is thought to be an "amorph," i.e., a completely inactive gene; thus group O cells contain unaltered H substance. Anti-H is found in the serum of some subjects of group A_1; and in all subjects of the "Bombay" phenotype.

The Bombay Phenotype

The Bombay phenotype, which is very rare, is produced by interaction between the ABO genes and a rare mutant at a different locus. Its name comes from its discovery in 1952 in Bombay when two men needed transfusion, one after stab wounds and the other after a railway accident. Their cells were not agglutinated by anti-A, anti-B or anti-H; their serum contained all three antibodies.

Figure 9–1 is a pedigree showing the Bombay phenotype. In this family a woman whose red cells were grouped as O produced an AB child. Family studies led to the conclusion that the woman's "true blood group" was B, but she could not form the B antigen. (It is noteworthy that her parents were consanguineous.) Most persons possess a gene *H* (*HH* or *Hh*) required for the development of the H substance from a precursor. Persons of the Bombay phenotype are homozygous for an inactive allele (amorph) *h*. When H is not formed, the enzymes determined by *A* and *B* genes have no substrate on which to act, so that *hh* persons cannot make the A or B antigen even if they have the *A* or *B* gene.

Secretion of Blood Group Substances

The A, B and H antigens of the ABO blood group system occur in other tissues as well as on the red cell, and in most people are secreted in a

TABLE 9–5 ABH ANTIGENS IN SECRETORS AND NONSECRETORS

Blood Group	Antigens of Secretors Red Cells	Saliva	Antigens of Nonsecretors Red Cells	Saliva
O	H	H	H	—
A	A	A and H	A	—
B	B	B and H	B	—
AB	A and B	A, B and H	A and B	—

water-soluble glycoprotein form in a number of body fluids as well as in an alcohol-soluble glycosphingolipid form on the red cells. The ability to secrete these substances is determined by a secretor gene, *Se*; its allele, *se*, has no known function. Secretors of H, A and B are either *SeSe* or *Sese*; nonsecretors (about 22 percent of the population) are *sese*. The ABH antigens on the red cells and in the saliva of secretors and nonsecretors are shown in Table 9–5.

Genetic Pathways of Synthesis of ABO Blood Group Substances

The biosynthesis of the ABO blood group substances has been unraveled chiefly by the work of Watkins and Morgan. Only one of their many papers on the subject is cited in our references (Morgan and Watkins, 1969). They have studied the biosynthesis of blood group substances in secretions, but it is assumed that red cell antigens are synthesized in a similar process.

In its basic structure a secreted blood group substance is a macromolecule (more precisely, a family of closely related macromolecules) with a polypeptide core to which are attached a large number of sugar chains. The antigenic specificity of the molecule is conferred by the terminal sugars (Fig. 9–2). These sugars are added to the core by the action of specific transferases, which presumably are the direct products of the *H, A* and *B* genes. The steps may be summarized as follows:

SUGAR TRANSFERASES

Gene	Transferase	Acceptor	Sugar Added*	Blood Group Substance
H	H-transferase	Precursor	L–fucose	H
h	—	—	—	Unchanged precursor
A	A-transferase	H substance	GalNAc	A
B	B-transferase	H substance	D–Gal	B
O	—	—	—	Unchanged H

*Abbreviations: GalNAc, N–acetylgalactosamine, D–Gal, D–galactose

The relationship between the blood group genes, their products and the blood group substances is shown schematically in Figure 9–3.

Landsteiner's discovery of the ABO blood groups in 1900 and Garrod's work on inborn errors of metabolism (dating from about 1902) both coincided closely in time with the rediscovery of Mendel's laws. For many years the two

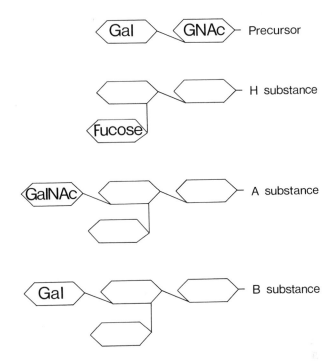

Figure 9–2 Diagrammatic representation of the terminal sugar sequences of the polysaccharide chains of secreted blood group substances. H specificity results from the addition of L-fucose to a precursor chain. H substance is converted into A substance by the addition of an N-acetyl-D-galactosamine (GalNac) residue or into B substance by the addition of a D-galactose residue. Each biosynthetic step is catalyzed by a specific transferase enzyme.

types of genetic variation seemed to be quite different; blood groups involved antigenic differences and came in a variety of normal forms, whereas inborn errors were concerned with rare variants in enzymes. We now see that in both, the basic mechanism is a gene mutation causing synthesis of a variant enzyme. The difference in frequency is a consequence of the relative biological fitness of the normal and mutant forms. We will return to this topic in a later chapter (Chapter 15).

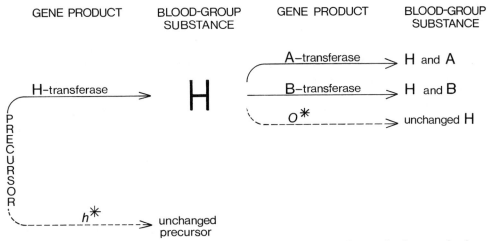

Figure 9–3 Diagrammatic representation of the genetic pathways leading to the biosynthesis of the H, A and B blood group substances. *Genes h and O are amorphs, i.e., they have no detectable effect on the precursor. (Based on the work of Morgan and Watkins and of Ceppellini.)

The Lewis Blood Group System and Its Relationship to ABO and Secretor

In the Lewis blood group system there are two known antibodies, anti-Le^a and anti-Le^b, which define two antigens, Le^a and Le^b. In most whites the red cells are either Le(a+b−) or Le(a−b+). Le(a−b−) is quite rare in whites but relatively common in West Africans. The Le(a+b+) phenotype is not observed in adults.

Unlike other blood group systems, the Lewis groups did not at first lend themselves to a simple genetic explanation, because of their complicated interactions with the ABO and secretor systems. Secretors (*SeSe* or *Sese*), as previously noted, secrete ABH substances in water-soluble form. Persons whose red cells are Le(a+b−) *never* secrete ABH substances in the saliva, persons whose red cells are Le(a−b+) *always* do so, and persons whose red cells are Le(a−b−) are *usually* secretors.

The interpretation is that the Lewis substances are not primarily red cell antigens but antigens of the body fluids, and are only secondarily absorbed onto the red cells from the plasma. The presence of the Le^a and Le^b antigens in the serum and saliva is governed by a pair of allelic genes, *Le* and *le*. The interactions of the Lewis and secretor genes produce the following *red cell* phenotypes:

Lewis Genotype	Secretor Genotype	Red Cell Phenotype	Approximate Frequency (whites)
LeLe or *Lele*	*SeSe* or *Sese*	Le(a−b+)	0.69
LeLe or *Lele*	*sese*	Le(a+b−)	0.26
lele	*SeSe, Sese* or *sese*	Le(a−b−)	0.05

The interaction of the four loci, H, secretor, Lewis and ABO can now be summarized (Table 9–6). Note that though the H and ABO loci are involved in the synthesis of both the secreted substances (water-soluble glycoproteins) and the red cell substances (alcohol-soluble glycolipids), the Lewis and secretor loci are involved only in the synthesis of the glycoproteins, not the glycolipids, and the products of their interaction with the H and ABO loci are present on the red cell only because they have been absorbed from the plasma.

THE MNSs BLOOD GROUP SYSTEM

After the discovery of the ABO blood group system in 1900, no other blood group antigens were found until 1927, when Landsteiner and Levine used a different approach. They injected human blood into rabbits, which developed antibodies against it. The "immune serum" formed by the rabbits

TABLE 9-6   PHENOTYPES PRODUCED BY INTERACTION OF *H, SECRETOR,*
                        LEWIS AND *ABO* GENES

Genotype			Antigens of Red Cells			Antigens of Secretions		
H	*Se*	*Le*	ABH	*Le^a*	*Le^b*	ABH	*Le^a*	*Le^b*
HH or *Hh*	*SeSe* or *Sese*	*LeLe* or *Lele*	+	−	+	+	+	+
"	*sese*	*LeLe* or *Lele*	+	+	−	−	+	−
"	*SeSe* or *Sese*	*lele*	+	−	−	+	−	−
"	*sese*	*lele*	+	−	−	−	−	−
hh	*SeSe* or *Sese*	*LeLe* or *Lele*	−	+	−	−	+	−
hh	*sese*	*lele*	−	−	−	−	−	−

+ indicates antigens present; − indicates antigens absent. The A and B antigens present, determined by the alleles present at the ABO locus, are not shown individually in this table.

could distinguish between different human red cell samples. The antigens so recognized were called M and N. The P blood group system was found in the same series of experiments.

As originally described, the MN groups were models of genetic simplicity. They appeared to depend upon a pair of codominant alleles, *M* and *N*, roughly equal in frequency, which produced three genotypes, *MM, MN* and *NN*, and three corresponding phenotypes, M, MN and N.

Blood Group (Phenotype)	Genotype	Reaction With		Approximate Frequency (European)
		Anti-M	Anti-N	
M	*MM*	+	−	0.28
MN	*MN*	+	+	0.50
N	*NN*	−	+	0.22

The Ss subdivisions of the MN groups, which make the genetic pattern less simple, were discovered some 20 years later when the first example of anti-S was recognized in Sydney. It now appears that combinations of MN and Ss are inherited as units, i.e., *MS, Ms, NS* and *Ns*. The genetic interpretation of these combinations is not yet clear; the situation is somewhat similar to that for the Rh groups (see below). Rarely in blacks and frequently in Congo Pygmies, red cells have been found that lack both S and s.

The MN groups have little importance in blood transfusion or in maternal-fetal incompatibility. Their major significance in medical genetics is that their relative frequencies and codominant pattern of inheritance make them useful in the solution of identification problems.

Many population surveys show a greater number of MN heterozygotes than would be expected on the basis of the Hardy-Weinberg equilibrium. The reason for this excess is not known; there is no obvious heterozygote advantage that would explain it.

THE Rh BLOOD GROUP SYSTEM

The Rh blood groups rank with the ABO groups in clinical importance, because of their relation to hemolytic disease of the newborn (HDN) and their importance in transfusion. The Rh antigen was so named because Rhesus monkeys were used in the experiments in which it was discovered. Anti-Rh reacts with the red blood cells of about 84 percent of white North Americans, who are thus Rh positive; the remaining 16 percent are Rh negative.

The Rh system is genetically complex, but as an introduction it can be simply described in terms of a single pair of alleles, *D* and *d*. Rh positive (Rh+) persons are *DD* or *Dd,* and Rh negative (Rh−) ones are *dd*. A *dd* woman may form anti-D when pregnant with a *Dd* fetus, and any *dd* person can form anti-D in response to transfusion with Rh+ blood. As described below, there are many other specificities in the Rh system, but the great majority of cases of HDN recognized clinically are caused by anti-D.

To describe the genetics of the Rh system more completely, it is necessary to introduce two separate sets of symbols: the CDE gene notation of Fisher and Race and the R gene notation of Wiener.

According to Fisher and Race, the Rh blood groups are determined by a series of three closely linked genes, *C, D* and *E*, with allelic forms *c, d* and *e*. Antibodies directed against the antigens determined by all of these except *d* are known, but anti-d has never been detected. There are eight possible combinations of these six genes, which are haplotypes comparable to the HLA haplotypes but are more often called gene complexes. Their frequencies are very different in different parts of the world. The highest frequency of Rh− individuals is found in European populations, especially in the Basques, where it reaches 30 percent. Close to 100 percent of Orientals and North American Indians are Rh+. The following are the frequencies of the eight most common Rh gene complexes in an English population.

CDe	0.41	*cdE*	0.01
cde	0.39	*Cde*	0.01
cDE	0.14	*CDE*	low
cDe	0.03	*CdE*	low

Sometimes these genes are listed in the order DCE, since there is some suggestion that this is the order of the three genes on the chromosomes.

Each person has two Rh gene complexes, making the total number of diploid combinations of the eight gene complexes listed above 36. Additional alleles (*C*ʷ, *D*ᵘ and so forth) are known, making the total well over 100. Rh− persons are *dd* irrespective of the presence of *C* and *E*, but because of the comparative rarity of *Cde, cdE* and *CdE* as compared with *cde* the great majority of Rh− persons are *cde/cde*.

Wiener, one of the co-discoverers of the Rh system, did not accept the three-allele theory, but preferred to think of Rh as a single locus with a large series of alleles. His notation for the eight gene complexes listed above is as follows.

Fisher and Race Notation	Wiener Notation
CDe	R^1
cDe	R^0
cDE	R^2
CDE	R^z
Cde	r'
cde	r
cdE	r''
CdE	r^y

The three most frequent genotypes in North American whites are:

CDe/cde	R^1r	0.33
CDe/CDe	R^1R^1	0.17
cde/cde	rr	0.15

In Africa cDe and cde both occur with high frequency and other combinations are all uncommon. It is speculated that all the other types might have arisen from cDe by mutation followed by recombination, as shown in Figure 9–4.

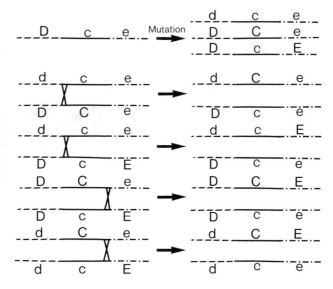

Figure 9–4 Diagrammatic representation of the possible origin of the Rh gene complexes by mutation and recombination. Note that to generate the very rare CdE combination, one of the two types undergoing recombination must itself be a recombinant.

Unusual Rh Genotypes

Among the rare variants of Rh, there is one in which the Rh antigens are completely absent. This is Rh_{null}, first found in an Australian aborigine, which is sometimes associated with a hemolytic anemia known as Rh_{null} disease. In another very rare variant, D is present but the C and E series of antigens are missing. In individuals with either of these variants, incompatibility problems are likely to arise in transfusion or pregnancy.

Hemolytic Disease of the Newborn

One of the major contributions of genetics to medicine has been in the discovery of the Rh system and its role in hemolytic disease of the newborn (HDN). In HDN, the life span of the fetal red cells is shortened by the action of specific antibodies formed by the mother against antigens of the child and transmitted to the fetus by placental transfer. The disease thus begins *in utero* and continues as long as some maternal antibody remains in the infant, which is up to three months after birth. Thus, though its basis is a genetically determined antigenic difference between mother and child, HDN is an acquired hemolytic anemia and must be clearly distinguished from hereditary anemias such as those caused by G6PD deficiency or hereditary spherocytosis.

There are two main types of HDN: one due to Rh incompatibility, when the mother is Rh− and the fetus is Rh+, and the other due to ABO incompatibility, when the mother is O and the fetus A. Most cases of HDN recognized clinically are due to Rh incompatibility; ABO incompatibility is difficult to diagnose but tends to be so mild as to require no treatment. HDN is only rarely caused by incompatibility for other blood group systems, such as Kell or Duffy. In a white population, HDN occurs approximately once in 100 births.

Mechanism of Hemolytic Disease of the Newborn

Normally during pregnancy small amounts of fetal blood cross the placental barrier and reach the maternal blood stream. When this happens in an Rh− mother with an Rh+ fetus, the Rh+ fetal cells may stimulate the formation of anti-Rh, which is then transferred to the fetal circulation (Fig. 9–5), where it attaches to the red cell membranes.

In the fetus, red cells that are heavily "coated" with anti-Rh are rapidly removed from the circulation. The fetus becomes anemic and responds by releasing large numbers of erythroblasts (nucleated immature red cells) into its blood (thus accounting for one of the names sometimes used for this disease, *erythroblastosis fetalis*). Hydrops, a consequence of the anemia, may cause intrauterine death.

After birth, the rapid destruction of red cells produces a large amount of bilirubin, which causes jaundice to appear during the first day of life. If hyperbilirubinemia is not promptly prevented by replacement transfusion, in which the infant's cells are replaced by Rh− cells, deposition of unconjugated bilirubin in the brain (kernicterus) may produce cerebral damage, which few

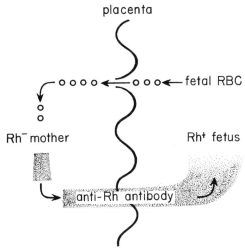

Figure 9–5 Diagrammatic representation of the immunization of an Rh-negative woman by cells from her Rh-positive fetus. The mother is stimulated to form anti-Rh, which passes to the fetal blood stream and causes hemolysis of the fetal red cells.

infants survive. Those who do survive may have high-frequency deafness, mental retardation or the athetoid type of cerebral palsy.

Factors Influencing the Development of Rh Hemolytic Disease

HDN can develop only when the fetus has inherited from the father an antigen that the mother lacks, but the number of genetic opportunities for HDN is far in excess of the number of cases that actually develop. The probability that a given infant will develop HDN depends upon a number of factors.

1. The father as well as the mother may be Rh−, so the fetus can only be Rh−.

2. The father may be heterozygous and transmit to the fetus his d rather than his D allele. Calculating from Hardy-Weinberg considerations, one can estimate that even when her husband is Rh+, the chance that an Rh− woman is carrying an Rh+ fetus is only 60 percent.

3. The fetus may be protected because the mother is O and the fetus non-O (i.e., A or B). The protection afforded by incompatibility on the ABO blood group system has long been recognized. If the mother is group O she has naturally occurring anti-A and anti-B, and it is believed that these antibodies will destroy any A or B fetal red cells in the maternal circulation before they have an opportunity to stimulate anti-Rh. The probability that a mother is O and has an non-O fetus is about 15 percent. (The frequency of group O times the frequency of non-O alleles = $0.46 \times 0.32 = 0.15$).

4. Usually an Rh negative mother does not become immunized until she has had at least one Rh-incompatible pregnancy or transfusion. In fact, with modern treatment she may never become immunized, since primary immunization to the Rh antigen on fetal cells can now be prevented in nearly all cases by treatment of the Rh− mother of an Rh+ fetus with anti-Rh immune globulin, prepared from people who have become sensitized by transfusion or incompatible pregnancy and have the capacity to form potent antibodies.

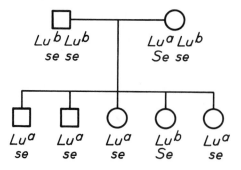

Figure 9–6 A pedigree demonstrating Lutheran-secretor linkage, described in the text. From Sanger and Race, Heredity 12:513–520, 1958.

(This capacity itself appears to be genetically determined.) Because abortion and amniocentesis for prenatal diagnosis have been incriminated as means of Rh sensitization, nonimmunized Rh— women who undergo amniocentesis or who abort should receive Rh immune globulin as a prophylactic.

THE LUTHERAN BLOOD GROUP SYSTEM

The particular genetic interest of the Lutheran blood groups is that they provided the first example of autosomal linkage and crossing over in man, the Lutheran-secretor linkage. A pair of codominant alleles, Lu^a and Lu^b, determine the chief blood groups within the Lutheran system, Lu^a being much less common (0.04) than Lu^b (0.96). Figure 9–6 illustrates the Lutheran and secretor genotypes in a family described by Sanger and Race (1958). The father is homozygous for Lu^b and se, and the mother is heterozygous at both loci; this is the classical "back-cross" mating, in which the children's phenotypes reveal the genotype of the heterozygous parent. Only the maternal contribution to the children's genotypes is shown, since the father can transmit only Lu^b and se. In this family, the genes Lu^a and se are obviously on one maternal chromosome and Lu^b and Se on the other. This particular pedigree does not show crossing over between the two loci, but other family studies have shown that the recombination frequency is about 10 percent.

The myotonic dystrophy locus is in the same linkage group. This linkage relationship can theoretically be exploited in suitable families for the prenatal diagnosis of myotonic dystrophy, because it is possible to determine the secretor phenotype of the fetus from amniotic fluid. (A secretor fetus has H substance in the amniotic fluid.)

THE KELL BLOOD GROUP SYSTEM

The Kell system was originally described in terms of two alternative phenotypes, Kell positivē (K+) and Kell negative (K—), though like other blood group systems it has become more complicated as new antibodies have been found to detect antigens other than K and k. Kell positive people make up 2 to 9 percent of various populations, with lower frequency in blacks than in whites. The Kell system is sometimes implicated in hemolytic disease of the newborn, if a K− mother forms anti-K in response to fetal K antigens. This is one possible cause of a sequence of stillbirths. As in Rh hemolytic disease, protection is afforded if the mother and fetus are ABO-incompatible.

It has recently been found that the precursor of the Kell antigen is determined by an X-linked locus, Xk, which is very near to Xg but not identical to it.

THE DUFFY BLOOD GROUP SYSTEM

At the Duffy locus there are two known alleles, Fy^a and Fy^b, and a third has been postulated on the basis of family studies.

Genotype	Phenotype	Approximate Frequency (whites)
$Fy^a Fy^a$ $Fy^a Fy$	Fy(a+b−)	0.20
$Fy^a Fy^b$	Fy(a+b+)	0.46
$Fy^b Fy^b$ $Fy^b Fy$	Fy(a−b+)	0.33
$FyFy$	Fy(a−b−)	0.001

The Fy (a−b−) phenotype, or in other words the $FyFy$ genotype, is very common in Central and Western Africa and in American blacks (40 percent or more $FyFy$), but very rare in whites (about 0.1 percent). Recently it has been reported that the antigens Fy^a and Fy^b may act as red cell receptors for malarial parasites. Persons of the Fy(a−b−) phenotype might therefore have a degree of resistance to malaria, and this selective advantage might account for the high Fy frequency in Africa. In whites, the frequency of the Fy gene is about 0.03 ($\sqrt{0.001}$), and this means that a relatively large number of people are heterozygous Fy^aFy or Fy^bFy. This can lead to errors in family testing. For example:

In the above pedigree, failure to recognize that the father is Fy^aFy rather than Fy^aFy^a, and that the child is Fy^bFy rather than Fy^bFy^b, could lead to the false interpretation that the man is not the biological father of the child.

The distinction of the Duffy locus as the first to be assigned to any autosome has been mentioned earlier. Once Duffy had been assigned, markers were rapidly added to chromosome 1 (see Fig. 11–5; Chapter 11).

THE DIEGO BLOOD GROUP SYSTEM

The major interest of the Diego system is that the Di^a antigen appears to be a specific marker for Orientals. Its frequency varies from about 36 percent in some South American Indian tribes to 10 percent in Japanese to practically 0 percent in most other groups. Surprisingly Canadian Eskimos, though undoubtedly of Mongolian origin, are entirely Di(a−). An antithetical antigen Di^b has also been identified.

Genotype	Phenotype	Approximate Frequency Mongolian	Other
Di^aDi^a Di^aDi	Di (a+)	0.10	0
$DiDi$	Di (a−)	0.90	1.00

THE Xg BLOOD GROUP SYSTEM

The Xg blood group's special importance, which has been mentioned earlier, is that its locus is on the X chromosome and appears not to participate in X inactivation like other X-linked loci. The antibody that first defined the Xg^a antigen was found in a man who had been transfused many times because of a familial vascular disorder. Only a small number of other examples of anti-Xg^a have been found so far. The genotypes and phenotypes of the Xg system are shown on page 74, where they are described as examples of X-linked inheritance. The phenotype frequencies are:

Genotype		Phenotype	Approximate Frequency (whites)
Males	Xg^aY XgY	Xg(a+) Xg(a−)	0.67 0.33
Females	Xg^aXg^a Xg^aXg	Xg(a+)	0.89
	$XgXg$	Xg(a−)	0.11

The Xg locus has been useful in mapping the X chromosome, as shown by a recent tentative map (Fig. 9–7). It is also an informative marker in some cases of X chromosome aneuploidy (Chapter 7) and a useful illustration for calculation of the population frequency of X-linked genes (p. 292).

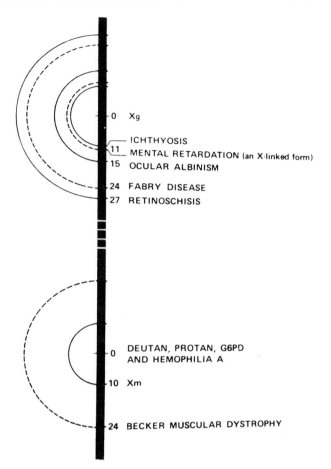

Figure 9-7 A tentative map of the human X chromosome, based on linkage studies in human families. The semicircles are used to indicate that though the distance of a given locus from a marker (Xg or the cluster of loci that includes G6PD) is known, the position of the loci relative to one another has not been established. Solid lines, established linkage; interrupted lines, very probable linkage. From McKusick, *Mendelian Inheritance in Man. Catalogs of Autosomal Dominant, Autosomal Recessive and X-Linked Phenotypes.* 5th ed., Johns Hopkins University Press, Baltimore, 1978. After Race and Sanger, *Blood Groups in Man.* Blackwell Scientific Publications, Oxford, 1975.

ADDITIONAL INDEPENDENT BLOOD GROUP SYSTEMS

Several other blood group systems have been defined or are in the process of definition. Table 9–1 lists nine of these, which will not be discussed individually here: P, Cartwright (Yt), Auberger (Au), Dombrock (Do), Colton (Co), Sid (Sd), Scianna (Sc), Rodgers (Rg), Chido (Ch) and I. Of these, the most useful as a genetic marker is Dombrock, because approximately 64 percent of North European whites are Do(a+).

THE "USEFULNESS" OF BLOOD GROUP SYSTEMS AS GENETIC MARKERS

The blood groups and other polymorphic characters are useful in a number of types of genetic problems. Disputed paternity may be the most common problem, but it is not the only one. Occasionally, newborns are accidentally interchanged in hospital nurseries, or one or both sets of parents suspect there has been an interchange. The proportion of extramarital children, mainly but not exclusively in firstborns, is high enough to permit misinterpretation of family data in genetic studies, so confirmation of pedigree information by blood grouping may be necessary, particularly if the condition under study is rare or the other findings (in carrier tests, for example) seem anomalous. Determination of twin zygosity usually requires objective tests, and blood groups are the most convenient. These and several other types of problems can often be solved by blood groups. If all the known blood group systems are considered, several million different phenotypes can be defined, and the most common of these (in England, which is better studied than most countries in this respect) has a frequency of one in 1000 in females, one in 1,429 in males. The combination is: O, Ms/Ns, P, CDe/cde, Lu(a—b+), K—k+, Le(a—b+), Fy(a+b+), Jk(a+b+), Yt(a+b—), Au(a+), Do(a+b+), Co(a+b—), Sc1,—2 and Xg(a+).

A polymorphism's "usefulness" is its probability of distinguishing between two random individuals, and usefulness is measured as one minus the sum of the squares of the phenotype frequencies, which is, of course, the chance that two random samples will not have the same phenotype. By this rule, the order of usefulness of 14 blood group systems is as follows:

MNSs	83.6 percent	
Rh	80.5 "	
ABO	67.2 "	(using subgroups A_1 and A_2)
Duffy	62.7 "	
Kidd	62.5 "	
Dombrock	61.8 "	
Xg (males)	43.9 "	
Lewis	43.3 "	
P	33.4 "	
Xg (females)	19.0 "	
Kell	16.3 "	
Colton	15.8 "	
Yt	14.9 "	
Lutheran	14.1 "	

Some antigens are particularly useful in making ethnic distinctions. The antigen V of the Rh system occurs in 27 percent of American blacks, but only in 0.5 percent of whites. Fy(a—b—) is found in over 40 percent of American blacks and Js^a of the Kidd system in 20 percent, but neither occurs in whites. Di(a+) is virtually confined to Orientals. As this kind of information grows, it can give objective evidence about population roots and migrations.

BLOOD GROUP CHIMERAS

Monozygotic twins frequently have a common prenatal circulation, but dizygotic twins, even when they have fused placentas, do not usually develop anastomoses between their circulations. It is therefore very rare indeed in man (though quite usual in cattle) for any mixture of blood to occur between dizygotic twin fetuses. Only about 12 cases have been recorded in which one or both members of a dizygotic twin pair have a mixture of two kinds of blood. In the first such case to be described, a woman was observed to have a mixture of A and O cells. It was later found that she was a twin whose twin brother had died 25 years earlier. A person of this kind, with blood cells of two different genotypes derived from two different zygotes, is called a **blood group chimera.** (Dispermic chimeras, probably usually formed by fusion of DZ twin zygotes or early embryos, are also known.)

In blood group chimeras, the exchange of hematopoietic cells takes place in fetal life, prior to the development of immune competence. The foreign cells do not stimulate the host to form antibodies against them, so they are accepted in the host's bone marrow, where they can persist and multiply throughout life. Since only hematopoietic cells and not germ cells have been exchanged, the host can, of course, transmit to his offspring only his own blood group genes (i.e., his "true" blood group).

GENETIC MARKERS IN SERUM

The technique of starch gel electrophoresis, introduced by Oliver Smithies in 1955, has led to the discovery of many polymorphisms of great interest to geneticists and biochemists. In this technique, starch gel is used as a supporting matrix in which the protein components of a sample of serum or a tissue extract can be separated in an electrically charged field. The proteins in the sample migrate at different rates, determined chiefly by their electrical charge. Electrophoresis of a serum sample, followed by protein staining, reveals a pattern of bands in which the different serum proteins assume characteristic positions, as shown in Figure 9–8. When the usual pattern has been established, variants can be detected by differences in their speed of migration and thus in their position in the electrophoregram, or occasionally by the complete absence of one component.

When a mutation results in an altered polypeptide, the alteration may bring about a difference in net electrical charge and thus a difference in

Figure 9–8 Diagrammatic representation of the proteins of normal serum separated by starch gel electrophoresis.

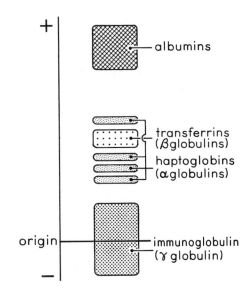

electrophoretic pattern. Only those variants that have a different charge can be detected electrophoretically, and it is believed that only one-third of base substitutions in a gene would result in a detectable variant (Harris, 1971).

Several other ·supporting media for electrophoresis, such as acrylamide gel and agarose gel, are now available, but starch gel is still widely used.

In this section, only two of the known polymorphisms of serum are described. One of these is the haptoglobin polymorphism, which was discovered by Smithies in the first application of the technique of starch gel electrophoresis and has proved to be of particular genetic significance because it illustrates a new mechanism of genetic variation. The second is the protease inhibitor (Pi) polymorphism, which has growing clinical significance.

HAPTOGLOBIN

Haptoglobins are α globulins with the property of binding hemoglobin. There are three main types in man, Hp 1–1, Hp 2–1 and Hp 2–2, determined by two allelic genes, Hp^1 and Hp^2.

Genotype	Phenotype
Hp^1Hp^1	Hp 1–1
Hp^1Hp^2	Hp 2–1
Hp^2Hp^2	Hp 2–2

Hp^1 has two subtypes: Hp^{1F} (fast) and Hp^{1S} (slow). Since Hp^{1F}, Hp^{1S} and Hp^2 are allelic, several phenotypes are possible.

The three common haptoglobin types form distinct patterns with starch gel electrophoresis. Hp 1–1 forms a single zone, and Hp 2–2 forms a series of zones such as one might expect to find with a mixture of polymers of a basic unit similar to Hp 1–1. Hp 2–1 has a polymeric series different from Hp 2–2 and unlike a mixture of Hp 1–1 and Hp 2–2 (Fig. 9–9).

The polypeptides corresponding to Hp^{1F} and Hp^{1S} differ in only one amino acid, but Hp^2 is responsible for a different polypeptide that resembles a theoretical fusion of parts of two Hp¹ chains. Smithies et al. (1962) proposed that this similarity resulted from *nonhomologous crossing over* between a pair of chromosomes, so that part of one Hp^{1F} gene has been added to part of an Hp^{1S} gene (Fig. 9–10). The principle of nonhomologous crossing over with resulting partial gene duplication could be of major evolutionary significance, since it permits formation of "new" genes without actual chemical change, and it could allow parts of two different genes to combine so that the resulting single polypeptide may have properties formerly divided between two polypeptides. This unusual crossover event seems to have happened only once, perhaps quite early in human evolution. The Hp^2 allele seems to be in the process of replacing Hp^1 as the common allele at the locus (Fig. 9–11). It is useless to speculate about what its selective advantage might be, when the physiological role of haptoglobin is poorly understood and its total or near-total absence (quite common in some black populations) does not seem to be deleterious.

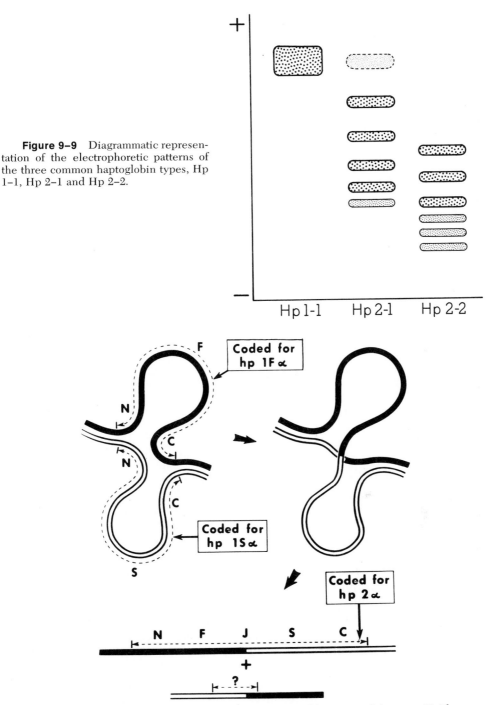

Figure 9–9 Diagrammatic representation of the electrophoretic patterns of the three common haptoglobin types, Hp 1–1, Hp 2–1 and Hp 2–2.

Figure 9–10 Diagrammatic representation of the possible origin of the gene Hp^2 by nonhomologous crossing over in an $Hp^{1F} Hp^{1S}$ heterozygote. The chromatid from one parent (shown in solid black) contains the code for the hp 1F α-polypeptide, and the chromatid from the other parent (shown by a double line) contains the allelic code for hp 1S. The locus is shown on each chromosome by a dotted line. Nonhomologous crossing over between these two alleles produces a "new" gene, Hp^2, coding for the hp 2 α-polypeptide. The new gene contains almost the whole of the Hp^1 and Hp^2 genes, joined at point J. A gene with a deletion corresponding to the duplication is also formed. From Smithies, Connell and Dixon, Nature 196:232–236, 1962.

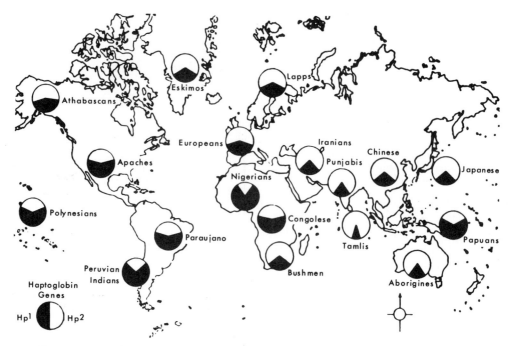

Figure 9-11   Geographic distribution of the *Hp¹* and *Hp²* alleles. From Kirk, *The Hapto-globin Groups in Man.* Karger, Basel, 1968, and Harris, *The Principles of Human Biochemical Genetics.* 2nd ed., North-Holland Publishing Company, Amsterdam and London; Elsevier-North Holland, New York, 1975, by permission.

THE PROTEASE INHIBITOR (Pi) SYSTEM

Alpha-1 antitrypsin is a serum protein of the α_1 fraction that can neutralize the activity of trypsin and other proteolytic enzymes. It is one of several protease inhibitors in serum. The *Pi* locus is highly polymorphic, and its mutant alleles give rise to varying degrees of reduced activity. The identification is complicated, requiring either electrophoresis in one direction followed by "crossed electrophoresis" in agarose containing anti-α1AT antibodies, or isoelectric focusing (Fig. 9–12).

Here we will mention only two alleles, the common allele *Pi*ᴹ and the deficiency allele *Pi*ᶻ. *Pi*ᶻ*Pi*ᶻ homozygotes have alpha-1 antitrypsin deficiency, which is associated with severe lung and liver disease in childhood. *Pi*ᴹ*Pi*ᶻ heterozygotes and persons with other genotypes giving rise to moderate reduction of alpha-1 antitrypsin activity appear to have a greater risk of impaired lung function due to smoking than do ordinary *Pi*ᴹ*Pi*ᴹ homozygotes. Alpha-1 antitrypsin deficiency is, of course, not the only cause of pulmonary emphysema and impaired lung function, but it is one of the genetic factors that confers genetic susceptibility upon some members of the population. Other associations of the *Pi* polymorphism with disease are under study.

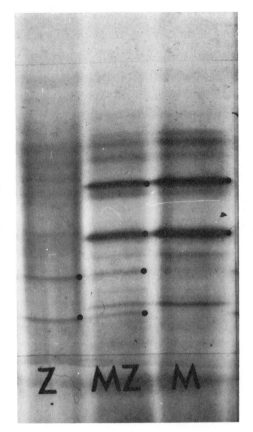

Figure 9–12 Genetic variants (Pi types) of alpha-1 antitrypsin, as separated by isoelectric focusing in acrylamide gel (LKB pH 4–5). Dots indicate major bands for each type. Deficient type is Pi type Z (genotype ZZ). Photograph courtesy of D. W. Cox.

RED CELL ACID PHOSPHATASE

Red cell acid phosphatase is an enzyme which, despite its name, is not restricted to the red cell, though it may be the only acid phosphatase present there. It is presented here because it is not only a commonly used genetic marker of blood, but is also an example of quantitative variation resulting from a simple underlying system of genetic variation and may be a model for many other systems in which quantitative variation exists.

The basic data are presented in Table 9–7. There are three main alleles that together give rise to six phenotypes. (Other less common variants are omitted.) The different variants have different means and ranges of enzymatic activity, though there is considerable overlap. In the general population the range of activity is wide and the distribution unimodal.

As Table 9–7 shows, the activity of the different phenotypes appears to support the concept that each allele is responsible for a certain average activity (ACP^A 62 units, ACP^B 94 units and ACP^C 118 units).

TABLE 9-7   RELATIONSHIP OF RED CELL ACID PHOSPHATASE ACTIVITY TO GENOTYPE*

Genotype	Mean Activity
$ACP^A ACP^A$	122
$ACP^B ACP^A$	154
$ACP^B ACP^B$	188
$ACP^C ACP^A$	184
$ACP^C ACP^B$	212
$ACP^C ACP^C$	No data

*Data from Spencer et al., 1964. Nature 201:299–300.

In Figure 9–13, the distribution of red cell acid phosphatase in the general population and the distribution for each phenotype are plotted. It is apparent that the supposedly unimodal curve is actually a composite of the sums of the various individual distributions.

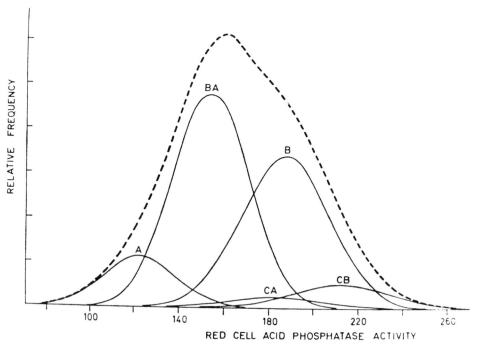

Figure 9–13   Distribution of red cell acid phosphatase activities in the general population (dotted line) and in the separate phenotypic classes (solid lines). Note that the apparent continuous distribution represents the sum of several separate distributions. From Harris, *The Principles of Human Biochemical Genetics*. 2nd ed., North-Holland Publishing Company, Amsterdam and London; Elsevier-North Holland Inc., New York, 1975, by permission.

GENERAL REFERENCES

Giblett, E. R. 1969. *Genetic Markers in Human Blood.* Blackwell Scientific Publications, Oxford.

Harris, H. 1975. *The Principles of Human Biochemical Genetics.* 2nd ed. North-Holland Publishing Company, Amsterdam and London; Elsevier North-Holland Inc., New York.

Morgan, W. T. J., and Watkins, W. M. 1969. Genetic and biochemical aspects of human blood group A-, B-, H-, Le^a- and Le^b-specificity. Br. Med. Bull. 25:30–34.

Race, R. R., and Sanger, R. 1975. *Blood Groups in Man.* 6th ed. Blackwell Scientific Publications, Oxford.

PROBLEMS

1. An Rh-negative woman has an Rh-negative child.
   a) What are the possible genotypes of the father?
   b) What proportion of children born to Rh-negative women are Rh negative?

2. An AB mother has an AB child.
   a) What are the possible blood groups of the father?
   b) What proportion of matings of AB women could theoretically produce an AB child?

3. A woman of blood group B, N has a child of blood group O, MN. She states that a certain man of blood group A, M is the father of the child.
   a) Can the man be excluded as the father on this evidence?
   b) If the man and woman are both Rh positive and the child is Rh negative, can the man be excluded as the father?
   c) If the man is now found to be a secretor, whereas the mother and child are nonsecretors, can he be excluded?
   d) If the putative father's blood groups were A, N, Rh negative, what would your conclusion be?

4. Assign each of the following children to the right set of parents.

Children	Parents
O	AB × O
A	A × O
B	A × AB
AB	O × O

5. A woman who in infancy was affected with hemolytic disease of the newborn marries a man who did not have hemolytic disease but had several older sibs who were affected.
   a) Give the probable genotypes of the couple and their parents.
   b) If they have several children, what is the chance that they will have a child with hemolytic disease?

10
SOMATIC CELL GENETICS

An entirely new approach to human genetics has evolved in recent years through technical advances in the study of cultures of somatic cells that allow them to be manipulated much as cultures of microorganisms have been used by molecular geneticists. The range of problems that can be investigated by the techniques of somatic cell genetics is broad; it encompasses the analysis of the expression of mutant cells in culture, gene mapping, the study of differentiated function and malignancy. Findings from studies in somatic cell genetics have already been mentioned elsewhere in this book. Here we will briefly describe the background and methodology of such studies and some of the kinds of new information they provide.

There are many advantages to studying genetic disorders in cell culture rather than in living patients. Obviously, if a cell line from a patient with a rare disorder or unusual karyotype can be established when the patient is available, it can be studied later at any convenient time. Established cell lines can be revived after having been frozen in liquid nitrogen for long periods. They can be mailed to specialized laboratories or stored in a central depository for use by many different investigators. Cell lines can also be used for a variety of experimental procedures that might raise both practical and ethical problems if they were attempted with living patients.

The usefulness of cultured somatic cells for research in medical genetics depends upon three main factors:

1. The feasibility of growing cells in long-term culture.

2. The possibility of identifying and characterizing genetic differences in cultured cells.

3. The possibility of performing some type of genetic analysis with the cells.

Much of the success of somatic cell genetics has come about because each of these prerequisites has been achieved.

CELL CULTURES

Techniques for cell culture have taken a long time to develop and to be widely used, partly because of the demanding nutritional requirements of cultured cells and partly because successful maintenance of a culture for a long period requires elaborate precautions to avoid contamination. Only a few types of cells can be cultured successfully; highly differentiated tissues, such as muscle, usually deteriorate very rapidly in culture. The kinds of cells most frequently used in cultures are the following:

1. **Peripheral lymphocytes** in short-term culture. These cells do not meet the requirement of long-term survival, as they persist in culture for only 72 hours or so and are never available in the quantities needed for most biochemical assays. However, they are used so extensively for chromosome analysis and in immunogenetics that they cannot be overlooked.

2. **Fibroblasts.** Fibroblasts are the most useful cell type for genetic studies. They are cultured from small explants of skin or other tissues, set up in culture vessels in nutrient media and maintained under closely controlled environmental conditions. The cells, which have a characteristic spindle shape (Fig. 10–1), grow in monolayers and must be subcultured at frequent intervals because overcrowding leads to contact inhibition and cessation of growth. Usually they have normal karyotypes, though clones of cells with abnormal karyotypes arise in culture with a fairly high frequency, perhaps reflecting loss of rigid control over the stability of the karyotype in culture. Though fibroblast cultures have a fairly long life span, they undergo senescence after 50 to 60 generations in culture, the duration of survival being longer in cultures set up from a young donor than in cultures from an older person.

3. **Permanent lines** of transformed cells. Some cell lines undergo malignant transformation in culture, either spontaneously or experimentally by viral transformation, and other cell lines have been established directly from malignant tissue. These lines do not have the property of contact inhibition and do not undergo senescence. They have unstable karyotypes. The best-known example is the HeLa cell line, established many years ago from a patient with cervical carcinoma. The HeLa cell line is heteroploid, with 63 to

Figure 10–1 Fibroblast culture under low-power magnification. Fibroblasts are spindle-shaped and grow as a monolayer.

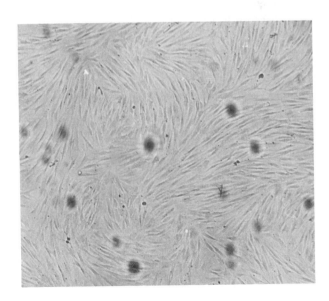

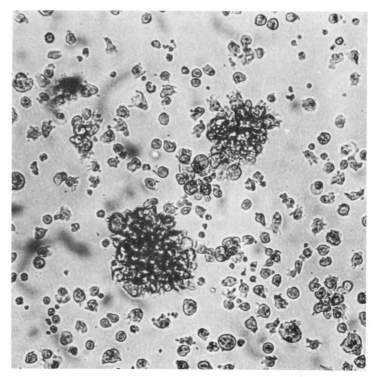

Figure 10–2 Lymphoblastoid cells in culture, under high-power magnification. The cells grow in suspension and typically form small clumps.

65 chromosomes in its various sublines or even within a line. The growth properties of HeLa cells are so exceptional that a few years ago it was found that cell lines in many laboratories had become contaminated by HeLa cells in the course of laboratory procedures and that the HeLa cells had overgrown and replaced the original lines.

4. **Lymphoblastoid cell lines.** Though cultures of peripheral blood lymphocytes do not ordinarily persist in culture, they can be induced to do so if. they have been infected (perhaps transformed) by Epstein-Barr virus, the cause of infectious mononucleosis. Lymphoblastoid cell lines have several advantages; they are easier to manipulate experimentally than fibroblasts, because they grow in small clumps in suspension rather than in monolayers (Fig. 10–2), and they do not undergo senescence. A disadvantage is that the karyotype is somewhat unstable, and specific types of karyotypic alteration develop within the cultures over a period of time.

5. **Amniotic fluid cell cultures.** Samples of amniotic fluid obtained from a pregnant mother by amniocentesis contain cells derived from the fetal skin and mucous membranes and from the inner surface of the amnion. These cells of fetal origin include a small fraction that have at least some growth potential in culture, though as a rule amniotic fluid cell cultures do not grow vigorously or for many generations. These fetal cells may be epithelial cells or fibroblasts. They are used to determine the fetal karyotype or to look for a specific genetic marker (or absence of such a marker), for example, the pres-

ence or absence of an enzyme determined by a gene for which both parents are heterozygous.

PHENOTYPIC EXPRESSION IN CELL CULTURE

Medical cytogenetics rests on the principle that the karyotype of cultured cells is, with rare exceptions, characteristic of the organism from which the culture was derived. A fortunate aspect of somatic cell genetics is that this principle holds with even greater accuracy for many genetic markers. Though fibroblasts and lymphoblasts do not express the whole range of structural genes in the genome from which they are derived, both do express a large number of genes concerned with metabolic functions. Many normal and variant enzymes that are expressed in blood or other tissues and can be examined by electrophoretic techniques, enzyme assays or other methods are also expressed in cultured cells. Thus cultured cells can be used in the investigation of metabolic disorders and of individual variation in many genetic markers. They have been especially useful in prenatal diagnosis of genetic defects.

Unfortunately, many enzymes and other proteins expressed in specialized cell types in the intact organism are not expressed in cultured cells. Consequently, cell culture has been less valuable so far in the analysis of genetic variants of enzymes characteristic of liver, muscle or other specialized tissues than in the analysis of enzymes of more widespread distribution.

SOMATIC CELL HYBRIDIZATION

The technique of somatic cell hybridization is based on the finding that cells of different origin grown together in culture will occasionally fuse with one another (Barski et al., 1960). Though the frequency of spontaneous fusion is low, it can be greatly enhanced by the addition of inactivated Sendai virus or polyethylene glycol to the culture medium. Using Sendai virus to improve the likelihood of cell fusion, Harris and Watkins (1965) showed that hybrids could be obtained between human and mouse cells, and it is now known that hybridization can be accomplished both between cells of widely different species and between different cell types from members of the same species.

When cell fusion does take place, a further problem is how to select the fused cells from the background of other cells (unfused cells, or fused cells of identical instead of different origin) in which they are growing. In the earliest experiments, only hybrid cells with growth advantages over the parental cells could be isolated, but a major improvement in technique came with the invention of the HAT technique (Littlefield, 1964). This method is shown schematically in Figure 10–3.

HAT SELECTION

The HAT technique is so named because it makes use of hypoxanthine, aminopterin and thymidine. Purine analogues such as 8-azaguanine (8AZA)

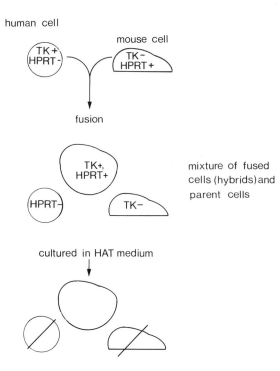

Figure 10–3   Schema of the HAT selection technique described in the text.

are toxic when converted to nucleotides by the action of the enzyme hypoxanthine-guanine phosphoribosyl transferase (HPRT). Cells lacking HPRT (HPRT−) are resistant to 8AZA. Cells from patients with Lesch-Nyhan syndrome are naturally HPRT−, and mutants with the same deficiency arise occasionally in culture. Either of these can be selected from a mixed culture because they resist 8AZA and survive, whereas HPRT+ cells are killed. Similarly BUdR (bromodeoxyuridine), an analogue of thymidine kinase (TK), will select for TK− cells in culture. The drug aminopterin inhibits the synthesis of both purine and thymidine. Thus cells that lack either HPRT or TK fail to grow if aminopterin is present in the culture. As illustrated in Figure 10–3, if an HPRT− human cell line and a TK− mouse cell line are fused and plated in HAT medium, both the parent lines are killed and only the fused cells, which have mouse HPRT and human TK, are capable of survival.

CLONING

When hybrid cells have been selected by their survival, they are grown as clones for further study. A clone is a cell line derived from a single cell. All the cells of a clone have the same karyotype (though the clone must be monitored regularly to ensure that the karyotype remains the same; see below). They also have the same active X chromosome, as would be predicted on the basis of the Lyon hypothesis.

CHROMOSOME LOSS IN HYBRID CELLS

A further discovery of crucial importance to the usefulness of somatic cell hybridization came about when it was found that in human-rodent hybrid cells, the human chromosomes are selectively lost, rapidly at first and then more gradually (Weiss and Green, 1967). The reason for this phenomenon is still not known, but it is a rule that in any interspecific hybrid, one or the other chromosome set is preferentially lost. Thus it is possible to analyze man-mouse hybrids that have the full set of mouse chromosomes but different complements of human chromosomes, or perhaps even a single remaining human chromosome. Though this is not genuine genetic segregation, it does give the opportunity of performing experiments that allow genetic conclusions to be drawn, and it has been extensively applied to map human enzyme genes to their respective chromosomes.

The assignment of human genes to chromosomes by means of somatic cell hybridization makes use of the electrophoretic differences that exist between human and mouse enzymes. For example, mouse G6PD can readily be distinguished from human G6PD by electrophoresis; in an electrophoregram from a hybrid culture, the presence of the human G6PD band confirms the presence of the human X chromosome (Fig. 10–4).

The first gene to be assigned to its chromosome on the basis of genetic analysis in somatic cells was the TK gene. This work was part of the same study that demonstrated the sequential and preferential loss of human chromosomes from hybrid lines. The mouse cell line used was TK–, so with

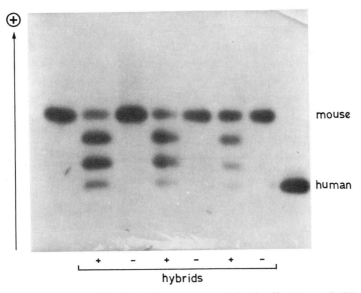

Figure 10–4 Expression of G6PD in human-mouse hybrid cells. Mouse G6PD is at the left, human G6PD at the right, with G6PD from six human-mouse hybrid lines between them. Of the six hybrid lines, three have the human enzyme and three lack it. The intermediate bands seen in the hybrids result from fusion of mouse and human enzyme subunits. Photograph courtesy of S. Povey.

selection in HAT medium only those hybrids possessing human TK survived. Eventually it was apparent that the only chromosome consistently present in the surviving lines was chromosome 17, which must therefore have the TK locus.

LINKAGE ANALYSIS

An enzyme locus can be assigned to a specific chromosome when the enzyme phenotype is retained when that chromosome is retained, and lost when it is lost. The following example (Creagan and Ruddle, 1977) is a simplified illustration. Clones of somatic cell hybrids are selected in which there are certain combinations of human chromosomes, as shown. The three clones are examined for the expression of a given phenotype. If the phenotype is expressed in all three clones, its locus is on chromosome 1; if it is expressed in clones A and B but not C, its locus is on chromosome 2; and so on.

HUMAN CHROMOSOMES

	1	2	3	4	5	6	7	8
Clone A	+	+	+	+	−	−	−	−
Clone B	+	+	−	−	+	+	−	−
Clone C	+	−	+	−	+	−	+	−

Stocks of hybrid clones covering all of the 24 different human chromosomes (22 autosomes, X and Y) are available, so it is usually unnecessary to make new hybrids to look for linkages. Instead, the existing panels can be used to find the chromosomal locus of any enzyme that remains unmapped. Probably most of the genes that can be assigned to human chromosomes by the somatic cell hybridization techniques have already been located (see linkage map, Chapter 11). A number of general observations are emerging: for example, enzymes in the same metabolic pathway are scattered, not clustered, on the map; loci for lysosomal enzymes are also scattered, and genes for mitochondrial and soluble forms of the same enzyme are separated; the gene map is closely similar in man and chimpanzee, but not in man and mouse; and X-linked genes are, without exception, conserved throughout the mammals.

Somatic cell hybridization gives a different type of information from that learned from classical linkage studies in families. It shows that certain loci are syntenic, but not how far apart they are, which is the first thing learned in classical genetic studies. Nothing is known about the genetics of many of the assigned loci except their chromosomal location and the fact that they differ in rodent and man.

SELECTIVE SYSTEMS OTHER THAN HAT

HAT selection has been very widely used in somatic cell genetics, but a number of other methods for the selection of fused cells or for the identifica-

tion of mutants have been devised. One convenient trick is the use of peripheral lymphocytes as the human parent in hybrids, since the very short life
span of peripheral lymphocytes means that they will soon be lost.

Most experiments in mutagenesis make use of a selective system designed to separate a particular kind of mutant from the pool. For example,
addition of the cardiac glycoside ouabain to the pool will result in the selection of mutants that are resistant to this substance, which inhibits ATPase
(adenosine triphosphatase) on the cell membrane. Chu and Powell (1976)
have reviewed the techniques available for selection of cultured somatic
cells; though many such techniques exist, further improvements are needed.
At present the lack of suitable selection techniques is a barrier to many kinds
of manipulations that would be of interest to somatic cell geneticists.

OTHER APPLICATIONS OF SOMATIC CELL GENETICS

Genetic Transfer by Individual Chromosomes

Attempts have been made to fractionate metaphase chromosomes and to
use the separate fractions to transfer limited numbers of genes from one
organism to another. If individual human chromosomes are fractionated and
added to a culture of HPRT− mouse fibroblasts, they are phagocytized. If this
is followed by HAT selection, some of the survivors can be shown to have
human HPRT but no recognizable human chromosomes. If these surviving
cells are then hybridized with yet another HPRT− cell line derived from the
hamster and selected again in HAT medium to identify which mouse chromosome has incorporated the human HPRT gene, it is found that the integration
is random (though closely linked human genes tend to be integrated *en bloc*).
TK, ouabain resistance and a few other genes have also been transferred.

Chromosome transfer is a potentially powerful technique for manipulating cells, which may be able to provide much information about the organization of genes and the regulation of their expression.

DIFFERENTIATED CELL FUNCTION

So far, most of the studies in somatic cell genetics have involved the study
of genes normally expressed by the relatively undifferentiated kinds of cells
used in the hybridization experiments. The expression of differentiated function has received less attention, though it is potentially very important and
some progress has been made.

A significant observation is that in hybrids between mouse hepatoma
cells (which manufacture albumin) and human leucocytes (which do not make
albumin), *human* albumin is made (Darlington et al., 1974). The human gene
for serum albumin is activated in the hybrid cell. Unfortunately, reactivation
of inactive genes has not been the rule in cell hybrids. In a hybrid between a
hamster melanoma line (synthesizing melanin) and a mouse fibroblast line,
pigment synthesis is extinguished in the hybrid even when all the hamster

chromosomes are present, though extinction can be overcome by using a tetraploid melanoma line. Again, when cells of the Friend erythroleukemia mouse line (which has the special property of synthesizing hemoglobin) are hybridized with a human fibroblast line, no hemoglobin of either the mouse or the human type is made, and hemoglobin messenger RNA is absent. Thus in this case the human fibroblasts suppress hemoglobin production at the mRNA level. Since we need to know much more about the regulation of differentiated function, it is disappointing that on the whole, suppression rather than expression seems to be the rule.

STUDIES OF MALIGNANCY IN SOMATIC CELL HYBRIDS

Somatic cell hybridization techniques have been applied to studies of malignant transformation. When normal and malignant mouse cells are hybridized, the hybrid cells are nonmalignant, until loss of a specific normal chromosome in a subclone restores malignancy to that clone. This observation has led to the suggestion that normal cells contain genes that can suppress the cells' ability to induce tumors. However, in other experiments in which human malignant cells are fused with normal mouse cells, the hybrid progeny express malignancy and are capable of inducing tumors in "nude" mice. ("Nude" mice are mutants that lack a thymus gland, lack cell-mediated immunity and consequently cannot reject transplants.)

It has been known for some time that malignant transformation can be induced in cell culture by addition of the virus SV40 (simian virus 40). Somatic cell hybridization has been used to investigate whether the transforming virus is integrated into a specific chromosome or chromosomes, or whether integration is random. Figure 10–5 outlines one such experiment. Human HPRT− cells transformed by SV40 are fused with TK− mouse cells, and HAT selection is used to pick out the hybrid clones. Among the hybrid clones, the antigen characteristic of SV40 is found only on hybrid cells that retain human chromosome 7. The obvious conclusion is that the SV40 virus has become integrated into chromosome 7 and nowhere else in the genome.

MUTAGENESIS IN CELL CULTURE

Somatic cells in culture undergo mutation at a low rate, and the rate can be increased by the use of mutagens. The usual method is to mutagenize cells by means of irradiation or a chemical mutagen and then pass them through a selective system that will allow survival of mutants of a particular type but not wild-type cells. For example, selection with 8AZA allows the survival of any HPRT− mutants, which are resistant to 8AZA, and selection with ouabain allows ouabain-resistant mutants to survive.

There are many problems in the interpretation of the results of such experiments. For example, most mutagens kill a large proportion of the cells in the culture, and the measurement of mutations in the survivors does not allow for this. Because many cells are lost, the growth rate may appear much

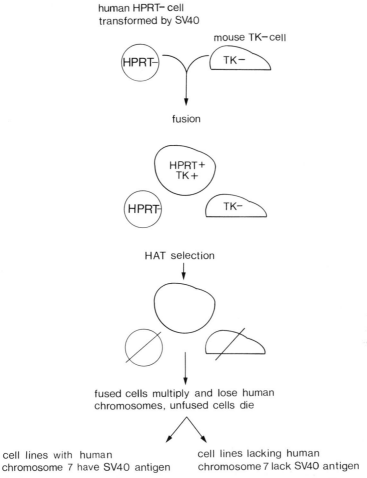

Figure 10–5   Schema of experiment to demonstrate that when human cells are transformed by SV40 virus, the virus is incorporated into human chromosome 7.

lower in mutagenized cultures than in controls during the early days of an experiment, and the rate of spontaneous mutation in the controls may then appear to be even higher than the rate of induced mutation in the mutagenized line. Metabolic cooperation can also lead to a problem in interpretation; if the cell density is high and many cells are in contact, even HPRT− cells may be killed by 8AZA.

When allowance is made for problems such as these, it is possible to get some idea of the mutation rate in culture. As an example, the mutation rate of resistance to 8AZA is about 10^{-4} per cell division in mutagenized cultures, as compared with about 10^{-8} spontaneously. It is hard to know whether 8AZA resistance truly represents mutation at the HPRT locus, since other types of change (for example, a mutation in permeability of the cell membrane to 8AZA) might also confer resistance. It is also puzzling that tetraploid and even octaploid cells seem to have much the same mutation rate as diploid cells. Thus it is hard to know whether resistance to 8AZA is a true mutation.

Mutation at the TK locus has been examined using BUdR resistance to select for mutants. Recall that the TK locus is on chromosome 17 and that absence of TK is a recessive mutation. If heterozygotes are used as the original cell line, an allelic mutation would result in TK− cells. In hamster and human cells, the spontaneous mutation rate of about 10^{-7} can be increased to about 10^{-4} by mutagens.

It is difficult to compare the mutation rate in cultured cells, which is expressed as the rate per cell per cell generation, with the mutation rate in individuals, which is expressed as the rate per gamete per generation. Given an estimated population mutation rate of 10^{-5} for HPRT deficiency, a rate of 10^{-8} in cell culture seems not unexpected.

COMPLEMENTATION

The technique of co-cultivating cells of different origins has been used to learn whether or not two different mutations are at the same locus or at different loci. Mutations at the same locus have defects in the same protein, so complementation does not occur. In contrast, if the mutations are at different loci, metabolic cooperation between cells of the two types in culture allows each to compensate for the deficiency of the other. The demonstration of complementation between Hurler and Hunter syndromes in fibroblast culture referred to earlier was the first application of this technique to somatic cell genetics, though it is widely used in molecular genetics. It has since been applied to the analysis of several groups of disorders (notably other mucopolysaccharidoses and xeroderma pigmentosum). Further extension of this kind of study should be rewarding.

GENERAL REFERENCES

Chu, E. H. Y., and Powell, S. S. 1976. Selective systems in somatic cell genetics. Adv. Hum. Genet. 7:189–258.

Creagan, R. P., and Ruddle, F. H. New approaches to human gene mapping by somatic cell genetics. In: Yunis, J. J. (ed.), 1977. *Molecular Structure of Human Chromosomes*. Academic Press, New York.

Croce, C. M., and Koprowski, H. 1978. The genetics of human cancer. Sci. Am. 238:117–125.

Siminovitch, L. 1976. On the nature of hereditable variation in cultured somatic cells. Cell 7:1–11.

PROBLEMS

1. A certain metabolic disorder is expressed in cultured fibroblasts. Cell cultures from six different patients are grown in all possible paired combinations, with the following results:

	A	B	C	D	E	F
A	−					
B	+	−				
C	−	+	−			
D	+	+	+	−		
E	+	−	+	+	−	
F	−	+	−	+	+	−

+ represents complementation in culture; − represents failure to complement.

a) How many complementation groups are there?
b) Assign each patient to a complementation group.
c) What do these results tell you about the genetics of the condition?

2. A series of mouse-human hybrid clones have the following human chromosomes:

Clone I   4, 7, 8, 10, 20, X
Clone II   1, 3, 7, 21, X
Clone III   9, 13, 19, 21, X
Clone IV   1, 4, 10, 19, X
Clone V   2, 3, 13, 20, X

Human enzyme A is present in clone V but absent in the other four clones. Human enzyme B is present in all five clones. Human enzyme C is present in clones III and IV but not in the other clones. All five clones lack human enzyme D. What conclusions can be drawn about the chromosomal location of these four enzymes?

11

LINKAGE

According to Mendel's law of independent assortment, genes that are not allelic assort independently of one another. **Linkage** is an important exception to this law; if two genes are linked, that is, if their loci are on the same chromosome and reasonably close together, they do not assort independently but are transmitted to the same gamete more than 50 percent of the time. It is only when the loci are on different chromosomes, or far apart on the same chromosome, that the law of independent assortment is followed.

The concept of linkage can be extended to encompass not only those genes that are so closely linked that they tend to remain together, but also other genes on the same chromosome, even when they are too far apart to show linkage in family studies. The term **syntenic** was coined by Renwick (1969) to refer to genes on the same chromosome.

FAMILY STUDIES

Classical family studies used to be the only source of information about human linkages, but until other methods were invented progress in mapping human genes was disappointingly slow. Apart from X-linkage, which is easy to identify by its genetic pattern, only three linkages came to light before about 1968; these were the linkage of the Lutheran blood group locus and the locus for the secretor trait (Mohr, 1951), the linkage of the Rh and elliptocytosis loci (Lawler, 1954), and the linkage of the ABO and nail-patella syndrome loci, which has been mentioned earlier (Renwick and Lawler, 1955). The rate of progress in mapping human genes has been transformed by somatic cell hybridization and other technical and intellectual advances, and about two hundred autosomal genes have now been assigned to their respective chromosomes.

Linkage has been referred to repeatedly in earlier sections. Here we will introduce the basic principles of linkage analysis and its cytological basis and will then review the newer techniques and the contributions they have made to the human linkage map.

The most useful kind of mating for the study of linkage is the so-called **double backcross,** in which one parent is doubly heterozygous (Aa Bb) and the other is homozygous for the recessive allele at each locus (aa bb). Matings in which the alleles at each locus are codominant can also be used; the point is to be certain what the parents' genotypes are. In either of these matings, the phenotypes of the offspring show what genes the heterozygous parent has transmitted to each child. If the A and B loci are not linked, the theoretical proportions of the four possible genotypes in the progeny are:

1/4 Aa Bb   1/4 aa Bb
1/4 Aa bb   1/4 aa bb

In a single human family there are too few children to allow deviations from these proportions to be identified. However, in a large enough series of families of the same kind, a significant deviation from the 1:1:1:1 ratio means that A and B are assorting nonrandomly or, in other words, that they are linked.

Genes may be linked in **coupling** or in **repulsion**; using an alternative terminology, we may say that genes may be linked in **cis** or **trans** configuration. Suppose we are interested in genes A and B of a person whose genotype is Aa Bb. There are two possible linkage phases of A and B, as follows:

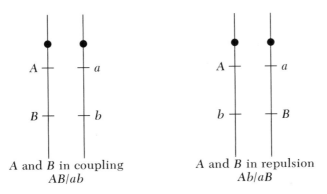

A and B in coupling   A and B in repulsion
AB/ab   Ab/aB

Note the symbols for the genotypes: AB/ab and Ab/aB. The slash here indicates that the genes A and B, or A and b, are on the same chromosome. Normally the two loci should be in equilibrium; that is, the relative proportions of AB and Ab should be determined only by the population frequencies of B and b. As an example, if the gene frequencies are:

A 0.90   B 0.60
a 0.10   b 0.40

the corresponding frequencies of the combinations are expected to be:

AB 0.54   aB 0.06
Ab 0.36   ab 0.04

If linkage equilibrium is not present, an explanation is looked for in terms of natural selection. Linkage disequilibrium exists for alleles at some of the HLA loci, and this phenomenon is thought to underlie the associations of HLA and disease (see Chapter 8).

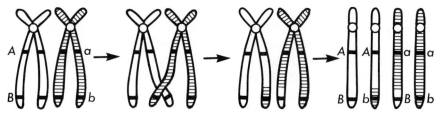

Figure 11-1   Crossing over and recombination in an *AB/ab* double heterozygote. At the end of meiosis, the gametes are of four types: two parental types, *AB* and *ab*, and two recombinants, *Ab* and *aB*.

If the linkage phase of genes *A* and *B* in the doubly heterozygous parent is known, it is easy to predict the expected genotypes of the offspring. For example, if the parents are *AB/ab* × *ab/ab*, the offspring are expected to have both trait A and trait B or neither; if the parents are *Ab/aB* × *aa/bb*, the offspring are expected to have trait A or trait B but not both. When exceptions occur, the cause is recombination between the A and B loci, and the exceptional offspring are **recombinants.**

The physical basis of recombination is the crossing over of chromosomal material that occurs in meiotic prophase and is evidenced by the chiasmata seen holding bivalents together at that stage (see Fig. 2–6). Figure 11–1 is a diagram of crossing over in an *AB/ab* double heterozygote, resulting in the production of four types of gametes: two parental types, *AB* and *ab*, and two recombinants, *Ab* and *aB*. In the following pedigree, II-5 is a recombinant. (Note that the term recombinant refers to either the gamete or the resulting individual.)

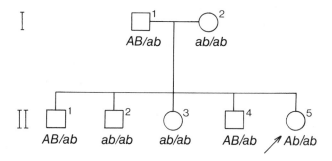

Although, as noted above, a single human family is rarely if ever large enough to give definitive evidence of linkage between two loci, in a series of families the following observations might be made:

Parents   *AB/ab* × *ab/ab*

Offspring 45% *AB/ab*   5% *Ab/ab*   5% *aB/ab*   45% *ab/ab*

Thus, among the offspring 90 percent have the parental combinations of genes A and B, and 10 percent are recombinants. The genes are said to show 10 percent recombination, and to be "10 units apart."

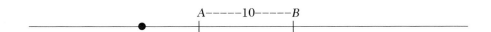

A third linked locus C can be added to this linkage map if its distance from the A and B loci can be measured. Consider the following findings:

A - C 13 percent recombination (13 units apart)
B - C 5 percent recombination (5 units apart)

The order of the genes must then be A - B - C, and C can be mapped as follows:

Note that the A - C distance as measured by percentage of recombination is less than the sum of the A - B and B - C distances. The reason is that **double crossovers,** which do not result in recombination, lead to an underestimate of the distance between A and C (Fig. 11–2).

The map unit is sometimes called the **centimorgan,** in honor of T. H. Morgan, the discoverer of genetic linkage (1911). Over short distances the percentage of recombination is equivalent to the centimorgan, but over longer distances it gives an underestimate, as described.

Linkage analysis is more complicated if the linkage phase of the genes under study is unknown, and usually in human genetics it is unknown. Sometimes, however, information from a three-generation family is helpful. Figure 11–3 illustrates two families, in each of which the mother is doubly heterozygous, carrying X-linked genes for forms of color blindness (cb) and hemophilia (h). Here the mother's father's phenotype tells the linkage phase of the two genes in the mother.

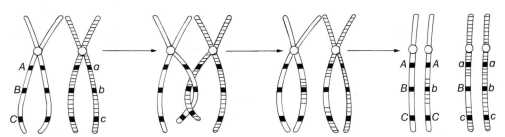

Figure 11–2   A double crossover between the A and C loci does not result in recombination between A and C or between a and c. When an intermediate locus B is included, the double crossover is recognizable.

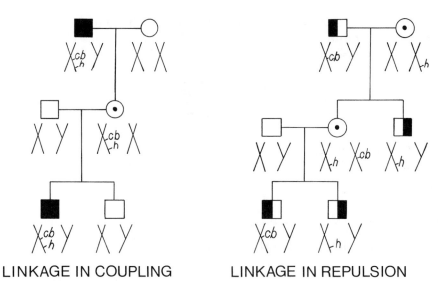

LINKAGE IN COUPLING   LINKAGE IN REPULSION

Figure 11–3 Linkage in coupling and in repulsion. In both pedigrees, the mother is heterozygous for hemophilia (h) and color blindness (cb). If these genes are linked in coupling and recombination does not occur, the sons are either doubly affected or normal. If linkage is in repulsion and there is no recombination, the sons have one condition or the other but not both.

Figure 11–4 is another example, showing a three-generation pedigree giving data for the nail-patella syndrome and the ABO blood groups and in which the doubly heterozygous mother, II-2, has obviously received both the A gene and the *nail-patella* gene from her father. Thus she has these two genes in coupling. One of her seven children, III-6, has inherited the A gene but not the *nail-patella* gene, so this child is a recombinant. The "recombination fraction" is then 1/7. Actually it is about 10 percent for a large series of families combined. It is important to recognize that if the mother's linkage phase was unknown, we would not know whether III-6 was a recombinant or whether the other six children were all recombinants.

Often in human linkage studies information can be obtained for only two generations. There is a special mathematical method of extracting linkage information in these cases, known as the **lod** (log odds) method. Without attempting to go into detail, we will simply note that the principle is to set up a series of theoretical recombination frequencies (0, 5 percent, 10 percent, and so on up to 50 percent, which is random assortment) and then calculate the odds for a particular pedigree on the basis of a given recombination fraction, as compared with the likelihood if the recombination fraction is 50 percent. If this ratio is expressed as a logarithm, information from different pedigrees can be combined by simple addition. The lod method is treated in detail in several texts, for example, in the text by Cavalli-Sforza and Bodmer cited in the General References of this chapter.

Until lately, the chance of finding that a rare human trait was linked to some other marker was slight, partly because of difficulty in finding enough families of the right kind to carry out the study and partly because the known markers were few, with much unmapped territory between. Now that the chromosome map has been so well surveyed, the probability of finding a linkage if a rare trait is tested against all the possible known markers has greatly improved.

OTHER METHODS OF DETECTING LINKAGE

Most of the progress made over the last 10 years or so in constructing the human gene map has come about through the development of new approaches other than classical family studies. These have almost all been mentioned, at least briefly, in other chapters, but a summary is in order here.

1. Combination of cytogenetic and genetic markers. This was the method used to map the Duffy locus to chromosome 1, the first assignment of a gene to a specific autosome.

2. Somatic cell hybridization. This technique can reveal to what chromosome a given gene can be assigned and, if a line with a translocation is used as the human parent, can even show the regional location of the gene (see Chapter 10).

3. Gene dosage methods. **Deletion mapping** is the absence of an expected phenotype together with loss of a specific chromosomal region. An example is the finding of an interstitial deletion of a short segment of the long arm of chromosome 13 in some sporadic cases of retinoblastoma; this may mean that there is a locus in that area for the retinoblastoma gene and its normal allele. In contrast to deletion mapping, **duplication mapping** requires finding the expression of three alleles (two from one parent), or 150 percent of the normal activity of a given enzyme, together with trisomy for a chromosome or duplication of a chromosome segment.

4. Hybridization of a specific type of RNA (or DNA) to a specific segment of a chromosome. This technique, known as *in situ* hybridization, involves heating DNA to cause the strands to separate, then allowing them to reanneal

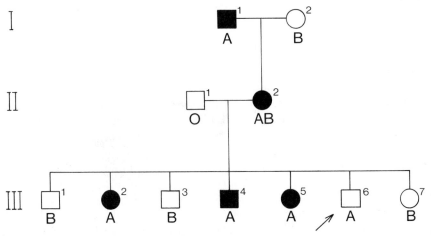

Figure 11–4 The ABO nail-patella linkage in a three-generation family. Black symbols: nail-patella syndrome. Letters below symbols: ABO blood groups. In this pedigree, gene *A* and gene *np* are in coupling, and III–6 is a recombinant.

at a lower temperature. If radioactively labelled RNA is present during the reannealing stage, it may bond with DNA that has a base sequence complementary to its own. After autoradiography, the radioactive RNA reveals the location of the complementary DNA sequence. This is how the genes for 18S and 28S ribosomal RNA have been localized to the short arms of the five pairs of acrocentric chromosomes (see Chapter 3). This technique depends mainly on the availability of pure messenger RNA of a specific type. It is being used to map the human α and β hemoglobin genes.

5. Molecular homology and gene fusion, as shown by the δβ polypeptide chain of the Lepore hemoglobins and the γβ chain of Hb Kenya (see Chapter 5). *abnormal* *produced by abnormal crossing over —* *does not involve homologous regions 2 chromosomes*

Several other experimental approaches or analyses of special situations have been used to add new information about linkages or to confirm gene assignments. They are beyond the scope of this chapter, but are described and documented in some of the references, especially Creagan and Ruddle (1977) and McKusick and Ruddle (1977).

THE HUMAN LINKAGE MAP

Figure 11–5 is a map of the human genome as known by early 1979. As it indicates, about 200 human gene loci have already been mapped, through the collaborative efforts of many investigators. The gene assignments include three that are particularly noteworthy: the ABO and Rh blood group loci and the major histocompatibility complex. Every chromosome has at least one marker, and chromosome 1 has more than 20. Figure 11–5 does not show all that is known about the regional location of genes within chromosomes, nor does it show several linkage groups that have not yet been assigned (including the Lutheran blood group-secretor trait linkage, the first autosomal linkage to be established through family studies). Though much detail remains to be filled in, the map already can contribute to our knowledge of human biology and evolution.

The direct clinical applications of information about the human gene map are relatively few but important or potentially important. Two major uses are evident:

1. Identification of the presence of a gene for a particular genetic disorder by its linkage to a known marker. This can, at least in theory, be done either prenatally or later. The possibility of using the linkage of the loci for the secretor trait and for myotonic dystrophy in prenatal diagnosis has been noted (see Chapter 9). There are several other linkages that might be exploited in a similar way. For example, the loci for classic hemophilia and G6PD are very closely linked, so in suitable families a fetus at risk of having hemophilia might be detectable. The G6PD locus has a low frequency of variants in many populations, so its usefulness as a marker in those populations is limited, but in other groups its variants are frequent, and in those groups it would be useful. For Huntington chorea, a close linkage might be useful in picking out the family members who have the Huntington gene, either before birth or later, but no such linkage is known, and the late onset age and diagnostic problems of Huntington disease hamper efforts to map it.

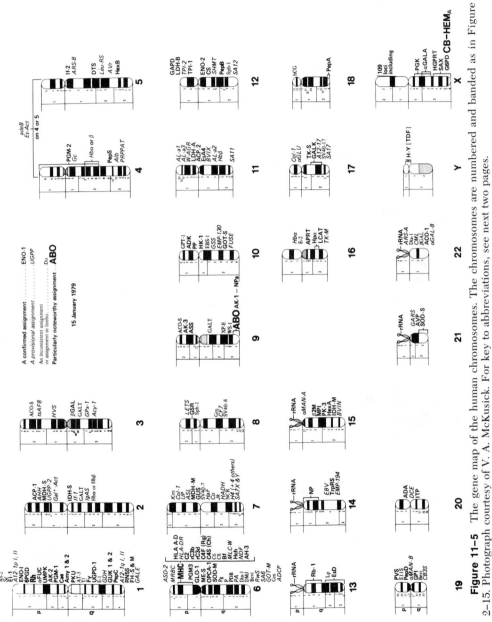

Figure 11–5 The gene map of the human chromosomes. The chromosomes are numbered and banded as in Figure 2–15. Photograph courtesy of V. A. McKusick. For key to abbreviations, see next two pages.

Key: # This symbol indicates that the locus so marked is the site of one or more mutations that "cause" disease. Liberties have been taken with the definition of both *cause* and *disease*. For example, the polio sensitivity locus and the Rh locus are marked. In some cases it is not certain that the "disease locus" is the one that has been mapped, e.g., acid phosphatase-2 may not be the locus of the mutation in lysosomal acid phosphatase deficiency.

ABO	= ABO blood group (chr. 9)
ACO-M	= Aconitase, mitochondrial (chr. 22)
ACO-S	= Aconitase, soluble (chr. 9)
ACP-1	= Acid phosphatase-1 (chr. 2)
#ACP-2	= Acid phosphatase-2 (chr. 11)
adeB	= Formylglycinamide ribotide amido-transferase (chr. 4 or 5)
#ADA	= Adenosine deaminase (chr. 20)
#ADCP	= Adenosine deaminase complexing protein (chr. 6)
ADK	= Adenosine kinase (chr. 10)
A12-1pI, II	= Adenovirus-12 chromosome modification site-1p I and II (chr. 1 k)
A12-1qI, II	= Adenovirus-12 chromosome modification site-1q I and II (chr. 1 q)
A12-17	= Adenovirus-12 chromosome modification site-17 (chr. 17)
#AH-3	= Adrenal hyperplasia III (21-hydroxylase deficiency) (chr. 6)
AHH	= Aryl hydrocarbon hydroxylase (chr. 2)
AK-1	= Adenylate kinase-1 (chr. 9)
AK-2	= Adenylate kinase-2 (chr. 1)
AK-3	= Adenylate kinase-3 (chr. 9)
AL	= Lethal antigen: 3 loci (a1, a2, a3) (chr. 11)
#Alb	= Albumin (chr. 4)
Acy-1	= Aminoacylase-1 (chr. 3)
Amy-1	= Amylase, salivary (chr. 1)
Amy-2	= Amylase, pancreatic (chr. 1)
#An-2	= Aniridia, type II Baltimore (chr. 1)
#ARS-A	= Arylsulfatase A (chr. 22)
#ARS-B	= Arylsulfatase B (chr. 5)
#APRT	= Adenine phosphoribosyltransferase (chr. 16)
#ASD-2	= Atrial septal defect, secundum type (chr. 6)
#ASL	= Argininosuccinate lyase (chr. 7)
#ASS	= Argininosuccinate synthetase (chr. 9)
#AT-3	= Antithrombin III (chr. 1)
AVP	= Antiviral protein (chr. 21)
AVr	= Antiviral state regulator (chr. 5)
Bevi	= Baboon M7 virus infection (chr. 6 or 19)
Bf	= Properdin factor B (chr. 6)
2M	= 2-Microglobulin (chr. 15)
BVIN	= BALB virus induction, N-tropic (chr. 15)
BVIX	= BALB virus induction xenotropic (chr. 11)
#C2	= Complement component-2 (chr. 6)
C4F	= Complement component-4 fast (chr. 6)
C4S	= Complement component-4 slow (chr. 6)
C6	= Complement component-6 (chr. 6)
#C8	= Complement component-8 (chr. 6)
#Cae	= Cataract, zonular pulverulent (chr. 1)
CB	= Color blindness (deutan and protan) (X chr.)
CB3S	= Coxsackie B3 virus susceptibility (chr. 19)
#CF7	= Clotting factor VII (chr. 8)
Ch	= Chido blood group (chr. 6) — same as C4S
#CML	= Chronic myeloid leukemia (chr. 22)
Co	= Colton blood group (chr. 7)
#Col-1	= Collagen I (α1 and α2) (chr. 7. and 17)
CS	= Citrate synthase, mitochondrial (chr. 12)
DCE	= Desmosterol-to-cholesterol enzyme (chr. 20)
#Dia-1	= NADH-diaphorase (chr. 6 or 22)
#DMJ	= Juvenile diabetes mellitus (chr. 6)
Do	= Dombrock blood group (chr. 1)
#DTS	= Diphtheria toxin sensitivity (chr. 5)
#EBS-1	= Epidermolysis bullosa, Ogna type (chr. 10)
#EBV	= Epstein-Barr virus integration site (chr. 14)
#E-1	= Pseudocholinesterase-1 (chr. 1)
#E-2	= Pseudocholinesterase-2 (chr. 16)
E11S	= Echo 11 sensitivity (chr. 19)
El-1	= Elliptocytosis-1 (chr. 1)
EMP-130	= External membrane protein-130 (chr. 10)
EMP-195	= External membrane protein-195 (chr. 14)
ENO-1	= Enolase-1 (chr. 1)
ENO-2	= Enolase-2 (chr. 12)
Es-Act	= Esterase activator (chr. 4 or 5)
EsA4	= Esterase-A4 (chr. 11)
EsD	= Esterase D (chr. 13)
FH-M	= Fumarate hydratase, mitochondrial (chr. 1)
FH-S	= Fumarate hydratase, soluble (chr. 1)
#FUC	= Alpha-L-fucosidase (chr. 1)
FUSE	= Polykaryocytosis inducer (chr. 10)
#Fy	= Duffy blood group (chr. 1)
Gal+-Act	= Galactose + activator (chr. 2)
#GAL A	= galactosidase A (Fabry disease) (X chr.)
GAL B	= -galactosidase B (chr. 22)
# GAL	= -galactosidase (chr. 3) (see also chr. 12 and 22)
#GALK	= Galactokinase (chr. 17)
#GALT	= Galactose-1-phosphate uridyltransferase (chr. 2, 8 or 9)
#GALE	= Galactose-4-epimerase (chr. 1)
#GLU	= -glucosidase (chr. 17)
GAPD	= Glyceraldehyde-3-phosphate dehydrogenase (chr. 12)
GAPS	= Phosphoribosyl glycinamide synthetase (chr. 21)
Gc	= Group-specific component (chr. 4)
GDH	= Glucose dehydrogenase (chr. 1)
GLO-1	= Glyoxylase I (chr. 6)
Gm	= Immunoglobulin heavy chain (chr. 8)
GOT-M	= Glutamate oxaloacetate transaminase, mitochondrial (chr. 6)
GOT-S	= Glutamate oxaloacetate transaminase, soluble (chr. 10)
#GPI	= Glucosephosphate isomerase (chr. 19)
GPT-1	= Glutamate pyruvate transaminase, soluble (chr. 10)
#GPx-1	= Glutathione peroxidase-1 (chr. 3)
#G6PD	= Glucose-6-phosphate dehydrogenase (X chr.)
#GSR	= Glutathione reductase (chr. 8)
GSS	= Glutamate-gamma-semialdehyde synthetase (chr. 10)
GARS	= Glycinamide ribonucleotide synthetase (chr. 21)

GUK-1 & 2 = Guanylate kinase-1 & 2 (chr. 1)
#GUS = Beta-glucuronidase (chr. 7)
H4 = Histone H4 and 4 other histone genes (chr. 7)
HADH = Hydroxyacyl-CoA dehydrogenase (chr. 7)
HaF = Hageman factor (chr. 7)
#Hbα = Hemoglobin αchain (chr. 16)
#Hbβ = Hemoglobin βchain (chr. 11)
hCG = Human chorionic gonadotropin (chr. 18)
#Hch = Hemochromatosis (chr. 6)
#HEM-A = Classic hemophilia (X chr.)
#HexA = Hexosaminidase A (chr. 15)
#HexB = Hexosaminidase B (chr. 5)
#HGPRT = Hypoxanthine-guanine phosphoribosyltransferase (X chr.)
HK-1 = Hexokinase-1 (chr. 10)
HLA (A–D)= Human leukocyte antigens (chr. 6)
HLA–DR = Human leukocyte antigen, D-related (chr. 6)
Hp = Haptoglobin, alpha (chr. 16)
#HVS = Herpes virus sensitivity (chr. 6)
H-Y = Y histocompatibility antigen (Y chr.)
IgAS = Immunoglobulin heavy chains attachment site (chr. 2)
If-1 = Interferon-1 (chr. 2)
If-2 = Interferon-2 (chr. 5)
IDH-M = Isocitrate dehydrogenase, mitochondrial (chr. 15)
IDH-S = Isocitrate dehydrogenase, soluble (chr. 2)
#ITP = Inosine triphosphatase (chr. 20)
Jk = Kidd blood group (chr. 7)
Km = Kappa immunoglobulin light chains, Inv. (chr. 7)
#LAP = Laryngeal adductor paralysis (chr. 6)
#LCAT = Lecithin-cholesterol acyltransferase (chr. 16)
LDH-A = Lactate dehydrogenase A (chr. 11)
LDH-B = Lactate dehydrogenase B (chr. 12)
LETS = Large, external, transformation-sensitive protein (chr. 8)
LeuRS = Leucyl-tRNA synthetase (chr. 5)
Lp = Lipoprotein—Lp (chr. 13)
MAN-A = Cytoplasmic D-mannosidase (chr. 15)
#MAN-B = Lysosomal D-mannosidase (chr. 19)
MDH-M = Malate dehydrogenase, mitochondrial (chr. 7)
MDH-S = Malate dehydrogenase, soluble (chr. 2)
ME-S = Malic enzyme, soluble (chr. 6)
MHC = Major histocompatibility complex (chr. 6)
MLC-W = Mixed lymphocyte culture, weak (chr. 6)
MPI = Mannosephosphate isomerase (chr. 15)
MRBC = Monkey red blood cell receptor (chr. 6)
MTR = 5-Methyltetrahydrofolate: L-homocysteine S-methyltransferase (chr. 1)
NCR = Neutrophil chemotactic response (chr. 7)
NDF = Neutrophil differentiation factor (chr. 6)
#NP = Nucleoside phosphorylase (chr. 14)
#NPa = Nail-patella syndrome (chr. 9)
#OPCA-1 = Olivopontocerebellar atrophy I (chr. 6)
P = P blood group (chr. 6)
PA = Plasminogen activator (chr. 6)

#PDB = Paget disease of bone (chr. 6)
PepA = Peptidase A (chr. 18)
PepB = Peptidase B (chr. 12)
PepC = Peptidase C (chr. 1)
PepD = Peptidase D (chr. 19)
PepS = Peptidase S (chr. 4)
Pg = Pepsinogen (chr. 6)
PGK = Phosphoglycerate kinase (X chr.)
PGM-1 = Phosphoglucomutase-1 (chr. 1)
PGM-2 = Phosphoglucomutase-2 (chr. 4)
PGM-3 = Phosphoglucomutase-3 (chr. 6)
6PGD = 6-Phosphogluconate dehydrogenase (chr. 1)
PRPPAT = Phosphoribosylpyrophosphate amidotransferase (chr. 4)
PK-3 = Pyruvate kinase-3 (chr. 15)
#PKU = Phenylketonuria (chr. 1)
PP = Inorganic pyrophosphatase (chr. 10)
#PVS = Polio sensitivity (chr. 19)
#RB-1 = Retinoblastoma-1 (chr. 13)
rC3b = Receptor for C3b (chr. 6)
rC3d = Receptor for C3d (chr. 6)
Rg = Rodgers blood group (chr. 6); same as C4F
#Rh = Rhesus blood group (chr. 1)
RN5S = 5S RNA gene(s) (chr. 1)
#RP-1 = Retinitis pigmentosa-1 (chr. 1)
rRNA = Ribosomal RNA (chr. 13, 14, 15, 21, 22)
#RwS = Ragweed sensitivity (chr. 6)
SA6 = Surface antigen 6 (chr. 6)
SA7 = Species antigen 7 (chr. 7)
SA11 = Surface antigen 11 (chr. 11)
SA12 = Surface antigen 12 (chr. 12)
SA17 = Surface antigen 17 (chr. 17)
SAX = X-linked species (or surface) antigen (X chr.)
Sc = Scianna blood group (chr. 1)
SHMT = Serine hydroxymethyltransferase (chr. 12)
SOD-M = Superoxide dismutase, mitochondrial (chr. 6)
SOD-S = Superoxide dismutase, soluble (chr. 21)
Sph-1 = Spherocytosis, Denver type (chr. 8 or 12)
SV40-7 = SV40—integration site-7 (chr. 7)
SV40-8 = SV40—integration site-8 (chr. 8)
SV40-17 = SV40—integration site-17 (chr. 17)
TDF = Testis determining factor (Y chr.)—probably same as H-Y
Tf = Transferrin (chr. 1)
TK-M = Thymidine kinase, mitochondrial (chr. 16)
TK-S = Thymidine kinase, soluble (chr. 17)
#TPI-1 & 2 = Triosephosphate isomerase-1 and -2 (chr. 12)
TrpRS = Tryptophanyl-tRNA synthetase (chr. 14)
tsAF8 = Temperature-sensitive (AF8) complementing (chr. 3)
UGPP-1 = Uridyl diphosphate glucose pyrophosphorylase-1 (chr. 1)
UGPP-2 = Uridyl diphosphate glucose pyrophosphorylase-2 (chr. 2)
UMPK = Uridine monophosphate kinase (chr. 1)
UP = Uridine phosphorylase (chr. 7)
#WS-1 = Waardenburg syndrome-1 (chr. 9)
#W-AGR = Wilms tumor—aniridia, ambiguous genitalia, mental retardation (chr. 11)
#XP-E = Xeroderma pigmentosum, Egyptian (chr. 9)

As mentioned above, the family information may be inadequate in a given instance, either because the parents' genetic makeup is unsuitable or because the linkage phase is unknown. Consider myotonic dystrophy: families in which the parent with myotonic dystrophy is heterozygous for the secretor trait and the normal parent is a non-secretor constitute only about 10 percent of the families at risk of producing children with this disease. The next requirement, to know whether the two dominant alleles are in coupling or in repulsion in the doubly heterozygous parent, cannot always be fulfilled even if the grandparents are available for study. Finally, the two loci are not very closely linked, so not infrequently either a normal fetus would be diagnosed as affected or an affected fetus would be diagnosed as normal.

2. Understanding the cause of malformations in chromosome abnormality syndromes. For example, most of the array of phenotypic abnormalities characteristic of Down syndrome may result from trisomy for just the terminal segment of the long arm of chromosome 21 rather than trisomy of the whole chromosome. A locus for superoxide dismutase (SOD) is in this area. It seems possible that a disturbance of superoxide metabolism resulting from excessive levels of SOD is a factor in the pathogenesis of Down syndrome.

Further progress in filling up the linkage map will probably come about through a combination of classical family studies and laboratory methods. Family studies are particularly useful in determining gene order and map distances, as well as in assigning genes that are not expressed in cell culture. Concurrently with the human linkages, those of other mammals are being worked out, and a new dimension is being added to our knowledge of evolutionary relationships.

GENERAL REFERENCES

Cavalli-Sforza, L. L., and Bodmer, W. F. 1971. *The Genetics of Human Populations.* W. H. Freeman and Company, Publishers, San Francisco.

Creagan, R. P., and Ruddle, F. H. New approaches to human gene mapping by somatic cell genetics. In; Yunis, J. J., ed., 1977. *Molecular Structure of Human Chromosomes.* Academic Press, New York.

McKusick, V. A., and Ruddle, F. H. 1977. The status of the gene map of the human chromosomes. Science 196:390–405.

Renwick, J. H. 1971. The mapping of human chromosomes. Ann. Rev. Genet. 5:81–120.

PROBLEMS

1. Duchenne muscular dystrophy (DMD) and the deutan type of red-green color blindness (cb) are both X-linked. A woman whose father is color blind has five sons: the first and fourth have DMD and cb, the second has cb only, and the third and fifth have DMD only.

a) Sketch the pedigree.
b) Which sons are recombinants? (In your answer, ignore the probability that the mother carries a new mutation for DMD.)
c) Are the loci closely linked?
d) Now assume that the mother is a carrier of a new DMD mutation. What conclusions can be drawn about the recombinant sons and about the closeness of the linkage?

2. The ABO and nail-patella loci are about 10 map units apart on chromosome 9. (As in many other linkages, the recombination frequency is higher in females than in males, but this point can be ignored in your answer.) In a large series of families in which one parent has blood group AB and the nail-patella syndrome (with the A and *np* alleles in coupling) and the other parent is group O and normal, what phenotypes and in what proportions are expected in the offspring?

3. In mouse-human cell hybrids, all clones with human chromosome 17 have human thymidine kinase (TK).
a) Name another human enzyme that is also formed in such cell hybrids.
b) If the only representative of human chromosome 17 present is 17q− and TK is synthesized but the other enzyme is not, what conclusion could be drawn?

4. What percentage of the normal level of superoxide dismutase is expected in a child with translocation Down syndrome? With monosomy 21?

5. The loci for myotonic dystrophy and the secretor trait are linked. A man with myotonic dystrophy who is a secretor is the son of a man with myotonic dystrophy who was a secretor and a woman who had no muscle disease but was a secretor. This man marries a woman who has no muscle disease and is a nonsecretor. Their 10 children include five with myotonic dystrophy who are secretors, one with myotonic dystrophy who is a nonsecretor and four with no muscle disease who are nonsecretors.
a) Sketch the pedigree, showing the relevant genotypes.
b) Does it support linkage of myotonic dystrophy and the secretor trait?
c) What percentage of the children are recombinants?
d) A son in this family marries a normal woman who is a nonsecretor. Comment on the possibility of prenatal detection of myotonic dystrophy in their children.

12

MULTIFACTORIAL
INHERITANCE

In Chapter 1, the three types of genetic disorders have been listed and briefly characterized, and in later chapters single-gene traits and syndromes caused by chromosome abnormalities have been treated in more detail. We now turn to the third major type of genetic disorder, the type in which inheritance is **multifactorial**. Multifactorial inheritance is less familiar and more difficult to analyze than the other types of inheritance, but it accounts for much of the normal variation in families, as well as for a large number of common disorders, including many common congenital malformations. Multifactorial traits tend to cluster in families, but their genetic patterns are not clear-cut in individual families. However, if a series of families is studied, certain common characteristics can be seen. In this chapter the basis of multifactorial inheritance is described, the concept of a **threshold trait** is introduced, and the distinguishing characteristics of multifactorial inheritance are given, chiefly as they apply to some common congenital malformations.

A multifactorial disorder is one which is "determined by a combination of genetic and environmental factors," without reference to the nature of the genetic component(s). The term **polygenic** is often used interchangeably with "multifactorial," but should be reserved for use in a more restricted sense for conditions "determined by a large number of genes, each with a small effect, acting additively" (Fraser, 1976). In actual experience, it is often hard to judge whether or not environmental factors are concerned or whether all the genes determining a trait have small and additive effects. For example, more complex causes of genetic variation may be present, such as several genes with larger effects. However, the family patterns of many disorders follow the expectations for multifactorial inheritance, and for simplicity we will omit other terms.

NORMAL VARIATION

Many traits have a unimodal distribution in the population. This type of distribution is characteristic of multifactorial inheritance, but by no means restricted to it; as described earlier (Fig. 9–13), red cell acid phosphatase

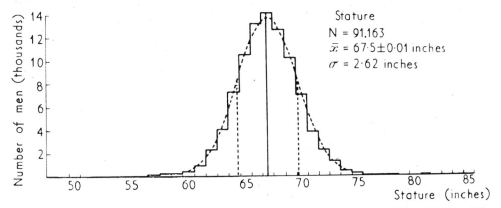

Figure 12-1 Distribution of stature in a sample of 91,163 young English males in 1939. Mean (x̄) and standard deviation (σ) are indicated. A normal curve with the same mean and standard deviation is also shown. From Harrison et al., *Human Biology.* 2nd ed. Oxford University Press, 1977, by permission. Data from Martin, Physique of young adult males, HMSO, 1949.

activity is determined by combinations of three alleles at a single locus but appears to have a unimodal distribution. However, there are a number of normal characteristics that are distributed unimodally and have family patterns characteristic of multifactorial inheritance. Here we will mention three of these traits: stature, total fingerprint ridge counts (described in Chapter 17) and intelligence as measured by the intelligence quotient.

A population distribution of stature is shown in Figure 12–1. Stature is a multifactorial trait in which environmental and genetic factors are both important. Stature distributions usually show a slight hump at the lower end of the range, accounted for by people who are exceptionally short, usually because of some gene with a major effect on stature (achondroplasia, for example). A few people at the upper end of the range owe their lofty stature to Marfan syndrome or to an XXY or XYY karyotype, but the conditions causing increased stature seem to be too few and too rare to affect the curve.

To explore how stature is inherited in families, the most useful measure is the **correlation coefficient**, described in any standard textbook of statistics. If mating is random, in the sense that mates are chosen without regard to genotype, and if environmental differences are not involved, the correlation between parent and child, or between sibs, is 0.50. In more general terms, the correlation between relatives is proportional to their genes in common (Fisher, 1918). **Genes in common** are those genes inherited from a common ancestral source.

At this point more terminology must be introduced. Relatives may be classified as **first-degree, second-degree**, and so on in terms of how many steps apart they are in the family tree. Relatives of the same degree have the same proportion of genes in common. For multifactorial inheritance, all relatives within the same degree may be considered together. Table 12–1 lists the closest degrees of relationship, the chief examples of relatives of each degree and the proportion of genes each has in common with a given propositus.

Note that the principle of genes in common also applies to single-gene

TABLE 12–1   DEGREES OF RELATIONSHIP AND GENES IN COMMON

Relationship to Propositus	Proportion of Genes in Common with Propositus
MZ twin	1
First-degree relative (parent, DZ twin or other sib, child)	1/2
Second-degree relative (grandparent, uncle or aunt, half-sib, nephew or niece, grandchild)	1/4
Third-degree relative (great-grandparent, great-grandchild, first cousin and so forth)	1/8

traits; that is, the probability that two relatives share the same gene, inherited from the same source, is the same as the proportion of all their genes that they have inherited from the same source.

Table 12–2 lists the correlation coefficients of the three traits mentioned above. For all three, the correlation coefficient is close to the theoretical value. The correlation between relatives is not always as close as this for metrical traits, because it can be affected by numerous factors, especially by dominant rather than additive genes, by environmental variation and by nonrandom mating.

As Table 12–2 shows, the correlation coefficients are particularly close to the theoretical values for fingerprint total ridge count (TRC), defined in Chapter 17. Mating is likely to be random with respect to TRC (unless TRC is associated with some other characteristic, such as ethnic origin, for which mating is nonrandom). TRC is a particularly good example of a metrical trait determined by multifactorial inheritance.

REGRESSION TO THE MEAN

Regression toward the mean is a common observation in family studies of metrical traits. Consider, as an example, the correlation of height in mother

TABLE 12–2   CORRELATION BETWEEN RELATIVES FOR CERTAIN HUMAN TRAITS

Trait	Relationship	Correlation Coefficient[*]	Genes in Common[**]
Stature	Sib-sib	0.56	0.50
	Parent-child	0.51	0.50
Fingerprint total ridge count	MZ twins	0.95	1.00
	DZ twins	0.49	0.50
	Sib-sib	0.50	0.50
	Parent-child	0.48	0.50
	Parent-parent	0	0
Intelligence (IQ)	Sib-sib	0.58	0.50
	Parent-child	0.48	0.50

[*]Data from Bodmer and Cavalli-Sforza, 1976, and from Holt, 1961, 1968.
[**]Proportion of genes in common equals theoretical correlation coefficient.

and daughter. A daughter receives half her genes from her mother. Theoretically, for a series of mothers of given height, the range of stature in the fathers is the same as the population distribution. Thus the average height of the daughters should be midway between the mother's height and the population mean.

In everyday terms, the law of regression to the mean accounts for the observation that parents who are especially "extreme" in terms of their location in the normal curve are likely to have children who, on average, are less unique than the parents. This can be seen, for example, at both ends of the range for stature or for intelligence. Especially brilliant parents tend to have children less outstanding than themselves though still above average in intelligence, whereas dull normal parents find that their children, on the average, rank higher than themselves, though still below average.

This example ignores the effect of major deleterious genes on intelligence. Penrose (1963) observed that whereas parents of moderately retarded children tend to have below-average intelligence, parents of severely retarded children have a normal range of intelligence. Moderate retardation is usually multifactorial, whereas severe retardation usually has a single major cause, as in phenylketonuria or Down syndrome.

HERITABILITY

To try to separate the relative roles of genes and environment for multifactorial traits, the idea of **heritability** has been developed. Heritability is defined as the proportion of the total phenotypic variance of a trait that is caused by additive genetic variance. (Students of statistics will recall that variance is a measure of dispersion about the mean, that is, a measure of how much an individual value is likely to differ from the typical value, the mean.) In general terms, heritability is a measure of whether the role of genes in determining a given phenotype is large or small. There are standard methods for its estimation, if the frequency of a trait is known in the population and in relatives, or in MZ as compared with DZ twins. The methodology is described by Falconer (1967), by Bodmer and Cavalli-Sforza (1976) and in many other references.

THRESHOLD TRAITS

Many common congenital malformations result from a failure in a developmental process, so that subsequent developmental stages cannot be reached. The rate of development is determined by many factors, some genetic and others environmental; in other words, the underlying components of the rate may be visualized as a continuous distribution. If a process fails to reach a certain point at an appropriate time, a malformation may result; if so, the continuous variable is now made up of normal and abnormal classes, separated by a threshold (Fig. 12–2).

The idea of a threshold effect was originally developed by Sewall Wright (1934) in connection with a study of the inheritance of polydactyly in guinea

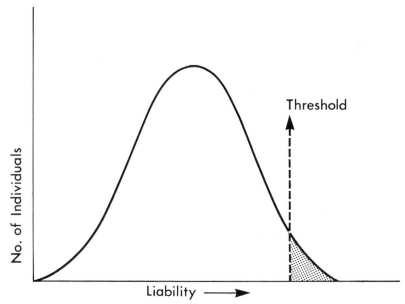

Figure 12-2 Quasicontinuous variation. Liability to a given trait is normally distributed, but the population is divided into normal and abnormal classes by a threshold.

pigs. Usually the guinea pig has three toes, but occasionally a four-toed one is born. Wright found that single-gene inheritance could not explain the transmission of polydactyly and decided that it must result from a threshold effect in a trait determined by many genes. The group as a whole had continuously distributed liability to the trait, but the threshold separated it into two classes, one with polydactyly and one without. Later Gruneberg (1952), working with another type of congenital malformation in another animal, the mouse, coined the term **quasicontinuous variation** for this kind of inheritance.

A number of human congenital malformations have been shown to fit the family patterns expected of multifactorial inheritance with a threshold. The method of analysis is simply to collect information on the frequency of the malformation in the general population and in different categories of relatives. From this information, the **empiric risk** of the condition can be estimated. The empiric risk is based solely on past experience, without knowledge of the nature of the genetic and environmental factors in the pathogenesis of the malformation.

CLEFT LIP AND CLEFT PALATE

Cleft lip with or without cleft palate is etiologically distinct from cleft palate alone (Fogh-Anderson, 1942; Fraser, 1970). Both categories are heterogeneous and include: cleft lip and/or palate as part of the syndromes determined by single mutant genes (about 30 known); cases which are part of chromosomal syndromes (especially trisomy 13); cases resulting from teratogenic agents (the rubella virus, thalidomide, some anticonvulsants); and forms that appear in nonfamilial syndromes. When all these have been removed, the remaining cases, the great majority, are in the multifactorial group.

Cleft Lip With or Without Cleft Palate (CL or CLP). This disorder originates as a failure of fusion of the frontal prominence with the maxillary process, at about the seventh week of development. About 60 to 80 percent of those affected are males. There is wide variation in frequency in different ethnic groups: one per 1000 in whites, 1.7 per 1000 in the Japanese and 0.4 per 1000 in American blacks. When neither parent is affected, the recurrence risk in subsequent sibs of an affected child is about 4 percent, but after two affected children it is 9 per cent (Curtis et al., 1961). The risk varies with the severity of the defect; for example, the recurrence risk in the combined data of several studies cited by Fraser (1970) is as follows:

Cleft lip and palate, bilateral	5.6 percent
Cleft lip and palate, unilateral	4.1 percent
Cleft lip, unilateral, without cleft palate	2.6 percent

Cleft Palate (CP). Failure of fusion of the secondary palate has a population incidence of one in 2500 and is more common in females than in males. There is little ethnic variation in incidence. The recurrence risk in sibs is only about 2 percent, apparently regardless of family history, though more data are needed. Microforms of CP include bifid uvula and congenital palatopharyngeal incompetence, a condition in which the individual speaks as though she has a cleft palate, although there is no cleft.

Cleft palate has been carefully studied in the mouse. It is clearly multifactorial, and the threshold is easy to define. If the palate is to fuse normally, the palate shelves must move from a vertical position on either side of the tongue to a horizontal position above it. If the shelves move into position too late to meet and fuse, a cleft results. Many factors contribute to the underlying continuous variation of liability: the time the shelves move to the horizontal position, the width of the head at that time, the size and position of the tongue, the size of the shelves themselves and so forth.

PYLORIC STENOSIS

Congenital pyloric stenosis is five times as common in boys as in girls (males, five per 1000; females, one per 1000) and has a very unusual distribution in families (Table 12–3). Carter (1964, 1969, 1976) showed that the familial incidence can be explained if pyloric stenosis is a threshold trait (Fig. 12–3).

If the underlying liability to a trait is continuous, but there is a lower threshold in males than in females, then affected females are more "extreme" in range than are affected males, as shown by the relative area of each curve beyond the threshold. The more extreme the parent, the higher above the

TABLE 12–3   THE INCIDENCE OF PYLORIC STENOSIS IN OFFSPRING OF
PATIENTS AS COMPARED WITH THE INCIDENCE IN THE
GENERAL POPULATION*

Relatives	Risk	Multiple of Population Risk
Sons of male patients	1 in 18	× 11
Daughters of male patients	1 in 42	× 24
Sons of female patients	1 in 5	× 40
Daughters of female patients	1 in 14	× 70

*From data of Carter, 1969.

population mean is the mean of the offspring; in other words, offspring of affected females have a higher mean liability to pyloric stenosis than do offspring of affected males. However, the threshold is lower for males than for females, so whether the condition has been inherited from the father or from the mother, more sons than daughters are affected. The greatest risk (about 20 percent) is for sons of female patients. The same principles and risk figures apply to sibs as to offspring, since both offspring and sib have half their genes in common with the propositus.

CRITERIA FOR MULTIFACTORIAL INHERITANCE

Three characteristics of multifactorial inheritance have already been mentioned:

1. The correlation between relatives is proportional to the genes in common. Exceptions may occur if mating is nonrandom, if environmental variation is present, or if dominant rather than additive genes are involved.

2. On the average, if mating is random, if environmental differences are unimportant, and if the genes concerned are additive, the mean for the offspring is midway between the parental value and the population mean.

3. If a threshold trait is more frequent in one sex than the other, the recurrence risk is higher for relatives of patients of the less susceptible sex.

There are several other criteria for multifactorial inheritance with a threshold:

4. If the population frequency is p, the risk for first-degree relatives is $\sqrt{p}$ (Edwards, 1960). This is not true for single-gene traits. Figure 12–4 is a graph of the proportional change in risk to sibs with changing population incidence; it shows that the range for multifactorial traits is quite different from the range for single-gene traits (Newcombe, 1964). The lower the population risk, the greater the relative increase in risk for sibs. For example, neural tube defects (anencephaly and spina bifida) have a population frequency of about one in 130 and a recurrent risk of one in 20 in South Wales (a sevenfold increase), but

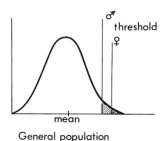

General population

Figure 12–3 An interpretation of the sex ratio and familial distribution of pyloric stenosis in terms of multifactorial inheritance, with a more extreme threshold in females than in males. In the normal population and among the relatives of affected individuals, more males than females are affected but affected females have more affected relatives of both sexes than do affected males. For further discussion, see text. Based on Carter, 1964. In: *Congenital Malformations: Papers and Discussions Presented at the Second International Conference* (M. Fishbein, ed.), International Medical Congress, p. 311.

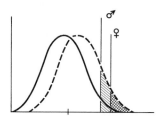

First–degree relatives of males

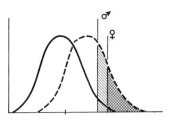

First–degree relatives of females

in London the population frequency is about one in 350 and the recurrence risk about one in 23, a fifteenfold difference (Carter, 1976).

Nora (1968) compared the incidence of several types of congenital heart defects in the general population and in first-degree relatives of patients and found close agreement with this rule (Table 12–4). However, only rather large differences in population frequency make an appreciable difference to the risk in relatives.

5. The recurrence risk is higher when more than one family member is affected. The risk figures for cleft lip and palate in sibs have already been mentioned. For single-gene defects, in contrast, the risk to subsequent sibs is the same regardless of the number of affected children in the sibship.

6. The more severe a malformation, the higher the recurrence risk. The data for cleft lip and palate given above bear this out. The threshold model would predict that the more extreme an individual is in the range of liability, the greater the severity of the malformation and the greater the risk that his offspring would fall beyond the threshold.

7. The risk to relatives declines sharply with increasingly remote degrees of relationship. Typical figures for several conditions are given in Table

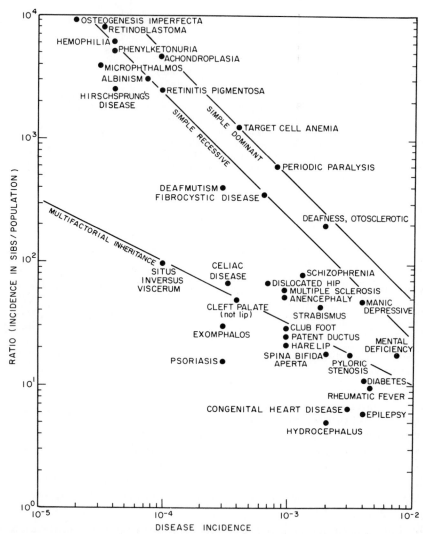

A.E.C.L. Ref. # A-2923-E

Figure 12–4 Comparison of the risk in sibs of propositi with single-gene traits and multifactorial traits. On a log scale, the incidence of each disease shown is plotted against the ratio of the sib incidence to the population incidence. Single-gene and multifactorial traits fall into two distinct clusters. From Newcombe, 1964. *In Congenital Malformations: Papers and Discussions Presented at the Second International Conference* (M. Fishbein, ed.), International Medical Congress, p. 347.

TABLE 12-4   FREQUENCY OF SIX COMMON CONGENITAL HEART
DEFECTS IN SIBS OF PROBANDS

Anomaly	Per Cent Frequency in Sibs*	Expected Frequency†
Ventricular septal defect	4.3	4.2
Patent ductus arteriosus	3.2	2.9
Tetralogy of Fallot	2.2	2.6
Atrial septal defect	3.2	2.6
Pulmonary stenosis	2.9	2.6
Aortic stenosis	2.6	2.1

*Data from Nora, 1968.
†$\sqrt{p}$, where p = population frequency of the specific defect.

12-5.   In contrast, for autosomal dominants the risk declines by 50 percent with each degree of relationship removed from the proband.

8. In twin studies, if the concordance rate in MZ pairs is more than twice as high as the rate in DZ pairs, the trait cannot be a simple dominant; if it is more than four times as high the trait cannot be a simple recessive either. The possibility of multifactorial inheritance must then be considered.

9. Although parental consanguinity is usually taken as a strong indicator of autosomal recessive inheritance, it can also signify that multiple factors with additive effects are involved. If recessive inheritance can be ruled out (by showing that the parent-offspring correlation is similar to the sib correlation), multifactorial inheritance is the only other possibility. A further point is that for multifactorial inheritance, the risk to subsequent sibs is higher when the parents are consanguineous than when they are unrelated, but for autosomal recessive inheritance the risk is the same whether or not the parents are consanguineous.

OTHER FACTORS AFFECTING RECURRENCE RISK

In addition to the factors mentioned above, several others can affect the recurrence risk. If an affected child has a large number of normal sibs already, the chance of recurrence is less than if there are no sibs. If the malformation

TABLE 12-5   FAMILY PATTERNS FOR SOME CONGENITAL MALFORMATIONS*

	Cleft Lip with or without Cleft Palate	Talipes Equino- varus	Congenital Dislocation of the Hip (Males Only)	Congenital Pyloric Stenosis (Females Only)
General population	0.001	0.001	0.002	0.005
First-degree relatives	×40	×25	×25	×10
Second-degree relatives	×7	×5	×3	×5
Third-degree relatives	×3	×2	×2	×1.5

*Data from Carter, 1969.

TABLE 12–6 EXAMPLES OF RECURRENCE RISKS (PERCENT) FOR CLEFT LIP ± CLEFT PALATE AND NEURAL TUBE MALFORMATIONS[*]

	Cleft Lip ± Cleft Palate	Anencephaly and Spina Bifida
No sibs affected		
Neither parent affected	0.1	0.3
One parent affected	3	4.5
Both parents affected	34	30
One sib affected		
Neither parent affected	3	4
One parent affected	11	12
Both parents affected	40	38
Two sibs affected		
Neither parent affected	9	10
One parent affected	19	20
Both parents affected	45	43
One sib and one second-degree relative affected		
Neither parent affected	6	7
One parent affected	16	18
Both parents affected	43	42
One sib and one third-degree relative affected		
Neither parent affected	4	5.5
One parent affected	14	16
Both parents affected	44	42

[*]Data based on Bonati-Pellié and Smith, 1974. J. Med. Genet. 11:374–377.

has previously appeared in both paternal and maternal relatives, the recurrence risk is appreciably higher. Computer programs to derive risk estimates for three common congenital malformations, taking into account a number of different family histories, have been prepared (Bonati-Pellié and Smith, 1974). Table 12–6 gives some of the recurrence figures estimated by computer.

CONCLUSION

The multifactorial/threshold model has not been accepted without criticism. In brief summary, the criticisms are based on the difficulty of testing some of the underlying assumptions and the fact that the data will fit other models, especially if heterogeneity or reduced penetrance is invoked. However, multifactorial inheritance does explain many of the formerly puzzling features of the family distributions of a number of disorders, and has stimulated research into the nature of the underlying factors that contribute to their causation.

GENERAL REFERENCES

Carter, C. O. 1965. The inheritance of common congenital malformations. Prog. Med. Genet. 4: 59–84.

Carter, C. O. 1976. Genetics of common single malformations. Br. Med. Bull. 32:21–26.

Edwards, J. H. 1969. Familial predisposition in man. Br. Med. Bull. 25:58–64.

Fraser, F. C. 1976. The multifactorial/threshold concept: uses and misuses. Teratology 14:267–280.

PROBLEMS

1. For several different congenital malformations, family and population studies have led to the following conclusions:

   For malformation A, the recurrence risk in sibs of affected individuals is the same as the recurrence risk in offspring, 10 percent. The risk in nieces and nephews is 5 percent, and in first cousins the risk is 2.5 percent.
   a) Is this more likely to be an autosomal dominant trait with reduced penetrance or a multifactorial trait?
   b) What other information might support your conclusion?

2. In a certain clinic, prenatal diagnosis by amniocentesis is available to any woman whose fetus has a risk of 1 percent or higher of a neural tube malformation (anencephaly or spina bifida). Which of the following women would qualify? Arrange your answers in order of risk.
   a) A woman who has had a child with anencephaly and no other children.
   b) A woman who has had a child with anencephaly and four normal children.
   c) A woman who has had two anencephalic offspring.
   d) A maternal aunt of an anencephalic.
   e) A wife of a paternal uncle who has a sib with spina bifida.

3. A baby girl has pyloric stenosis. What is the recurrence risk for a) her brother? b) her sister?

4. A series of children with a particular type of congenital malformation includes both male and female affected members. In all cases, the parents are normal. How would you determine that the malformation is more likely to be autosomal recessive than multifactorial?

5. A large sex difference in affected individuals is often a clue to X-linked inheritance. How would you establish that pyloric stenosis is multifactorial rather than X-linked?

13

GENETIC
ASPECTS OF
HUMAN
DEVELOPMENT

It is a matter of concern that although prenatal diagnosis, which was first introduced about 1967, is already widely used for prevention of the birth of genetically defective children, the genetics of human development is still very poorly understood. Of course, there are numerous practical difficulties that hamper progress in this area. Early human embryos and fetuses are simply not available in quantity or in good condition for laboratory studies, and there are serious ethical problems about the use of even those few that are available. Moreover, the approach of the experimental geneticist, who can breed specific mutants and obtain them for study at a sequence of stages, is entirely proscribed in humans. Consequently, much of our knowledge of development, normal and abnormal, has come from studies in other animals.

The human embryo develops from the zygote by a long and complex process involving cell **proliferation, differentiation** of cells into a variety of specialized types, and **morphogenesis**, the development of shape. Development requires differential gene action, and although we can describe at least some of the different patterns of gene activity that occur, we really know nothing at all about the mechanism by which at programmed times particular genes are switched on in specific tissues, while other genes are activated in other tissues, even though the genome is the same in virtually all cells of the organism. The process must not only allow the embryo to undergo a sequence of changes to achieve its eventual form, but must also allow it to continue to function while the changes are going on — it is business as usual during alterations.

Earlier mention (Chapter 3) was made of the Jacob-Monod model for regulation of gene activity. The operon model that regulates gene activity in prokaryotes is not known in man (at least, not yet), and the regulation of gene activity in man and other eukaryotes remains a mystery. In this chapter we will describe some of the progress that has been made in the understanding of gene activity during development, and we will return to the problem of congenital malformations, which has already been considered in Chapter 12.

GENE ACTION DURING DIFFERENTIATION

During the development of a tissue, changes go on in its enzyme profile. New enzymes put in their appearance at specific times, and many alter their isozyme pattern in different tissues in a regular and apparently regulated way as development proceeds (Edwards et al., 1977). However, the means whereby regulation is effected remains unknown. One of the first enzymes to be studied in terms of its developmental changes, and still one of the most instructive examples, is lactate dehydrogenase.

LACTATE DEHYDROGENASE

Lactate dehydrogenase (LDH) is a cellular respiratory enzyme that is widely distributed in the vertebrates and in some invertebrates. The LDH molecule is made up of four polypeptide subunits of two different kinds, A and B, encoded by two separate structural genes. The two kinds of chains combine randomly to form tetramers, in five different combinations: four A chains, three A and one B, two of each, one A and three B's, or four B's. The A and B subunits are often called the M (for skeletal muscle) and H (for heart) subunits respectively, because the LDH tetramer of skeletal muscle is composed of four A subunits and that of cardiac muscle is composed of four B subunits. The five different tetramers (isozymes) are distinguished by starch gel electrophoresis (Fig. 13–1).

Isozyme	*Subunit Composition*	*Speed of Migration*
LDH-1	B_4	Most rapidly migrating band
LDH-2	A_1B_3	
LDH-3	A_2B_2	
LDH-4	A_3B_1	
LDH-5	A_4	Most slowly migrating band

The isozyme pattern of a given tissue is determined by the random combination of the A and B subunits present in the tissue. If the subunits are present in equal amounts, the ratio of the various isozymes is 1:4:6:4:1 (the coefficients of the expansion of $[\frac{1}{2}A+\frac{1}{2}B]^4$). If the proportions are unequal, the isozyme pattern changes accordingly. The pattern characteristic of a given tissue shows progressive changes during fetal and postnatal development.

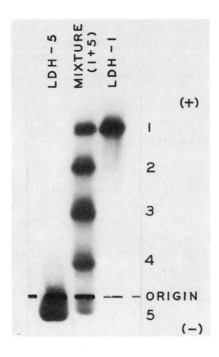

Figure 13–1 Lactate dehydrogenase isozymes as distinguished by starch gel electrophoresis. A mixture of LDH-1 and LDH-5 will produce all five isozymes. For further discussion, see text. From Markert, Science 140:1329–1330, 1963. Copyright © 1963 by the American Association for the Advancement of Science.

A third LDH locus determines a C subunit, which is restricted to sperm and its precursors and forms a separate isozyme, a tetramer of four C subunits, called LDH-X.

The reasons for isozymic differences are unknown, but presumably the differences are related to the functions of the tissues where specific isozyme patterns occur. Not much is known about the effect of mutant genes on the development of isozyme patterns, because so far it has been possible to study these isozymes only in terms of their normal variants.

THE DEVELOPMENT OF HEMOGLOBIN

The synthesis of hemoglobin in red cell precursors begins early in fetal development with the production of transitory embryonic hemoglobins and then of fetal hemoglobin, which is composed of two α and two γ globin chains, the γ chain being unique to fetal hemoglobin. During late fetal life and in the first few months after birth, the production of γ chains declines until fetal hemoglobin comes to represent only about 2 percent of the total hemoglobin, a level that is maintained thereafter. Meanwhile, the synthesis of β chains has begun, and it increases steadily until almost all the hemoglobin present is of the adult type, $\alpha_2\beta_2$. Concurrently, in late fetal life and after birth, δ chain synthesis has begun, though the hemoglobin of which it forms a part (Hb A_2) is never more than a small proportion of adult hemoglobin. These changes are shown in Figure 13–2. The mechanism of regulation of chain production is

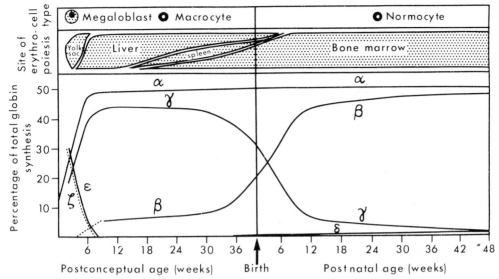

Figure 13–2 Development of erythropoiesis in the human fetus and infant. Types of cells responsible for hemoglobin synthesis, organ(s) involved and types of globin chain synthesized at successive stages (as understood at present) are shown. Redrawn from Wood, Haemoglobin synthesis during human fetal development. Br. Med. Bull., 32:282–287, 1976. Reproduced by permission of the Medical Department, The British Council. After Huehns and Shooter, 1965, and Kleihauer, 1970.

unknown; in fact, since hereditary persistence of fetal hemoglobin is a benign disorder, we have little idea of the reason for the switch to Hb A, and still less, any idea why Hb A_2 is also made. Nevertheless, the precision of the change-over indicates that a regulatory process is concerned.

CONGENITAL MALFORMATIONS

A congenital malformation is a malformation present at birth; the term "congenital" neither connotes nor excludes genetic etiology. Congenital malformations are often referred to as birth defects, though broadly speaking birth defects also include biochemical abnormalities manifest at or near the time of birth, whether or not they are associated with dysmorphism. In this section we will consider only traits that are malformations in the anatomical sense, determined by some disorder of morphogenesis.

Morphogenesis is an elaborate process during which many complex interactions must be performed in orderly sequence. We have little knowledge of the genes and gene products that bring about morphogenesis, though at times we can see their "negatives" in the consequences of mutations that interfere with normal development. For example, we know at least 10 mutant genes that affect the normal differentiation of the digits; if these are not allelic, there must be at least 10 gene loci responsible for the normal development of digit number and shape.

THE EFFECT OF GENES ON DEVELOPMENT

As we have seen in earlier chapters, dysmorphism can arise by a variety of mechanisms. In single-gene disorders, such as achondroplasia, a single major error in the genetic blueprint underlies the abnormality (or abnormalities, since pleiotropic effects are characteristic of genes acting during development). In chromosome disorders malformations are virtually always present and usually severe; they may be visualized as the result of the developing organism's effort to cope with the partially duplicated or incomplete blueprints it has been provided with. In multifactorial inheritance there may be no major error or really incorrect instructions, but a number of minor faults that combine to produce a defective end-product, in much the same way that a number of small defects in material and workmanship can result in a car that is a "lemon." In multifactorial inheritance one or more of the multiple factors may be environmental, but there are situations in which a fault of the environment alone is enough to cause a malformation, or more often a malformation syndrome. We must therefore define a fourth kind of causative agent, an environmental teratogen.

TERATOGENS

A teratogen is any agent that can produce a malformation or raise the incidence of a malformation in the population. Most known teratogens are infectious agents, radiation, or drugs. Some examples are given in Table 13-1.

Probably the best-known teratogen is thalidomide, which gained deserved notoriety when it was found to cause major limb malformations (phocomelia) and other malformations in children whose mothers had been given

TABLE 13-1 SOME TERATOGENIC AGENTS AND THEIR EFFECTS ON HUMAN DEVELOPMENT

Agent	Examples	Resulting Phenotypes
Viruses	Rubella	Congenital heart defects, cataracts, deafness
	Cytomegalovirus	Mental retardation
Radiation		Microcephaly
Drugs	Thalidomide	Phocomelia
	Aminopterin	Various physical malformations, including cranial dysplasia, upswept scalp hair, broad nasal bridge and low-set ears
	Progestin	Masculinization of female fetuses
	Alcohol*	Growth retardation, mental retardation, microcephaly and short palpebral fissures
	Anticonvulsants*	Retardation of physical and mental development, ocular hypertelorism, low-set ears and hypoplastic nails and phalanges

*Malformations occur in some but not all fetuses of mothers exposed to this agent.

the drug thalidomide during early pregnancy. During the short period in the early 1960's when this drug was used in pregnancy as a tranquilizer and anti-emetic, *all* the mothers who received the drug at a specific stage of early pregnancy, when differentiation of the embryo was proceeding rapidly, produced babies with some degree of phocomelia. Thus it appears that susceptibility to teratogenesis from thalidomide is a uniform trait of the human species that unfortunately is not shared by the experimental animals in which the drug was tested.

Thalidomide remains the only example of a teratogen for which the introduction of a drug led to a dramatic rise in the incidence of a specific type of malformation, and withdrawal of the drug was immediately followed by virtual disappearance of the malformation. None of the many other teratogens that have been examined, or suspected, in "epidemics" of malformations has been shown to have a definite cause-and-result connection with a given type of abnormality.

In the analysis of teratogenicity, four factors must be considered:

1. **Time of exposure** to the teratogen. This point has been mentioned above in connection with thalidomide. Another example is rubella embryopathy; after maternal rubella in pregnancy, if the infection occurs in the first four weeks of development, 50 percent of the infants have major malformations, in the third month 17 percent, and thereafter close to zero (Rhodes, 1961). Teratogens have their effect during the stage of active differentiation of an organ or tissue. Because a woman may not even realize she is pregnant at this early stage, it is wise for her to avoid any possibly teratogenic drugs until her pregnancy status has been determined.

2. **Dosage.** There is little information on dosage in human studies, but work with experimental animals has shown the expected dose-response relationship. The dose affecting the fetus is of course partly determined by the mother's ability to metabolize the chemical, so the maternal genotype can be a factor in the response.

3. **Genotype of the embryo.** Thalidomide is an example of a teratogen to which all humans are susceptible, but there are many examples in experimental animals of genetic differences in response to a teratogen. An instructive example of the influence of the fetal genotype on the teratogenicity of a drug is the interaction of the mouse mutant "luxoid" (which affects limb development) and the drug 5-fluorouracil. Polydactyly and limb reduction can result either from the homozygous luxoid genotype or from a high dose of the teratogen. Only a slight abnormality (polydactyly of the hind foot) results from the heterozygous genotype or from a low dose of the drug. But in combination, the heterozygous genotype and the low drug dosage produce a severe effect. There may be analogous instances in man, but we have no examples.

4. **Genotype of the mother.** As mentioned above, the mother's ability to detoxify certain drugs is a possible factor in determining their teratogenicity to her fetus. At present, anticonvulsant drugs used in the control of epileptic seizures and maternal alcohol ingestion are both under active investigation as possible causes of congenital malformations. Both maternal use of anticonvulsants and maternal ingestion of alcohol have been indicted in connection with the birth of infants with a combination of dysmorphic features, prenatal and postnatal growth retardation and retarded mental development, but only a

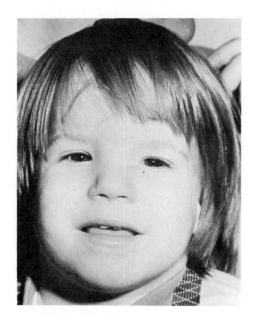

Figure 13–3 Facial appearance of a child aged two-and-one-half years, born to a chronic alcoholic mother. Note particularly the short palpebral fissures and mid-face hypoplasia. From Smith, *Recognizable Patterns of Human Malformation.* 2nd ed., W. B. Saunders Company, Philadelphia, 1976, p. 337.

minority of users of anticonvulsants or of alcohol during pregnancy have children with such problems. So far almost nothing is known about the possible effect of timing of administration, dosage, or maternal or fetal genotype on the expression of these malformation syndromes. A child with the features of fetal alcohol syndrome is shown in Figure 13–3.

The incidence of congenital malformations is somewhat increased in the offspring of diabetic mothers, but it is not clear whether it is the diabetes itself or inadequate management of it that causes these occasional malformations. Obviously this is a field requiring further work.

NEURAL TUBE DEFECTS

Anencephaly and spina bifida are neural tube defects that occur together frequently in families and are considered to have a common origin. They are noteworthy because they show great variation in incidence in relation to geography (more than 50 times as common in Belfast as in Lyons), socioeconomic level (lower levels more susceptible), season of conception and birth (autumn conceptions more susceptible) and parity (first births more susceptible than later births). The variation in incidence has led to an extensive search for environmental teratogens, but so far no agent has been unequivocally proven to be responsible.

In anencephaly the forebrain, midbrain, overlying meninges, vault of the skull and skin are all absent, and the survival after birth is a few hours at most. About two-thirds of the affected infants are female. Spina bifida is failure of fusion of the arches of the vertebrae, typically in the lumbar region. There are various degrees of severity, ranging from spina bifida occulta, in which the defect is in the bony arch only, to spina bifida aperta, often associated with meningocele (protrusion of the meninges) or meningomyelocele (protrusion of neural elements as well as meninges). The incidence is a little higher in girls than in boys. A child with an open neural tube defect is shown in Figure 13–4.

Many infants born with spina bifida are so severely abnormal that they do not survive, and others have minimal defects, but there is an intermediate group in which, even if surgical intervention is successful, the child is left with partial or complete paresis of the lower limbs and fecal and urinary incontinence. Thus at present there is much controversy over the ethical dilemma of whether to provide active treatment or withhold treatment if the child could not survive without surgery.

Anencephaly and spina bifida, like many other common congenital malformations, have multifactorial inheritance. The recurrence risks (which apply to the probability of having a child with either type of defect) have been summarized in Chapter 12.

Spina bifida and anencephaly can be diagnosed *in utero* by the presence of an elevated level of alpha fetoprotein in the amniotic fluid. Alpha fetoprotein may also be elevated in maternal serum, and methods are under development to allow screening of all pregnant women for elevated serum alpha fetoprotein levels, to be followed if indicated by amniocentesis for assay of alpha fetoprotein in amniotic fluid and by ultrasonography and possibly fetoscopy. Thus it seems that we may be on the verge of a breakthrough in prevention of the birth of children with either of these lethal or disabling defects.

DEVELOPMENTAL ANTIGENS AND DIFFERENTIATION

In the mouse, there is a complex gene locus known as the *T* locus which plays a very important part in early development and has a number of unusual

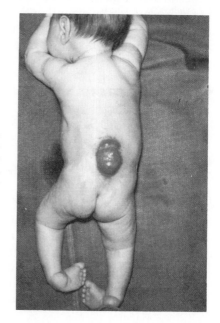

Figure 13–4 An infant with an open neural tube malformation and neurological involvement of the lower limbs. From Moore, *The Developing Human: Clinically Oriented Embryology.* 2nd ed., W. B. Saunders Company, Philadelphia, 1977, p. 352. Photograph courtesy of D. Parkinson.

properties. It is linked to the H-2 locus and may have an evolutionary relationship to it. One allele at this locus, the T allele, when in heterozygous combination with the wild-type allele (+) causes a shortened tail. TT is lethal. There are numerous other alleles (t^1, t^2 and so on) that may cause taillessness in combination with T, or may be lethal or semilethal at some specific embryonic stage or at birth. They usually restrict crossing over in the T–H-2 region of their chromosome, an effect that suggests that there has been a deletion. They often produce distorted segregation ratios, that is, $+t$ or Tt males transmit the t allele to more than 50 percent of their progeny; this mechanism serves to preserve the t alleles in wild populations, in spite of their deleterious effects.

It has recently been discovered that the wild-type allele determines an antigen of the cell membrane that is found only in early embryos, on sperm and on mouse teratocarcinoma cells, but not on adult cells. The same antigen is present in other mammalian embryos, or in other words it has been conserved throughout mammalian evolution. When the antigen is altered or missing, as in mice with mutant T or t alleles, the consequence is embryonic maldevelopment of a specific type, which for different mutants may be as diverse as failure of implantation, failure of formation of mesoderm, or defective formation of the spine. It has been postulated that the mutant alleles may produce defects of the cell membrane that in turn may prevent cell-cell recognition and thus interfere with normal differentiation (Bennett, 1977).

Does the human embryo have a system of developmental antigens homologous with those determined by the T-locus? There is some evidence that it does. The evolutionary conservation of the normal antigen suggests that there is a human locus for it. If so, it might be linked to the HLA locus, since the T locus is linked to the H-2 locus. To explore this possibility, research has begun into the question of the linkage of liability to spinal malformation (spinal bifida and related defects) and HLA type. Though not conclusive, the first reports suggest that a loose linkage does exist. Additional studies may confirm this, but at least it seems clear that the wild-type allele at the T locus determines an antigen on the cell surface that plays an important part in the development of mammalian embryos, including human ones.

MOUSE CHIMERAS

It is possible to make chimeric mice by fusion of cleavage-stage mouse embryos and even by injection of single cells from one mouse embryo to another at the blastocyst stage. The technique is delicate and is not always successful, but if the embryos have different genes for coat color it is easy to recognize the chimeras a week or so after birth by their patchy coats (Fig. 13–5). Chimeras derived from two fused embryos and thus having four parents are very useful experimental tools for the analysis of mammalian development. Chimeras have even been made from mouse teratoma cells combined with genetically different blastocysts. (A teratoma is an embryonic tumor.) These chimeras have been shown to be capable of forming reproductive cells of the same genotype as the tumor "parent," thus demonstrating that the tumor cell has not lost the capacity to differentiate into normal body tissues.

Figure 13–5 A "four-parent" chimeric mouse. The patchy coat shows the coat phenotypes of both parental lines. From Mintz, Gene control of mammalian pigmentary differentiation. I. Clonal origin of melanocytes. Proc. Natl. Acad. Sci. 58:344–351, 1967, by permission.

When three genetically different embryos combine to form a chimera, which consequently has six parents, studies of these triple chimeras suggest that adults are derived from three and only three cells in the inner cell mass of the blastocyst.

GENERAL REFERENCES

Heinonen, O. P., Sloane, D., and Shapiro, S. 1977. *Birth Defects and Drugs in Pregnancy.* Publishing Sciences Group, Inc., Littleton, Massachusetts.

Markert, C. L., and Ursprung, H. 1971. *Developmental Genetics,* Prentice-Hall, Englewood Cliffs, New Jersey.

Moore, K. L. 1977. *The Developing Human: Clinical Oriented Embryology,* 2nd ed. W. B. Saunders Company, Philadelphia.

Shepard, T. H. 1976. *Catalog of Teratogenic Agents.* 2nd ed. The Johns Hopkins University Press, Baltimore and London.

Smith, D. W. 1976. *Recognizable Patterns of Human Malformation.* 2nd ed. W. B. Saunders Company, Philadelphia.

Warkany, J. 1971. *Congenital Malformations: Notes and Comments.* Year Book Medical Publishers, Inc., Chicago.

14

ELEMENTS
OF
MATHEMATICAL
GENETICS

Though not all problems in medical genetics require a mathematical approach, there are areas that cannot be handled in any other way. A medical geneticist need not be an expert mathematician (though many are), but he or she should at least be aware of the pitfalls that await mathematical incompetents and the importance of obtaining expert assistance with the statistical aspects of research problems, from the planning stage through the analysis of the observations.

Special mathematical techniques are necessary in human genetics because of the special difficulties of dealing with human material. If human families were large enough for Mendelian ratios to be approximated (say, with 12 or more sibs), many of the problems resulting from ascertainment bias would disappear. Other complications would vanish if geneticists lived longer than the subjects of their research and could extend their observations over several generations. (Here pediatricians have an advantage over other clinicians, since they normally see two generations of a family at a time.) Most important of all, the impossibility of arranging test matings makes it necessary to accumulate the results of nature's own experiments, a sometimes laborious process that adds to the statistical difficulties and may make statistical methodology unavoidable.

PROBABILITY

Genetic transmission from parent to child can be compared with tossing coins. In both, there are two alternative possibilities: the coin can turn up heads or tails, and the parent can transmit one or the other of the two alleles at a locus. In either case, the outcome is determined by chance.

Statisticians define probability, the mathematical expression of chance, as the ratio of the number of occurrences of a specified event to the total number

274

of all possible events. The specified event may be that a tossed coin turns up heads. There are two possible events, heads and tails, so, for a single toss:

$$\frac{\text{Possibility of heads}}{\text{Total number of possibilities}} = \frac{1}{2}$$

The probability of tossing heads is $\frac{1}{2}$.

Similarly, the probability that a child whose father is of blood group AB will receive the A allele rather than the B allele is:

$$\frac{\text{Possibility of } A}{\text{Total number of possibilities}} = \frac{1}{2}$$

The probability of A is $\frac{1}{2}$.

PROBABILITY OF COMBINATION OF TWO INDEPENDENT EVENTS

A child receives his genotype from two parents, and the probability of receiving certain combinations of alleles from the two parents can be simulated by tossing two coins at once. The probability of heads is 1/2 for each coin, and the probability of heads turning up on both coins is $1/2 \times 1/2 = 1/4$. Similarly, the chance that a child, both of whose parents are of blood group AB, will receive the A allele from each parent is $1/2 \times 1/2 = 1/4$.

There are many genetic applications of this simple but important principle. Perhaps one of the most useful to the physician has to do with sex ratios in human families. The probability of a boy as the first-born child is 1/2. (Actually it is not exactly 1/2, but this is close to the precise figure and simplifies the arithmetic.) The next child is an "independent event" since sperm, like coins, have no memory; so the probability that the second-born child will be another boy is again 1/2. The probability that the first two children in a family will both be boys is $1/2 \times 1/2 = 1/4$. This procedure can be continued to obtain the probability of sons in any number of successive births; for example, the probability of a sequence of six sons is $(1/2)^6$ or 1/64.

THE BINOMIAL DISTRIBUTION

The seventeenth century mathematician Jakob Bernouilli was a member of a family noted for its many brilliant mathematicians, but his genetic significance as an example of hereditary genius is outweighed by his contribution to genetic theory: the binomial distribution (Bernouilli distribution). Though the binomial formula was originally discovered by Sir Isaac Newton in 1676, it was Bernouilli who gave the first rigorous proof of it.

Suppose the problem to be solved is not the probability of a boy at each of two births, but the probable distributions of boys and girls in two births. This is an extension of the problem of the probability of coincidence of random events.

If the probability of a boy is 1/2 and the probability of a girl is also 1/2, then the distribution of families with no boys, one boy and two boys among all two-child families is given by the expansion of $(1/2 + 1/2)^2$. In general terms: when there are two alternative events, one having a probability of p and the other having a probability of $q = 1 - p$, then in a series of n trials, the frequencies of the different possible combinations of p and q are given by the expansion of the binomial $(p + q)^n$.

Thus, for two-child families:

$$(p + q)^2 = p^2 + 2pq + q^2$$

$$p = q = \frac{1}{2}$$

$$p^2 = \text{families of 2 boys} = \frac{1}{4}$$

$$2pq = \text{families of 1 boy and 1 girl} = \frac{1}{2}$$

$$q^2 = \text{families of 2 girls} = \frac{1}{4}$$

For three-child families:

$$(p + q)^3 = p^3 + 3p^2q + 3pq^2 + q^3$$

$$p^3 = \text{families of 3 boys} = \left(\frac{1}{2}\right)^3 = \frac{1}{8}$$

$$3p^2q = \text{families of 2 boys, 1 girl} = 3\left(\frac{1}{2}\right)^2\left(\frac{1}{2}\right) = \frac{3}{8}$$

$$3pq^2 = \text{families of 1 boy, 2 girls} = 3\left(\frac{1}{2}\right)\left(\frac{1}{2}\right)^2 = \frac{3}{8}$$

$$q^3 = \text{families of 3 girls} = \left(\frac{1}{2}\right)^3 = \frac{1}{8}$$

The difference between a simple multiplication of the separate probabilities and the use of the binomial expansion is that the binomial expansion includes all the possible combinations of the two events. For example, in three-child families there are eight possible sequences:

♂ ♂ ♂	♀ ♂ ♂
♂ ♂ ♀	♀ ♂ ♀
♂ ♀ ♂	♀ ♀ ♂
♂ ♀ ♀	♀ ♀ ♀

The probability of each sequence is 1/8, but among families with two sons and a daughter, the daughter might be born first, second or last; the three possible sequences account for the total 3/8 probability of having two sons and a daughter among three children. The expected distribution of males and females in families of one to five children is summarized in Table 14–1.

Pascal's triangle is a convenient way to arrive at the coefficients of the binomial expansion. Each new coefficient in the triangle is made up by adding the two coefficients nearest it in the line above; e.g., on the fourth line, the 3 of $3p^2q$ is obtained by adding the 1 of p^2 and the 2 of $2pq$.

n	Expansion
0	1
1	$1p + 1q$
2	$1p^2 + 2pq + 1q^2$
3	$1p^3 + 3p^2q + 3pq^2 + 1q^3$
4	$1p^4 + 4p^3q + 6p^2q^2 + 4pq^3 + 1q^4$
5	$1p^5 \times 5p^4q + 10p^3q^2 + 10p^2q^3 + 5pq^4 + 1q^5$

Note that the number of terms in the expansion is always $n + 1$. For families of size n, there can be 0, 1, 2 ... n sons; hence there are $n + 1$ classes of family. If there are four children in a family, there can be five different ♂:♀ distributions: 4:0, 3:1, 2:2, 1:3, 0:4.

Mathematicians may prefer to use the general term of the binomial expansion:

$$\frac{n!}{m!\,(n-m)!}\; p^m\, q^{n-m}$$

n = total number in the series
$n!$ ("n factorial") is $n(n-1)\,(n-2) \ldots 1$.
p = probability of a specified event
$q = 1 - p$ = probability of the alternative event
m = number of times p occurs (in other words, the exponent of p).

For example, the probability of having three sons and two daughters in a five-child family is:

$$\frac{5 \times 4 \times 3 \times 2 \times 1}{(3 \times 2 \times 1)(2 \times 1)} \left(\frac{1}{2}\right)^3\left(\frac{1}{2}\right)^2 = \frac{10}{32}$$

In the examples used so far, the values of p and q have been equal ($p = q = 1/2$), but the binomial distribution can be used for other values of p and q. The same method can be applied, for example, to give the distribution of a recessive trait in the progeny of two heterozygous parents, but now the probability of the specified event (that a specific child will be affected) is 1/4, and the probability that a specific child will be unaffected is 3/4. We will return to this topic in the following section when discussing tests of genetic ratios. Gene frequencies in populations also depend upon simple binomial distributions (see Hardy-Weinberg law, chapter 15).

TABLE 14-1 DISTRIBUTION OF BOYS AND GIRLS IN FAMILIES

Number of Children in Family	$(p+q)^n$	Distribution
1	$\left(\frac{1}{2}+\frac{1}{2}\right)^1$	$\frac{1}{2}(1\,\sigma)+\frac{1}{2}(1\,♀)$
2	$\left(\frac{1}{2}+\frac{1}{2}\right)^2$	$\frac{1}{4}(2\,\sigma)+\frac{1}{2}(1\,\sigma:1\,♀)+\frac{1}{4}(2\,♀)$
3	$\left(\frac{1}{2}+\frac{1}{2}\right)^3$	$\frac{1}{8}(3\,\sigma)+\frac{3}{8}(2\,\sigma:1\,♀)+\frac{3}{8}(1\,\sigma:2\,♀)+\frac{1}{8}(3\,♀)$
4	$\left(\frac{1}{2}+\frac{1}{2}\right)^4$	$\frac{1}{16}(4\,\sigma)+\frac{4}{16}(3\,\sigma:1\,♀)+\frac{6}{16}(2\,\sigma:2\,♀)+\frac{4}{16}(1\,\sigma:3\,♀)+\frac{1}{16}(4\,♀)$
5	$\left(\frac{1}{2}+\frac{1}{2}\right)^5$	$\frac{1}{32}(5\,\sigma)+\frac{5}{32}(4\,\sigma:1\,♀)+\frac{10}{32}(3\,\sigma:2\,♀)+\frac{10}{32}(2\,\sigma:3\,♀)+\frac{5}{32}(1\,\sigma:4\,♀)+\frac{1}{32}(5\,♀)$

FURTHER COMMENTS ON THE SEX RATIO

The sex ratio is defined as the ratio of the number of male births to the number of female births. Since sex is determined by whether the sperm contributes an X or a Y chromosome to the zygote, and since X-bearing and Y-bearing sperm are theoretically formed in equal numbers at meiosis, the expectation is that the primary sex ratio (ratio at fertilization) should be 1.00. However, in all parts of the world there is an excess of male babies, and currently in North America the secondary sex ratio (ratio at birth) is about 1.05 (105 boys per 100 girls). It has been thought that more males than females die before birth, but sex ratio studies in abortuses contradict this idea (Carr, 1971). After birth, males have a higher mortality rate than females. By about age 25 the sex ratio reaches 1.00, and thereafter the excess of females becomes more and more pronounced. The reason for the higher death rate in males is not known, but it may be due in large part to the fact that males have only one X chromosome, so that harmful X-linked recessive genes are exposed to selection in males but not in females.

What effect does parental preference for one sex or the other have on the sex distribution in families and in the population? As an extreme example, consider the situation in which parents either have no more children after they have had one daughter, or continue to have children until they have a daughter. In this case, all one-child families will consist of one girl; all two-child families will have a boy and a girl, in that order; all three-child families will have two boys and a girl, in that order; and so on. The distribution of the sexes within these families will therefore be quite different from the binomial proportions set out in Table 14–1. But since at each birth equal numbers of boy and girl babies are born, the overall sex ratio in the population remains unchanged.

Of course, the sex ratio could be drastically altered if parents were to exercise preference for one sex before birth, terminating pregnancies in which the fetus was of the unwanted sex. In general, since males are usually preferred, this would raise the sex ratio. The long-term effects on the population could be dramatic and perhaps unexpected. In a world in which population control seems to many thoughtful observers to be the major problem facing our species, perhaps there is something to be said for lowering the relative number of female births. The result, in the next generation, would be a corresponding decrease in the total number of births, since population growth depends not on the total population, but on the number of fertile females it contains.

Population data on sex distribution in sibships of different sizes fit the binomial expectation reasonably well. There are, however, a few possible or real exceptions, such as excesses of unisexual sibships and a preponderance of males in some pedigrees and of females in others. In experimental animals there are some unusual genetic mechanisms that upset the normal sex ratio, for example, a "sex ratio" gene in the fruit fly that leads to almost 100 percent female offspring. However, on the whole the obstetrician can safely tell prospective parents that there is an even chance that the baby will be of the desired sex.

TESTS OF GENETIC RATIOS

An essential first step in the analysis of a genetic disorder is the demonstration of its pattern of inheritance, since this may suggest its mechanism and perhaps give clues leading to its prevention. Sometimes inspection of a collection of pedigrees will make the pattern obvious, or tests for heterozygote detection may be applied, but even then mathematical tests may be required to alert the investigator to discrepancies caused by such factors as abnormal segregation ratios, different prenatal viability in the two sexes, unsuspected heterogeneity and so forth.

X-linked inheritance, especially X-linked recessive inheritance, can usually be recognized quite easily by the characteristic pedigree pattern and by the discrepancy in the sex ratio of affected individuals. Autosomal dominant inheritance can usually be recognized without difficulty because the trait appears in each generation; the most likely type of pattern to be confused with it is X-linked dominance, but the two types are immediately distinguishable if one looks only at the offspring of affected males; if an affected male has any affected sons or any normal daughters, the trait cannot be X-linked.

If penetrance is reduced or if expressivity is variable, it is sometimes difficult to distinguish between autosomal dominant and multifactorial inheritance. Here, the chief means of differentiation is by comparison of the incidence of the trait in relatives of various degrees. Fuller discussions of the criteria for the different patterns of inheritance have been given in earlier chapters.

BIAS OF ASCERTAINMENT

In pedigree analysis, the **propositus** (also known as the index patient or proband) through whom the pedigree is ascertained must be omitted from calculations of the proportion of family members affected. As a demonstration of the effect of this kind of bias, consider the sex ratio of the families of a class of medical students. This class, composed of 197 men and 53 women, reported the number of males and females in their sibships to be 376 males and 226 females. The sex ratio of 1.7 deviated significantly from the expected 1.0. But when the data were broken down separately for male and female students it was found that:

The 197 males had 132 brothers and 127 sisters.
The 53 females had 50 brothers and 46 sisters.

Thus the sex ratio in the sibs of the propositi was 182 males: 171 females or 1.06, which is not significantly different from the expected ratio. The discrepancy in the first results was caused by the inclusion of the propositi.

The same kind of bias can arise in genetic studies if the propositus is included. If as usually happens, a series of families is selected because there is at least one affected member in each family, the expected ratio of normal to abnormal members is not observed unless an appropriate correction is made.

TESTS FOR AUTOSOMAL RECESSIVE INHERITANCE

Pedigree analysis to prove autosomal recessive inheritance has special difficulties not encountered with other pedigree patterns. An autosomal recessive trait appears in one-fourth of the offspring of two carrier parents. This is a simple and well-known rule, but to prove the quarter ratio is not always easy. Since carrier parents are usually identifiable only because they have produced an affected child, collections of data on families with recessive traits usually consist of parent-child sets in which both parents are phenotypically normal and at least one child is affected. Thus, carrier parents go undetected if all their children are normal and these normal children are not enumerated. As a result, the proportion of affected children in the families that are actually observed is well above the expected one-fourth.

The number of families missed because they include no affected child decreases with family size. Seventy-five per cent of one-child families are not recognizable, but only 3 percent of the 12-child families are missed. For families of typical sizes, the correction is large enough to be important.

Types of Ascertainment

The terminology used here for the different types of ascertainment used, by necessity or by design, in genetic studies, is that of Morton (1959), Crow (1965) and Cavalli-Sforza and Bodmer (1971). The reader is cautioned, however, that some textbooks use different terminology or different definitions.

Complete and Incomplete Ascertainment

If every pair of parents heterozygous for the condition under study or a truly random sample could be identified and all their offspring could be included, ascertainment would be **complete.** This may be possible for autosomal dominants, in which most matings that produce affected offspring are those of a heterozygote and a normal homozygote, but complete ascertainment is rarely possible for autosomal recessives. In actual practice ascertainment is almost always **incomplete**, i.e., only those sibships that contain at least one affected child are identified.

Truncate, Single and Multiple Ascertainment

Given incomplete ascertainment, the next point to be determined is how the sibships for study have been selected from the population. Selection is **truncate** if every affected child in the population is included in the survey (or if the group studied is a true random sample of the total group of affected children). If the population has been fully screened, any one af- fected child is equally likely to be included regardless of the number of affected children in his sibship. Selection is **single** if there is almost no chance that any one sibship will be ascertained more than once. In this model, the probability that any one child will enter the series may be quite low; for example, if an affected child has only a 1 percent chance of being included and a 99 percent chance of being missed, then a family in which there are two affected children each of whom has a 1 percent chance of being included, still has only a 2 percent $(1 - 0.99^2)$ chance of entering the study and a 98 percent chance of being missed. However, in actual prac- tice, more often than either of these two extremes there is **multiple** ascer- tainment. Some sibships are ascertained more than once, through more than one of their affected members, and families with more than one affect- ed child have a higher chance than one-child families of being ascertained, though still less than a 100 percent chance. For example, assume that the probability of ascertaining a particular child is only 80 percent. Then 20 percent of all sibships with one affected child will be lost to the study, but only 4 percent (0.20^2) of sibships with two affected members and less than 1 percent (0.20^3) of sibships with three affected members are missed.

To illustrate possible situations in which truncate, single and multiple selection might apply, consider how a study of the genetics of cystic fibro- sis (CF) might be organized.

1. *Truncate Ascertainment.* Every child with CF in a given area might be independently identified, regardless of the number of affected sibs, age, clinical status or other factors that might affect the child's likelihood of being found. Total ascertainment could involve an exhaustive search through clinics, hospital records, school health records, private physicians' practices and so forth. If the search is successful in finding every case of CF in the community, ascertainment is truncate, and in the collected family data for sibships of a given size, those with one, two, three or more affected members are distributed binomially, except for the class with no affected members. (Thus the distribution is truncate, which means "shortened by cut- ting off a part." The cut-off part is the group of families with heterozygous parents but no affected children.)

2. *Single Ascertainment.* Suppose our study of the genetics of CF is limited to a survey of all the children entering Grade 1 in a given year. It is very unlikely that any sibship with affected members would be encoun- tered more than once. However, the more affected children a family con- tains, the more likely that one of them will be entering Grade 1. In this situation, single selection methods are appropriate.

3. *Multiple Ascertainment.* If the case load of a particular CF clinic at a pediatric hospital is the source of the data, a good many affected children in the community might be missed because they do not attend the clinic. Some of these might be treated by private physicians, others might be in a hospital for chronically ill children and still others might be managed at home without regular medical attention. Among the children attending the clinic, there would be some with sibs who also were attending; in other words, some sibships would be ascertained more than once, and the proba- bility that a particular family would be included would depend partly but

not entirely upon the number of cases of CF in the sibship. This is the multiple ascertainment situation, which requires special statistical techniques that are beyond the scope of this discussion.

Corrections for Bias of Ascertainment

The Proband Method of Weinberg

In 1912 Weinberg proposed a simple method of correction for ascertainment bias, by discarding the proband from the calculations and determining the incidence of the disorder in his or her sibs. The proband (propositus) is used only as the individual through whom the sibship has been ascertained. This is the kind of correction applied to the data on sex ratio in sibships given earlier. It is especially suitable to the single ascertainment situation but can be used for multiple ascertainment if each proband (not each affected person, but each person through whom a given sibship has been ascertained) is considered separately. The proband method is often useful for a preliminary analysis and may be sufficient for all practical purposes.

The Method of Discarding the Singles

Li and Mantel (1968) proposed a simple method of testing for recessive inheritance, which is appropriate to use when ascertainment is truncate. It is only necessary to count the total number of individuals in all the sibships (T), the number of affected individuals in all the sibships (R) and the number of "singles," that is, the number of patients who are the only affected members of their sibships (J), then to calculate using the following formula:

$$\frac{R - J}{T - J}$$

The value can be compared with the expected 0.25 with the help of tables given by Li and Mantel for calculation of variance and standard error.

The Apert or A Priori Method

If it is suspected that a specific disease is inherited as an autosomal recessive, one method of testing is to assume that recessive inheritance is present; in other words, to set up an *a priori* expectation of recessive inheritance, and then to test the data for agreement with this hypothesis. Since the probability that a family with two heterozygous parents will be missed varies with family size, the calculation must be made separately for each size of sibship. The method can be used for either truncate or single ascertainment, but the expected proportion of affected children is different in the two cases, as will be shown.

Truncate Ascertainment.   As an example, let us begin by considering 16 two-child families, each with both parents heterozygous. One-quarter of the children are expected to be affected. How will these eight children be distributed in their respective sibships?

One-fourth of the families will have an affected child at the first birth, and one-fourth at the second birth:

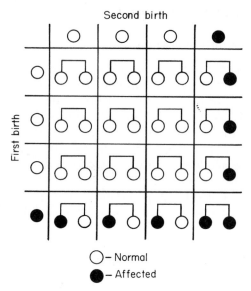

O – Normal

● – Affected

The chance distribution will produce 1 family with 2 affected children, 6 with 1 affected child each and 9 with no affected children (which will not be ascertained because they are not distinguishable from the rest of the population). The incidence of affected children in the observed families is 8/14, or 0.57.

Note the mathematical relationship: if the probability of a normal child is 3/4 and the probability of an affected one 1/4, then the distribution of the trait in two-child families is given as in the general situation, by the expansion of $(p + q)^n$. Here $p = 3/4$, $q = 1/4$ and $n = 2$.

Now consider 64 sibships of three children each, with heterozygous parents. The proportion of families with 1, 2 and 3 affected children can again be calculated from the expansion $(p + q)^n$, but now $n = 3$.

The population distribution of such families is:

$$\left(\frac{3}{4}\right)^3 + 3\left(\frac{3}{4}\right)^2 \left(\frac{1}{4}\right) + 3\left(\frac{3}{4}\right)\left(\frac{1}{4}\right)^2 + \left(\frac{1}{4}\right)^3$$

$$\frac{27}{64} + \frac{27}{64} + \frac{9}{64} + \frac{1}{64}$$

$\dfrac{27}{64}$ have no affected child and are not ascertained.

$\dfrac{27}{64}$ have 1 affected child

$\dfrac{9}{64}$ have 2 affected children

$\dfrac{1}{64}$ have 3 affected children

Since 27 of the 64 families are not ascertained, then among the 37 that can be studied the following numbers and proportions are found:

$$\frac{27}{37} \ (0.730) \text{ have 1 affected child}$$

$$\frac{9}{37} \ (0.243) \text{ have 2 affected children}$$

$$\frac{1}{37} \ (0.027) \text{ have 3 affected children}$$

The expected incidence of affected children in the ascertained sibships can now be calculated:

$$\frac{1}{3} \text{ of } 0.730 = 0.243$$

$$\frac{2}{3} \text{ of } 0.243 = 0.162$$

$$\frac{3}{3} \text{ of } 0.027 = 0.027$$

Expected proportion
  of affected children $= 0.432$

This method of correction of the expected proportion of affected children can be extended to sibships of any size (Table 14–2). When the expectations in a given sample have been calculated, the expected and observed numbers of affected sibs can be compared by the X^2 method (see later discussion).

Single Ascertainment. Whereas under truncate selection the assumption is that all sibships which have at least one affected child are included in the series, under single selection the assumption is that only a small proportion of the available sibships have been located. Under single selection, the probability that any one sibship will be included is low and directly proportional to the number of affected persons it contains.

Consider again the distribution of a recessive trait in three-child sibships. Since the probability that any sibship is included is now proportional to the number of affected sibs, each term of the expansion is multiplied by the number of affected members, and the expression becomes:

$$0(p)^3 + 1(3p^2q) + 2(3pq^2) + 3(q)^3$$
$$\text{or}$$
$$3p^2q + 6pq^2 + 3q^3$$

If we now divide by the common term $3q$, the distribution becomes

$$p^2 + 2pq + q^2$$

which is $(p + q)^2$. In other words, the distribution in the *ascertained* families is binomial, with $n =$ one less than the sibship size.

Table 14–2 summarizes the expected proportions of affected members under truncate and single selection for sibships of up to six members. Unfortunately, collections of data strictly suitable for analysis by either method are hard to acquire, since some intermediate situation is usually a better representation of reality. Other methods of dealing with more complicated situations have been devised, but these usually require considerable mathematical knowledge and the use of a computer. Differences in ascertainment are discussed here chiefly to alert the reader to the necessity of paying attention to how family data are collected.

THE X² TEST OF SIGNIFICANCE

Statistical tests of significance are used to determine whether a set of data conforms to a certain hypothesis. Textbooks of statistics and of genetics describe a variety of such tests and the circumstances under which they can appropriately be applied. One of the most useful is the X^2 (Chi square) test, which is applicable to many problems in genetics because it does not require that the data analyzed be more or less normally distributed, but only that the numbers in the different categories be known.

The calculation of X^2 is quite simple. If O is the observed number in each category and E is the number expected in that category on the basis of the hypothesis being tested, then X^2 is the square of the difference between O and E, divided by E, summed over all the categories; in other terms:

$$\Sigma \left[\frac{(O - E)^2}{E} \right]$$

The probability associated with a given value of X^2 can be obtained from X^2 tables originally prepared by Fisher (1944) and reprinted in many

TABLE 14–2 EXPECTED PROPORTIONS AND NUMBERS OF AFFECTED MEMBERS IN SIBSHIPS OF SIZES 1–6, UNDER TRUNCATE AND SINGLE ASCERTAINMENT

Sibship Size	Truncate Ascertainment Expected Proportion	Expected Number	Single Ascertainment Expected Proportion	Expected Number
1	1.0000	1.0000	1.0000	1.0000
2	0.5714	1.1428	0.6250	1.2500
3	0.4324	1.2972	0.5000	1.5000
4	0.3657	1.4628	0.4375	1.7500
5	0.3278	1.6390	0.4000	2.0000
6	0.2885	1.8246	0.3750	2.2500

textbooks. (Before using these tables, it is necessary to know the number of "degrees of freedom" available; the examples given subsequently should show how this can be determined.) A value of X^2 associated with a probability of less than 0.05 is considered to be "significant," i.e., to indicate a significant disagreement between the observations and the hypothesis being tested.

The hypothesis being tested may be a "null hypothesis," in the sense that the observed values do not differ from the expected, or that two variables being compared are not in any way associated. If the observed and expected values are significantly different, the null hypothesis is disproved.

Example 1. A series of patients with congenital pyloric stenosis includes 25 boys and five girls. Does this distribution differ from the normal sex ratio?

	Number Observed	Number Expected
Boys	25	15
Girls	5	15
Total	30	30

$$X^2 = \frac{(25 - 15)^2}{15} + \frac{(5 - 15)^2}{15} = 6.7 + 6.7 = 13.4$$

The expected values are calculated on the basis of 1:1 sex ratio. There is one degree of freedom: since the total must remain fixed at 30, only one of the two values can be freely assigned. Consulting a table of X^2, we see that for one degree of freedom, this value of X^2 would occur by chance with a probability of <0.01, i.e., less than once in 100 times. (The acceptable level of significance is $p = 0.05$.) Thus, it can be concluded that in this series the observed excess of boys is statistically significant.

Example 2. In a series of twins, at least one member of each pair having congenital dislocation of the hip, the defect was present in both members (concordant) in 34 of 86 MZ pairs and in 5 of 79 DZ pairs. Since congenital dislocation of the hip is about six times as frequent in females as in males, it is appropriate to compare like-sexed pairs rather than all DZ pairs.

	Concordant	Discordant	Total
MZ pairs	34	52	86
Like-sexed DZ pairs	5	74	79
Total	39	126	165

In this example, the expected proportion in each category can be calculated on the basis of the null hypothesis, which assumes that the degree of concordance is the same for MZ and DZ pairs. In the whole series, the proportion of MZ twins is $39/165 = 0.24$. The expected numbers then are:

$$MZ \text{ concordant } 0.24 \times 86 = 20.64$$
$$MZ \text{ discordant } 0.76 \times 86 = 65.36$$
$$DZ \text{ concordant } 0.24 \times 79 = 18.96$$
$$DZ \text{ discordant } 0.76 \times 79 = 60.04$$

$$X^2 = \frac{(34 - 20.64)^2}{20.64} + \frac{(52 - 65.36)^2}{65.36} + \frac{(5 - 18.96)^2}{18.96} + \frac{(74 - 60.04)^2}{60.04}$$

$$= 24.91$$

Since the marginal numbers in the table above must remain fixed, only one of the numbers of twin pairs can be filled in at random; thus there is one degree of freedom. Again, the X^2 table shows the calculated value of X^2 to be beyond the level of significance ($p < 0.001$). The conclusion is that the observed excess of MZ pairs is statistically significant. This suggests that congenital dislocation of the hip is in part genetically determined.

GENERAL REFERENCES

Cavalli-Sforza, L. L., and Bodmer, W. F. 1971. *The Genetics of Human Populations.* W. H. Freeman and Company, San Francisco.

Crow, J. F. Problems of ascertainment in the analysis of family data. In: Neel, J. V., et al., eds. 1965. *Genetics and the Epidemiology of Chronic Diseases.* U.S. Department of Health, Education and Welfare, Public Health Service Publication 1163, Washington, D.C.

Emery, A. E. H. 1976. *Methodology in Medical Genetics: An Introduction to Statistical Methods.* Churchill Livingstone, Edinburgh, London and New York.

Fisher, R. A. 1970. *Statistical Methods for Research Workers.* 14th ed. Oliver and Boyd, Edinburgh.

PROBLEMS

1. If a coin is flipped six times, what is the chance of getting:
   a) six heads? b) three heads and three tails?

2. In a four-child family, state the probability that:
   a) The fourth child will be a boy.
   b) All four children will be boys.
   c) At least one child will be a girl.

3. Two parents carry the same recessive gene for congenital deafness. What is the probability that:

a) Their first child will be deaf?
b) Their five children will all be normal?
c) Two of their five children will be deaf?

4. A man with Huntington chorea has three children. What is the chance that:
   a) None of the three will develop the disease?
   b) The first child will be affected and the next two normal?
   c) One child will be affected and two normal?

5. In a series of two-child families in each of which there is at least one child with cystic fibrosis, what proportion of all the children in the sibships will be affected:
   a) Under truncate selection?
   b) Under single selection?

15

POPULATION GENETICS

Population genetics is the study of the distribution of genes in populations and of how the frequencies of genes and genotypes are maintained or changed. Under certain ideal circumstances, genotypes are distributed in proportion to the gene frequencies in the population and remain constant from generation to generation. If gene frequencies and the corresponding genotype frequencies remained fixed, evolution could not proceed; hence population genetics encompasses factors concerned in human evolution.

The cornerstone of population genetics is the **Hardy-Weinberg law,** which is described in the next paragraphs. If a population is found not to be in Hardy-Weinberg equilibrium (that is, if the genotypes are not randomly distributed or if they change in proportion from generation to generation), there must be a reason. Factors known to disturb Hardy-Weinberg equilibrium include nonrandom mating, mutation, selection and migration.

GENE FREQUENCIES IN POPULATIONS: THE HARDY-WEINBERG LAW

The Hardy-Weinberg law is named for its discoverers, G. H. Hardy (an English mathematician) and W. Weinberg (a German physician), who independently defined it in 1908. It is reported that Hardy, when presented with the problem of why a dominant trait does not automatically increase until it replaces the recessive, worked it out immediately at the dinner table and thought it too trivial to publish; as it happens, he is remembered more for this piece of work than for his fundamental contributions to mathematical theory.

For any gene locus, in a population with random mating, the genotype frequencies are determined by the relative frequencies of the alleles of which they are composed. When we speak of population frequencies of genes, we have in mind a **gene pool** in which are collected all the alleles at that particular locus for the whole population. For an example using only two different alleles, we can think of the gene pool as a beanbag containing beans of two colors, white and black. (Population genetics is sometimes called "beanbag genetics".) The chance that in two draws a person will draw any one of the three possible combinations of beans (two white, two black or one of each color) depends on the frequency of each color in the bag. If p is the frequency of white beans and $q = 1 - p$ is the frequency of black beans, then the relative proportions of the three combinations are:

$$p^2 \text{ (2 white beans)} + 2pq \text{ (1 white, 1 black)} + q^2 \text{ (2 black)}$$

Returning to the example of the taster/nontaster traits used in Chapter 4, now let the white beans represent the allele T (taster) and the black beans the allele t (nontaster). The relative frequences of the genotypes for some arbitrarily selected values of p and q are shown below:

RELATIONSHIP OF GENE FREQUENCY TO GENOTYPE FREQUENCY

Gene Frequencies		Genotype Frequencies		
$p(T)$	$q(t)$	p^2 (TT)	$2pq$ (Tt)	q^2 (tt)
0.5	0.5	0.25	0.50	0.25
0.6	0.4	0.36	0.48	0.16
0.7	0.3	0.49	0.42	0.09
0.8	0.2	0.64	0.32	0.04
0.9	0.1	0.81	0.18	0.01
0.99	0.01	0.98	0.02	0.0001

If the different genotypes in a population are present in these proportions, the population is said to be in Hardy-Weinberg equilibrium. Note that this is a very simple application of the binomial expansion for $(p + q)^2$.

Hardy-Weinberg equilibrium can be disturbed, as noted above, by various factors, but it holds true for many genes, and its failure to hold true in a specific situation suggests the possibility of a factor that may be disturbing the equilibrium.

An important consequence of the Hardy-Weinberg law is that the proportions of the genotypes do not change from generation to generation. The next generation following random mating of a population in which the genotypes TT, Tt, tt are present in the proportions $p^2:2pq:q^2$, so that the mating types are in the proportions of the expansion $(p^2 + 2pq + q^2)^2$, is as follows:

FREQUENCIES OF MATING TYPES AND OFFSPRING

Mating Types	Frequency	TT	Offspring Tt	tt
$TT \times TT$	p^4	p^4		
$TT \times Tt$	$4p^3q$	$2p^3q$	$2p^3q$	
$TT \times tt$	$2p^2q^2$		$2p^2q^2$	
$Tt \times Tt$	$4p^2q^2$	p^2q^2	$2p^2q^2$	p^2q^2
$Tt \times tt$	$4pq^3$		$2pq^3$	$2pq^3$
$tt \times tt$	q^4			q^4

Sum of TT offspring $= p^4 + 2p^3q + p^2q^2 = p^2(p^2 + 2pq + q^2)$

Sum of Tt offspring $= 2p^3q + 4p^2q^2 + 2pq^3 = 2pq(p^2 + 2pq + q^2)$

Sum of tt offspring $= p^2q^2 + 2pq^3 + q^4 = q^2(p^2 + 2pq + q^2)$

The common factor $p^2 + 2pq + q^2$ can be dropped, and the proportions of the three genotypes are then seen to be $p^2:2pq:q^2$ as in the parental generation.

A widely used application of the Hardy-Weinberg equilibrium is the calculation of gene frequency and heterozygote frequency, if the frequency of the trait is known.

If the frequency of an autosomal trait $(q^2) = 1/10,000$

The frequency of q is $\sqrt{1/10,000} = 1/100$

The frequency of p is $1 - \dfrac{1}{100} = 99/100$

The frequency of heterozygotes is $2pq = 2\left(\dfrac{99}{100}\right)\left(\dfrac{1}{100}\right) \approx \dfrac{1}{50}$

Because p (the frequency of the normal allele) is usually close to 1, the heterozygote frequency is usually close to $2q$ (twice the gene frequency).

Frequencies of X-linked Genes

In Chapter 9, we have set out the relative frequencies of the X-linked Xg blood groups in males and females. These data will now be used to show how frequencies of X-linked genes relate to genotype frequencies.

Often the first clue that a trait is X-linked is its different frequency in males and females. For the Xg blood system, 67 percent of males are Xg(a+) and 33 percent are Xg(a−), whereas for females 89 percent are Xg(a+) and only 11 percent are Xg(a−).

Since males are hemizygous for X-linked genes, there are only two male genotypes:

Genotype	Phenotype	Frequency
Xg^aY	Xg(a+)	$p = 0.67$
XgY	Xg(a−)	$q = 0.33$

The frequencies of the two alleles are given directly by the frequencies of the two phenotypes.

Females have two X's, therefore two Xg alleles. Three genotypes are possible:

Genotype	Phenotype	Frequency
Xg^aXg^a	$Xg(a+)$	p^2
Xg^aXg	$Xg(a+)$	$2pq$
$XgXg$	$Xg(a-)$	$q^2 = 0.11$

$\left.\begin{array}{c} p^2 \\ 2pq \end{array}\right\} = 0.89$

From the male data, we see that $q = 0.33$. The square root of the frequency of the $Xg(a-)$ phenotype in females gives the same result ($\sqrt{0.11} = 0.33$). Substituting the male values of p and q, we find that $Xg(a+)$ females should have a frequency of 0.89, which is very close to actual observation.

Note that for deleterious X-linked recessives the ratio of affected males to affected females is q/q^2, or $1/q$. This relationship demonstrates that males are far more frequently affected.

For X-linked dominants, the ratio of affected males to affected females is

$$\frac{q}{2pq + q^2}$$

For rare traits (i.e., when q is very small) this ratio is very close to 1/2, which is to be expected since males have only one X (one "draw" from the gene pool) and females have two. For more common traits the ratio is greater than 1/2.

SYSTEMS OF MATING

Hardy-Weinberg equilibrium is maintained only if there is **random mating** (also known as panmixis) — that is, if any one genotype at a locus has a purely random probability of combining with any other genotype at that locus — so that the frequencies of the different kinds of matings are determined only by the relative frequency of the genotypes in the population. The random mating requirement is probably rarely fulfilled in practice. Within any population there are many subgroups differing genetically (for example, members of different ethnic groups each with characteristic gene frequencies) and nongenetically (for example, with respect to religion). Members of such subpopulations are more likely to mate within their own subgroup than outside it. If mating is nonrandom, it is said to be **assortative**, and it is a common observation that within any population mating tends to be positively assortative with respect to intelligence, stature, economic status and many other traits, some of which are determined genetically, or partly genetically. Assortative mating may be **positive**, as when there is a tendency for persons who resemble one another to marry more frequently than chance alone would indicate; or **negative**, in the sense that "opposites attract."

For some traits, it is unlikely that mating is assortative. Probably the blood groups have little primary effect upon random mating, though they may have a secondary effect, since their frequencies vary with ethnic background. The same is true of fingerprint ridge counts and many other traits which have no obvious phenotypic effect, or at least none that is expressed before the mating age.

Consanguineous mating is a special form of assortative mating. It can disturb Hardy-Weinberg equilibrium by increasing the proportion of homozygotes at the expense of heterozygotes. Of course, inbreeding alone does not affect the proportions of alleles in the next generation, but only their assortment into genotypes. However, it may expose recessive genes to selection and loss, and in this way can permanently alter the gene frequencies in the population.

In North America today, the incidence of consanguineous matings is lower than in older countries, probably partly because in this relatively newly settled area there are fewer cousins among whom to find a mate. The lower birth rate also reduces the number of cousins available. In many countries the improvements in communication of the last century have had the effect of breaking down isolates (enclaves within which the members intermarried much more frequently than they married outsiders). Nevertheless, there are many subgroups within which consanguineous mating is relatively common.

Of course, given true random mating, a certain small proportion of marriages would be consanguineous. For that matter, all marriages are at least distantly consanguineous. One can calculate the number of ancestors one had in previous generations (recommended as a better soporific than counting sheep). Estimating about three generations per century, at the time of the Declaration of Independence (1776) the young adults of today each had about 64 (2^6) ancestors, and these may well have been 64 different people; but at the time of the Magna Carta (1215) the number was about 2^{24}, or 17 million, and it is very unlikely that there were 17 million different people involved. Probably many people are homozygous for genes that have come down in different lines from one heterozygous ancestor. Few people can trace their ancestry in all its ramifications beyond the first few generations. French Canadians who can do this and whose progenitors have been in Canada since the early 1600s may find themselves to be inbred through 10 or more different lines of descent.

THE MEASUREMENT OF CONSANGUINITY

The measurement of consanguinity has some importance in medical genetics because consanguineous marriages have an above-average risk of producing offspring homozygous for some deleterious recessive gene. This risk is proportional to the closeness of the relationship of the parents concerned. For practical purposes, only first-cousin and second-cousin marriages occur often enough and carry a sufficiently increased risk to be of practical importance.

The coefficient of inbreeding (F) is the probability that an individual has received both alleles of a pair from an identical ancestral source, or the proportion of loci at which he is homozygous. In the accompanying sketch, IV–1 is the offspring of a first-cousin marriage. It has been shown that for any gene the father has, the chance that the mother also has it is one-eighth. For any gene he gives his child, then the chance that the mother has the same gene and will transmit it is $1/2 \times 1/8 = 1/16$. This is the coefficient of

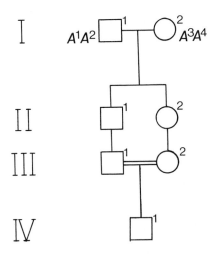

inbreeding for the child of first cousins. It signifies that he has a 1/16 chance of being homozygous at any one locus, or that he is homozygous at 1/16 of his loci. Alternatively, each of the four alleles in generation I has a 1/64 chance of being homozygous in IV-1, thus IV-1's probability of being homozygous is $4 \times 1/64 = 1/16$.

Some values of the coefficient of relationship and coefficient of inbreeding that might be encountered are listed in Table 15–1. In some inbred communities, the average coefficient of inbreeding may be as high as 0.03 or so.

MUTATION

New hereditary variations arise by mutation, and the new gene (or the individual who carries it) is called a **mutant.** Mutations are the only source of

TABLE 15–1 EXAMPLES OF COEFFICIENT OF INBREEDING (F) FOR SOME HUMAN POPULATIONS

Population	F Value
Canada:	
Roman Catholic	0.00004–0.0007
United States:	
Roman Catholic	0–0.0008
Hutterites	0.02
Dunkers (Pennsylvania)	0.03
Latin America	0–0.003
Southern Europe	0.001–0.002
Japan	0.005
India (Andhra Pradesh)	0.02
Samaritans	0.04

the material of evolution, upon which natural selection acts to preserve the fit and to eliminate the less fit.

In a sense, any change in the genetic material may be regarded as a mutation, but here we will follow the usage of Crow (1961) and define a mutation as a change that cannot be shown to depend upon a detectable chromosomal rearrangement or some sort of recombination mechanism. Though small chromosomal aberrations are impossible to distinguish from actual mutations in human material, the following discussion will be restricted as much as possible to "point mutations," i.e., mutations involving changes in base sequence in individual gene loci.

A mutation usually involves loss or change of the function of a gene. Since a random change is unlikely to lead to an improvement, most mutations are deleterious, though some confer an advantage and become established in the population by selection. The frequency of a gene in the population represents a balance between the mutation rate of the gene and the effectiveness of selection against it. If either the mutation rate or the effectiveness of selection is altered, the gene frequency changes.

Many mutations are lethal, leading to the death of the persons who receive them. A mutation that leads to sterility is just as lethal in the genetic sense as one that causes an early abortion. Others may be sublethal. Most of those actually observed in man are less severely detrimental, since a mutation cannot be identified at all unless it is compatible with life.

LOAD OF MUTATIONS

The load of genetic damage in man is not accurately known, but according to one estimate at least 6 percent of all persons born have some tangible genetic defect. At this rate, some 20,000 Canadian and 200,000 American babies are born each year with some sort of genetic handicap. Perhaps a quarter of all genetic defects are caused by single-gene mutations.

It is very difficult to know how many of these mutations are new and how many have been inherited from the previous generation, but Morton et al. (1956) have estimated that each individual in the population carries three to five **lethal equivalents.** A lethal equivalent is defined as one gene which, if homozygous, would be lethal, or two genes which, if homozygous, would be lethal in half the homozygotes and so on. One way of measuring lethal equivalents is by comparison of the mortality of the offspring of consanguineous and random marriages.

Even a slight heterozygote advantage can lead to the preservation and increase in the population of a gene that, when homozygous, is severely detrimental. The classic example is the resistance to malaria afforded by the sickle cell trait, described below. HbC, thalassemia, G6PD deficiency and the Fy allele of the Duffy blood group system are also thought to provide protection against malaria. It is suspected that a similar but unknown advantage may account for the high frequency of the cystic fibrosis (CF) gene, but there is no reasonable suggestion as to what the advantage might be. Human genes evolved under very different environmental conditions from those that exist today, and we can only speculate about what has made the CF gene so common in whites yet so rare in Orientals.

In experimental organisms, some genes that are detrimental when homozygous are transmitted through the sperm to far more than the usual 50 percent of the offspring of heterozygotes. This property, known as **meiotic drive,** is exemplified by the *t* alleles of the mouse, discussed earlier. A detrimental gene that confers an advantage on the gamete has a greatly enhanced chance of survival and increase.

MUTATION RATE

The spontaneous mutation rate varies for different loci, the limits of the observed range being about one in 10,000 (1×10^{-4}) and one in 1,000,000 (1×10^{-6}) per locus per gamete, with an average of about 1×10^{-5}. Some loci, especially the blood group loci, which have been studied in numerous families, are virtually never observed to mutate. At the opposite end of the range, the highest mutation rates measured in man are those of neurofibromatosis and Duchenne muscular dystrophy, both of which are in the range of 10^{-4}. Measurements of mutation rates in man are usually not highly accurate, but a value much greater than 10^{-5} may suggest that mutations at several different loci with similar phenotypic effects are being measured as though all occurred at a single locus, or that some other characteristic (reduced penetrance, for example) is interfering with the accuracy of the measurement.

For rare dominant mutations, the method of measurement is relatively straightforward. The number of cases, n, of a disorder that have been born to normal parents in a defined area over a defined time span and the total number of births in the same area during the same period, N, are determined. Then, since a new dominant trait requires a mutation in only one of two gametes, the mutation rate is $n/2N$. In practice, allowance must be made for unascertained cases, unclassified parents, reduced penetrance and, especially, genetic heterogeneity. If the same phenotype can be produced by mutation at more than one locus or by environmental factors, accurate measurement of the mutation rate is impossible.

The mutation rate of recessive genes is much more difficult to determine, because it is usually impossible to know whether the gene in question is a new mutation or has been inherited from a heterozygous parent. If there is a test for heterozygotes, the answer will be clear, but many recessives are absolutely recessive, with no demonstrable heterozygous expression. In any case, the probability is always in favor of inheritance rather than mutation, since the population frequency of recessive alleles is usually of the order of 0.01 to 0.03, whereas the frequency of mutation is not more than 0.0001. If heterozygote advantage is present, it may lead to a gross overestimate of the mutation rate.

For X-linked genes, the mutation rate is difficult to assess directly unless the trait determined by the mutation is a genetic lethal, in which case the frequency of the trait in males is one-third of the mutation rate (see later). Otherwise, it is often impossible to know whether a sporadic case of a disorder has been inherited from a heterozygous but phenotypically normal mother or whether it represents a new mutation.

One would expect that the older a parent is at the time of conception, the more likely he or she is to have accumulated mutations that the child might

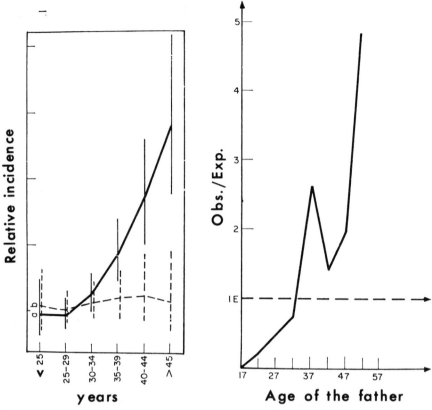

Figure 15-1 The effect of paternal age on the relative incidence of autosomal dominant mutations. Graph at left: (a) achondroplasia; (b) neurofibromatosis, tuberous sclerosis and osteogenesis imperfecta. Vertical bars are 95 percent confidence limits. Graph at right: Apert syndrome (acrocephalosyndactyly); line E is expected incidence if paternal age does not affect incidence. From Vogel, Mutations in man. In *Genetics Today*, Proceedings of the Eleventh International Congress of Genetics, Vol. 3, Pergamon Press, New York, 1963.

inherit. For fathers of children with new dominant mutations, this has been demonstrated. Fathers of new-mutant achondroplastic dwarfs are on the average a few years older than other fathers in the population (Fig. 15-1). The relationship does not hold for all dominant mutations and is not present, or at least not obvious, in older mothers.

ENVIRONMENTAL MUTAGENS

The mutation rate can be increased by a number of environmental agents, especially radiation and many chemicals. It has been very difficult to study the magnitude of such effects directly in man, and for the most part we have had to rely on experimental animals, especially fruit flies and mice. Methods are now being developed to assay for mutagenic effects in other systems, such as the Ames assay using microorganisms that are injected into mutagenized test animals, exposed there to the mutagen, then recovered for measurement

of their mutation rate (Ames et al., 1973). Another assay measures the incidence of abnormally shaped sperm in mice exposed to mutagens (Wyrobek
and Bruce, 1978). Both the Ames and the Bruce assays have been used to test
a variety of substances for mutagenicity.

Radiation

Ionizing radiation is by far the most potent mutagen known. The effect of
X-rays in increasing the rate of mutation was first observed by Muller (1927)
and earned him the 1946 Nobel prize. Man is exposed to a certain background
level of radiation from cosmic rays and other natural sources. This exposure
can be measured in terms of the roentgen (r), the unit of exposure dose, or the
rad, the unit of absorbed dose. (The units are very similar.) It is, of course,
only the dose absorbed by the gonad that is genetically significant. Natural
radiation provides perhaps 95 millirads of genetically significant exposure per
year, while man-made radiation (chiefly medical and dental X-rays) adds
another 40 or so mrads per year, bringing the total to about 135. Over an
estimated 30-year period to the end of the reproductive years, the total exposure is perhaps 3r.

A common measure of the effect of radiation in inducing mutations is the
"doubling dose," the dose that would be required to double the spontaneous
mutation rate. On the basis of mouse data this is thought to be in the range of
30r for man.

The gonadal exposure from diagnostic and therapeutic X-rays varies
greatly with the quality of the equipment used and the shielding provided,
therefore we will not attempt to provide an estimate of the gonadal dose in
different procedures. In mice, which appear to be similar to man in radiation
sensitivity, the relation between exposure and the number of mutations produced is linear, except possibly at very low intensity (Russell et al., 1959).
Thus it seems wise to reduce gonadal exposure to the lowest possible level in
persons who have not completed their reproductive years.

Evidence for the production of mutations by ionizing radiation in man has
been sought from a number of sources, but especially from studies of the
offspring of survivors of the atomic bombings of Hiroshima and Nagasaki
(Schull and Neel, 1958; Neel et al., 1974). Theoretically, radiation to the
parents could produce changes in the sex ratio, since X-linked lethal mutations might be induced. If only the father is irradiated, any radiation-induced
X-linked dominant mutations would pass only to his daughters, hence he
would be expected to have a deficiency of female children. If only the mother
is irradiated, her mutant X-linked recessive lethal genes would be expressed
in her sons, thus a deficiency of male children would be expected. Small
changes in the sex ratio of the offspring of irradiated women have been
observed, but with the numbers studied those changes are not significant.

Chemicals

Many chemicals are known to be carcinogenic (cancer-inducing), and
most of these are now known to be mutagenic as well. The first example,
found by Charlotte Auerbach in 1941, was mustard gas, an alkylating agent

that produces single-base substitutions in DNA. To be mutagenic, substances must permeate the nucleus and react with DNA, influencing either its stability or its replication. Thus the types of chemical mutagens known include for example, in addition to alkylating agents, deaminating agents, substances related to the dye acridine orange and base analogs, all of which are capable of interfering with DNA replication, though by different mechanisms.

Caffeine is a base analog and so may be a mutagen, but the importance of its mutagenic effect in mammals is still a matter of speculation.

SOMATIC MUTATIONS

Somatic mutation has already been mentioned (Chapter 10). Somatic cell mutations have, of course, no genetic consequences, but they may have significance for the individuals who carry them. One theory that has been advanced to explain the phenomenon of aging is that somatic mutations of many cells may impair their normal functioning. The role of somatic mutation in carcinogenesis is discussed in Chapter 18.

THE PROBLEM OF A SEX DIFFERENCE IN MUTATION RATES

For many years there has been a controversy over whether the mutation rate is the same in male and female germ cells. This question can be examined in X-linked mutations and has been tested for classic hemophilia and Duchenne muscular dystrophy. For classic hemophilia there is quite good evidence that the mutation rate is much higher in male than in female germ cells, and this possibility is borne out by the finding of late paternal age in the maternal grandfathers of patients. For Duchenne muscular dystrophy the situation is quite different; the mutation rate in male and female germ cells appears to be identical. Although the much greater number of DNA replications between zygote and gamete in males as compared to females certainly would explain an excess of mutations in males, at present no consistent difference has been found, and even when it does exist the magnitude of the effect seems smaller than would be expected. The whole question of a sex difference in mutation rates remains unresolved.

SELECTION

Darwin postulated natural selection as the factor of importance in evolution. In modern terms, survival of the fittest is interpreted as taking place through the action of selection upon new genotypes which have arisen by mutation or recombination.

In the biological sense, the term **fitness** has no connotation of superior endowment except in a single respect — the ability to contribute to the gene pool of the next generation. The many factors that affect fitness can operate at any stage of the life cycle, at least until the end of the reproductive period.

Perhaps selection can also operate after the reproductive period; there may be some selective value in a life span that lasts no longer than the productive years.

Dominant deleterious genes are openly exposed to selection, in contrast to autosomal recessive genes, most of which are hidden in heterozygotes. Consequently, the effects of selection are more obvious and can be more readily measured for dominants than for recessives.

SELECTION AGAINST DOMINANT MUTATIONS

A harmful dominant mutation, if penetrant, is expressed in any individual who receives it. Whether or not it is transmitted to the succeeding generation will depend on how deleterious it is. If it prevents reproduction and accordingly is not represented in the next generation, its relative fitness is zero. If it is just as likely as the normal allele to be represented in the next generation, its fitness is one. Most deleterious dominant mutations have a fitness value between zero and one. Many children with lethal multiple congenital defects may carry new dominant mutations, but it is impossible to be certain since these children do not reproduce.

If the mutation is deleterious but affected persons are fertile, they may contribute fewer than the average number of offspring to the next generation, i.e., their fitness may be reduced. The mutation will be lost through selection at a rate proportional to the loss of fitness of heterozygotes. The **coefficient of selection,** s, is the measure of the loss of fitness.

Fitness is measured as

$$f = 1 - s.$$

Achondroplastic dwarfs have about one-fifth as many children as do normal members of the population. Thus, their fitness is 0.20, and the coefficient of selection is 0.80. In the next generation, the frequency of the achondroplasia mutations passed on from the current generation is 20 percent. The remaining 80 percent are added through new mutation. The observed gene frequency in any one generation represents a balance between *loss* of alleles through selection and *gain* through recurrent mutation.

Selection against a dominant genotype can lower the frequency of the dominant gene precipitately. If no heterozygotes for the dominant gene for Huntington chorea reproduced, the incidence of the disease would fall in one generation to a level determined by the mutation rate, because the only chorea genes remaining in the population would be the new mutations.

SELECTION AGAINST RECESSIVE MUTATIONS

Selection against deleterious recessive genes is less effective than selection against dominants. Even if there is complete selection against homozygous recessives, it takes 10 generations to reduce the frequency from 0.10 to

0.05; and the lower the gene frequency, the slower the decline. Removing selection (for example, by successful medical management of children with cystic fibrosis or Tay-Sachs disease, so that they could survive and reproduce at a normal rate) raises the gene frequency just as slowly.

SELECTION AGAINST X-LINKED RECESSIVES

Deleterious X-linked recessive genes are exposed to selection in hemizygous males but not in heterozygous females. In X-linked recessives that are genetic lethals in the sense that affected males do not reproduce (or, in other words, have a fitness of zero) only the genes in the carrier females are passed on to the next generation. An important consequence is that one-third of all cases of the disorder are new mutants, born to genetically normal mothers who have no risk of having subsequent children with the same disorder. This topic is discussed further on page 341. In less severe disorders such as hemophilia, where fitness is reduced but is above zero, the proportion of new mutants is less than one-third.

Lately there have been suggestions that population data do not bear out this rule for Duchenne muscular dystrophy or for another much-studied X-linked recessive lethal, HPRT deficiency. At present it is not clear whether the unexpected findings have arisen through faulty ascertainment or inaccuracies in heterozygote tests for carriers, or whether a marked difference in mutation rate in males and females could account for the discrepancies. As noted above, there is no evidence that in Duchenne muscular dystrophy the mutation rate is different in the two sexes.

SELECTION AGAINST HETEROZYGOTES

If selection occurs at the gamete stage, single alleles are selected against. More frequently, selection takes place at the diploid stage and loss of one allele involves loss of its partner also. If the two alleles are different (that is, if the individual concerned is a heterozygote), the relative frequencies of the two alleles (or all alleles, if there are more than two possibilities) are altered, and they achieve a new balance at a different level but with a lower frequency of the rarer allele of the pair.

An example of selection against heterozygotes is provided by the Rh blood group system, though the removal of selection through improved medical management of Rh negative mothers, by administration of Rh immune globulin, will alter this picture for the future. Let us use a simplified terminology in which D represents any Rh positive gene complex and d any Rh-negative one. In a mating of a dd female with a DD or Dd male, a proportion of Dd infants will be lost or damaged because of Rh hemolytic disease of the newborn. Thus one D and one d allele are lost from the gene pool of the next generation. But since the frequency of D is higher than the frequency of d (about 0.60 as compared with 0.40), loss of the two alleles has a greater effect on the frequency of d than of D. Over many generations, the rarer allele should be lost. Why then is the frequency of d still so high? This intriguing question has no answer so far.

SELECTION FOR HETEROZYGOTES

Sickle cell anemia is a classic example of a situation in which the heterozygote is more fit in a particular environment than is either type of homozygote.

Recall that there are three genotypes with relation to normal adult hemoglobin and sickle cell hemoglobin. In certain parts of Africa the frequency of the HbS allele (specifically, the allele for an abnormal β globin chain) is higher than can be expected on the basis of recurrent mutation of an allele that, in homozygotes, has a fitness of zero. The situation in which two or more alleles at a given locus both occur with an appreciable frequency is known as a **polymorphism**, and the locus is said to be polymorphic.

In the case of the sickle cell polymorphism, a heterozygote advantage maintains the less favorable allele at its comparatively high frequency. Heterozygotes are resistant to the malaria organism *Plasmodium falciparum*. This is a parasitic protozoan that spends a part of its life cycle in the red cells of vertebrates, to which it is introduced by the bite of the vector, the *Anopheles* mosquito. Thus, in a malarial environment, normal homozygotes are susceptible to malaria and probably are almost all infected and relatively unfit; sickle cell homozygotes are selected against because of their anemia; but heterozygotes, whose red cells are not hospitable to the malaria organism and whose hemoglobin is quite adequate for normal conditions, are more fit and reproduce at a higher rate than either homozygote.

A selective advantage in one environment may not be advantageous, or may even be disadvantageous, in a different environment. Sickle cell heterozygotes do not have a particular advantage in North America, since malaria is not endemic. There is some evidence that having been removed from the circumstances that favored its maintenance, the frequency of the sickle cell gene is already dropping in North American blacks.

When selective forces operate in both directions, toward the preservation of an allele and toward its removal, a **balanced polymorphism** is said to exist. Removal of one of the selective pressures would be expected to allow a rapid fall in the frequency of the sickle cell allele. If malaria could be eradicated, there might be large changes in the frequency of the sickle cell gene and several other genes that appear to play a part in protection against malaria. Unfortunately, at present malaria seems far from being controlled.

POLYMORPHISM

Most of the discussion so far in this chapter has centered on mutations with clearly deleterious effects. However, as we have seen, there are numerous human mutations that are either genetically neutral or have such minor effects, in comparison with their alleles, that it is difficult to demonstrate selection against them. The loci at which these mutations are present are polymorphic. By a somewhat arbitrary rule, a polymorphic locus is one in which the most commonly occurring allele has a frequency of no more than 0.99, so at least 2 percent of the population must be heterozygous at that locus.

When Harris and his colleagues began to survey the human enzymes that could be examined by electrophoresis, an unexpectedly large number of variants came to light. In fact, about 30 percent of the enzyme and protein loci that have been surveyed have demonstrable polymorphism, and even this high figure must underestimate the true frequency. It seems to be the rule that many loci have two normal alleles each with an appreciable frequency as well as one or more rare alleles that can be accounted for by recurrent mutation.

What proportions of gene loci in a given individual are likely to be heterozygous? In man, recent data give an estimate of about 7 percent (Table 15–2), and additional diversity is being recognized as improved techniques are developed. For other species the estimate is much the same, ranging from 3 to 20 percent with a mean of about 6 percent. Thus it is clear that man and other animals have much diversity in their enzyme and protein makeup. The total number of different combinations of protein variants is so large that it is unusual for any two persons (except, of course, monozygotic twins) to have exactly the same set.

The extent of polymorphism has led to speculation about its significance in terms of evolution. The classical or "selectionist" view is that natural selection, acting upon mutations, is the primary factor in maintaining polymorphism. In some cases, a system with two common alleles is being maintained as a balanced polymorphism, with heterozygote advantage. In other cases, we may simply be observing an intermediate stage in the replacement of one major allele by another, as seems to be the case for the haptoglobins. Other special mechanisms could account for some polymorphisms; for example, two types of homozygotes might each be best fitted for a particular environment, and a cline with high frequencies of both alleles might develop between them. The major point is that selection in the Darwinian sense is operative.

The other view, chiefly associated with the name of Kimura, is that natural selection need not be the essential process in maintaining polymorphism, but that random genetic drift alone could account for the presence of two major alleles at a locus. One of the strong arguments in this "neutralist" or "non-Darwinian" view is that molecular evolution has proceeded at a more or less fixed rate for many proteins in widely different species over long periods of time, with such regularity that the rate of amino acid substitution is used as a measure of the evolutionary time sequence. In other words, in the neutralist view there is no need to postulate that one or the other allele is preferred or that the heterozygote is at an advantage. "Neutral" mutations could become fixed by chance.

The debate has had a stimulating effect on the study of the extent and significance of polymorphism. One observation, which supports the selectionist view, is that in the majority of enzyme polymorphisms the different genotypes have different enzyme activity. This would provide a reasonable basis for selection. The study of polymorphism is leading to deeper insight into the origin and maintenance of human diversity.

TABLE 15-2 ENZYME POLYMORPHISM FOUND BY ELECTROPHORESIS*

Number of loci screened	71
Number polymorphic	20
Percentage polymorphic	28.2
Average heterozygosity per locus	0.7

*Data of Harris and Hopkinson (1972) for Europeans

STABILIZING SELECTION

Population geneticists distinguish three main types of selection with respect to quantitative traits: stabilizing, directional and disruptive. The kinds of changes each of these types of selection brings about in regard to phenotypic distribution in the population are illustrated in Figure 15-2.

Stabilizing selection favors an intermediate optimum phenotype and selects against the **phenodeviants** at either extreme. The more extreme the deviant, the less its genetic contribution to the next generation. **Directional selection** is selection directed toward a new optimum, not at the mean of the population, in response to a new environmental challenge. **Disruptive selection,** rather than favoring any one phenotype, favors two quite different forms and selects against the intermediates; this may be viewed as directional selection of two separate subpopulations, in response to two different sets of environmental conditions.

Of the three types of selection, stabilizing selection appears to play the major role. It may be viewed as a tendency to maintain the status quo. For many human characteristics, an intermediate value is the favored optimum. An obvious example is birth weight, and here the effect of stabilizing selection in eliminating the more extreme phenodeviants has been clearly shown: babies much smaller or larger than the average are less likely to survive the perinatal period than those in the intermediate range.

There is not much evidence for directional selection within the human

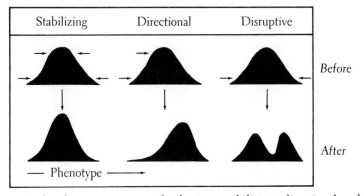

Figure 15-2 The three main types of selection: stabilizing, directional and disruptive. From Wallace, *Topics in Population Genetics.* W. W. Norton and Company, Inc., New York, 1968, p. 382.

species. For example, except for the last few generations, height appears to have changed little during the last 40,000 generations. Directional selection was basic to Darwin's concept of the mechanism of evolution, and it is much used by agricultural geneticists. Directional selection is a very powerful mechanism; many scientists and writers have speculated about what might happen to mankind if the techniques of cattle breeders were applied to our species.

MIGRATION

GENETIC DRIFT

Changes in gene frequencies may occur when new settlements are formed by members of an older group. There are two possible explanations. Perhaps migrants are themselves a separate subpopulation, differing from the population as a whole in energy, acquisitiveness, curiosity or whatever characteristics stimulate some members of a community to move away. Of more importance, however, is the fact that the gene frequencies of the migrants will probably not be representative of those of the population as a whole; and the smaller the group, the more likely that it will not accurately reflect the gene frequencies of the parent group.

One of the most striking demonstrations of drift is given by McKusick's studies of the Old Order Amish of Pennsylvania. Among the Amish, social custom has provided an excellent situation in which to see the effect of drift. The Amish intermarry, few if any genes being added to the population by marriage with outsiders. As a community grows, it is the custom for a few families to leave to set up a new colony elsewhere. Among one such group, by chance a founding father was heterozygous for a rare form of dwarfism with polydactyly known as Ellis-van Creveld syndrome. There are about 50 known cases of this rare disease among the 8000 living members of this group, and it is absent in other Amish communities descended from other ancestors. Thus, genetic drift can favor the establishment of genes that are not favorable or even neutral, but actually harmful. Another well-established example of random genetic drift is that of tyrosinemia in a remote area of Quebec (see Chapter 4). The socially isolated Amish and the geographically isolated French-Canadian groups illustrate the "founder principle" of Mayr (1963): if among a small number of founders who form a community there is a member with a rare recessive allele, the frequency of the allele is much higher within the community than outside it, and the small number of the original group of founders allows for a large effect of drift.

GENE FLOW

In contrast to the variations in gene frequencies in small populations resulting from genetic drift is the more gradual change in frequency in larger populations resulting from gene flow. A classic example is the steady change (cline) in the frequency of the B allele of the ABO blood group system from about 0.30 in Eastern Asia to about 0.06 in Western Europe (Fig. 15-3). The flow of white genes into American blacks is another good example; by several

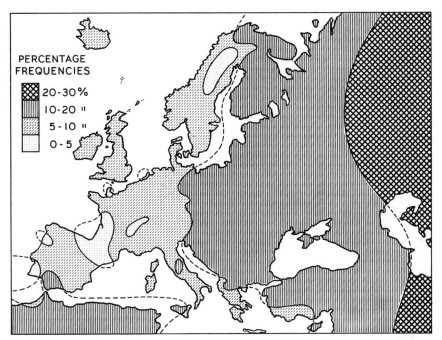

Figure 15-3 Cline of distribution of blood group B across Europe. After Mourant, *The Distribution of the Human Blood Groups.* Blackwell Scientific Publications, Oxford, 1954.

different measures such as comparison of the frequency of the strictly African R^0 allele in African and American blacks it has been shown that some 30 percent of the alleles now carried by American blacks are of white origin. More recently Reed (1969), using data for the "Caucasian" allele Fy^a of the Duffy blood group system and other alleles varying in frequency in blacks and whites, has found a lower proportion and differences in different geographic areas. He has calculated that the proportion of white genes in American blacks is higher in nonsouthern areas of the United States than in the south (22 percent in Oakland, California and 26 percent in Detroit, but only 4 percent in Charleston, South Carolina).

THE ORIGIN OF RACES

New varieties of living things evolve when new genes arise by mutation, are tried out in various combinations with other genes and in various environments, and are preserved. In man, many different subgroups have thus evolved, though all belong to a single species, *Homo sapiens*. These subgroups are commonly called races. A race is distinguished from other races by a gene pool in which gene frequencies (at a few or many loci) are characteristic of that population and of no other.

One of the most active areas of current research in human genetics is the accumulation of data on the frequency of many genetic markers in different populations. This is a matter of some urgency because, with improvement in communication and the consequent breakdown of isolates, few human populations will retain for many more years the characteristic gene frequencies they have acquired over the long time span of human evolution. The analysis of such data is the only way to obtain information about the evolution of human populations and the origin and maintenance of genetic polymorphisms in man.

GENERAL REFERENCES

Cavalli-Sforza, L. L., and Bodmer, W. F. 1971. *The Genetics of Human Populations.* W. H. Freeman and Company, San Francisco.

Harris, H. 1975. *The Principles of Human Biochemical Genetics.* 2nd ed. North-Holland Publishing Company, Amsterdam and London; Elsevier-North Holland, New York.

Kimura, M., and Ohta, T. 1971. *Theoretical Aspects of Population Genetics.* Princeton University Press, Princeton, New Jersey.

Mettler, L. E., and Gregg, T. G. 1969. *Population Genetics and Evolution.* Prentice-Hall Inc., Englewood Cliffs, New Jersey.

Morton, N. E. 1972. The future of human population genetics. Prog. Med. Genet. 8:103–124.

Wallace, B. 1968. *Topics in Population Genetics.* W. W. Norton and Company, Inc., New York.

PROBLEMS

1. In a certain population, three genotypes are present in the following proportions: AA 0.81, Aa 0.18, aa 0.01.
   a) What are the frequencies of A and a?
   b) What will their frequencies be in the next generation?
   c) What proportion of all the matings in this population are $Aa \times Aa$?

2. In screening programs to detect carriers of Tay-Sachs disease among Ashkenazic Jews, the frequency was found to be about 0.07 in Toronto and about 0.035 in Washington.
   a) For these two cities, calculate:
      i) the frequency of matings that could produce an affected child.
      ii) the incidence of Tay-Sachs disease (or, if prenatal diagnosis is used, affected fetuses) in the population.
   b) How can the difference in gene frequency between the two cities be explained?

3. In an isolated population of 800 individuals, all members are of blood group O. In another population, all are of blood group A. If 200 members of the second population are added to the first, what will the frequencies of the two blood groups be after a generation of random mating?

4. Two sisters marry two brothers. The son of one couple marries the daughter of the other couple. What is the coefficient of inbreeding of their children?

5. Monozygotic twin sisters marry dizygotic twin brothers. What is the coefficient of relationship of the children of one couple to the children of the other couple (i.e., what proportion of their genes are in common)?

6. a) In a population in which the incidence of brachydactyly is 1 in 10,000, what is the frequency of the gene for brachydactyly (an autosomal dominant)?
   b) In the same population, the frequency of phenylketonuria is 1 in 10,000. What is the gene frequency? The carrier frequency?
   c) The frequency of classic hemophilia in males is 1 in 10,000. What is the gene frequency? The frequency of heterozygous females?

7. If the blood group alleles M and N are equal in frequency, what are the proportions of the three possible MN blood groups?

16

TWINS IN
MEDICAL
GENETICS

Twins have a special place in human genetics. This is because diseases caused wholly or partly by genetic factors have a higher concordance rate in monozygotic than in dizygotic twins. Even if a condition does not show a simple genetic pattern, comparison of its incidence in monozygotic and dizygotic twin pairs can reveal that heredity is involved; moreover, if monozygotic twins are not fully concordant for a given condition, nongenetic factors must also play a part in its etiology. The importance of twin studies for comparison of the effects of "nature and nurture" was pointed out by Galton in 1875.

MONOZYGOTIC AND DIZYGOTIC TWINS

There are two kinds of twins, monozygotic (MZ) and dizygotic (DZ), or in common language, identical and fraternal twins. Monozygotic twins arise from a single fertilized ovum, the zygote, which divides into two embryos at an early developmental stage, i.e., within the first 14 days after fertilization. Because the members of an MZ pair normally have identical genotypes, they are like-sexed and identical with respect to such genetic markers as blood groups. They are less similar in traits readily influenced by environment; for example, they may be quite dissimilar in birth size, presumably because of differences in prenatal nutrition. Phenotypic differences between MZ co-twins may be produced by the same factors that cause differences between the right and left sides of an individual; for example, cleft lip may be bilateral or unilateral in an individual, and may be concordant or discordant in an MZ pair.

Dizygotic twins result when two ova, shed in the same menstrual cycle, are fertilized by two separate sperm. DZ twins are just as similar genetically as ordinary sib pairs, having, on the average, half their genes in common. Phenotypic differences between the members of a DZ pair reflect their genotypic dissimilarities as well as differences arising from nongenetic causes.

RELATIVE FREQUENCY OF MONOZYGOTIC AND DIZYGOTIC TWINS

There is a simple way of finding out how many of the twin births in a population are MZ and how many are DZ. MZ twins are always like-sexed, while approximately half the DZ twin pairs are boy-girl sets. Therefore, the total number of DZ pairs is twice the number of unlike-sexed pairs, and the number of MZ pairs can be found by subtracting the number of unlike-sexed pairs from the total number of like-sexed pairs.

$$\frac{\text{All twin pairs} - 2(\text{unlike-sexed pairs})}{\text{All twin pairs}} = \text{Frequency of MZ pairs}$$

For precision, a small correction is required because the sex ratio is not exactly 1:1, but the simple method described gives a good approximation. Among white North Americans, approximately 30 percent of all twins are MZ, 35 percent are like-sexed DZ and 35 percent are unlike-sexed DZ.

Comparison of the ratio of like-sexed to unlike-sexed pairs in populations with varying frequencies of twin births has shown that the proportion of MZ births relative to all births is much the same everywhere, about one in 300 births, but that the proportion of DZ births varies with ethnic group, maternal age and genotype (see below).

FREQUENCY OF TWIN BIRTHS AND HIGHER MULTIPLE BIRTHS

Among white North Americans, about one birth in 87 is a twin birth. Thus, two of every 88 babies — over 2 percent of the population — are born as twins. Triplets are born about once in $(87)^2$ births, quadruplets about once in $(87)^3$ births and so on, although, because of a much higher mortality, this mathematical relation (known as Hellin's law) is less closely approximated for mulitiple sets of four or more. Since twins have a higher infant mortality than singletons, the incidence of twins in the general population is somewhat lower than 2 percent.

The frequency of twin births varies with ethnic origin. In the United States the rate is higher in blacks than in whites. One of the highest frequencies (one twin birth in 20 to 30) is reported from Nigeria, and one of the lowest (one in 150) from Japan.

The frequency of DZ twins rises sharply with maternal age to age 35 and later declines, but the frequency of MZ twins is hardly affected by the mother's age. Scheinfeld and Schachter have calculated, for United States whites, the probability of producing MZ and DZ twins in relation to maternal age and parity (Table 16–1). These probabilities may be used, together with blood group data and other evidence, to help assess the probable zygosity of a twin pair.

The tendency for twins to run in families is well known. It is questionable whether there is a genetic factor in MZ twinning, but certainly a disposition to

TABLE 16-1 RELATIVE CHANCE THAT TWINS ARE MONOZYGOTIC IN
RELATION TO MATERNAL AGE AND PARITY°

Maternal Age	Birth Rank		
	1	2	3 or more
15–19	1.38	1.00	0.69
20–24	0.85	0.64	0.50
25–29	0.61	0.49	0.40
30–34	0.56	0.45	0.34
35–39	0.52	0.41	0.33
40–44	0.69	0.54	0.48

°The relative chance in favor of MZ twins is the ratio MZ frequency: DZ frequency. After data of Scheinfeld and Schachter, 1963.

DZ twinning is genetically determined. Whether the father plays any part in the occurrence of twinning among his children has been the subject of dispute for many years. The studies of Greulich (1934) seemed to show that the father's genotype was at least as important as the mother's in the production of DZ twins. This observation seemed inexplicable in view of the fact that the multiple ovulation which accounts for DZ twins has been considered to be a purely maternal event, related to the mother's FSH (pituitary follicle-stimulating hormone) level. The data of White and Wyshak (1964) from the genealogical records of the Mormon church show that female DZ twins produce twins at a rate of 17.1 sets per 1000 maternities, whereas wives of male DZ twins produce only 7.19. Sisters of DZ twins also produce twins at a high rate, but brothers do not. Thus the mother's genotype affects the frequency of DZ twins among her offspring, but the father's genotype has no effect. In other words, a disposition to multiple ovulation is an inherited trait and can be expressed only in females. The trait can be transmitted through males as well as through females; e.g., the brothers of DZ twins do not themselves produce a higher than average number of twins, but their daughters do.

Although there is thought to be no genetic factor in MZ twinning, there are some apparent exceptions, especially in families in which chromosomal abnormalities occur. Many authors have pointed out that twins are more common among individuals with chromosomal abnormalities and their families than in the general population. For example, Nance and Uchida (1964) noted an increased frequency of twins (presumably MZ) in 34 families in which there were 45, X aneuploids. The twinning rate was eight twin births in a total of 128 births, a fivefold to tenfold increase over the normal rate. Their data and those of other workers also show a slight increase in the frequency of twins in sibships of children with Down syndrome. Observations such as these suggest a common mechanism in the causation of twinning and chromosomal disorders.

When women are treated for sterility with massive doses of FSH to induce ovulation, they may produce several ova at once. Multiple births (quintuplets, sextuplets and even septuplets) have been reported following FSH treatment. These multiple sets are of course DZ.

DETERMINATION OF TWIN ZYGOSITY

It is often useful to know the zygosity of a twin pair (or of a higher multiple birth, in which the members may be an MZ set, all from separate zygotes, or a mixture). Accurate classification is a prerequisite if twins are to be used in research, and it may be medically useful, for example, if one twin develops diabetes, or needs a kidney transplant. Zygosity is determined on the basis of the type of placenta and fetal membranes or by looking for genetically determined similarities and differences in a twin pair.

PLACENTA AND FETAL MEMBRANES

A developing fetus is invested in two membranes: the inner, delicate **amnion** and the outer, thicker **chorion** attached to the placenta. If two fetuses are developing simultaneously in the same uterus, there are several variants of the placenta and fetal membranes. The most common types are shown in Figure 16–1 and summarized in Table 16–2.

A twin placenta is **monochorionic** if there is a single chorion, and **dichorionic** if there are two chorions (which may be secondarily fused where they meet). A monochorionic twin placenta may be either monoamniotic (with one amnion), or diamniotic (with two amnions).

Dizygotic twins have separate placentas, chorions, and amnions, but in about 40 percent of cases the two placentas and chorions are secondarily fused and superficially resemble a monochorionic placenta.

Monozygotic twins may have either dichorionic or monochorionic placentas, depending on the time in early embryonic development at which twinning occurred. In about 25 percent of cases, twinning of the embryo occurs before the third day, i.e., before the development of the chorion, so that two separate chorions are formed. There may be two separate placentas or, more commonly, a single secondarily fused placenta. Both these types of placentation also occur in DZ twins. In the majority of MZ twins, however, twinning occurs between the third and eighth day of development, by which time the differentiation of the blastocyst has already proceeded too far to allow duplication of the chorion. A monochorionic placenta (which is usually diamniotic) can be regarded as proof of monozygosity.

There are several varieties of monochorionic placentas, depending again on the stage at which twinning of the embryo occurred. Some of the distinguishable types, arranged in order from earlier to later time of twinning, are the following:

1. Each embryo has a separate amnion, and the two fetal circulations remain separate.

2. Each embryo has a separate amnion, but a common fetal circulation has developed by anastomoses of vessels in the placenta. This the most common type of monochorionic placenta.

3. Rarely, an even later twinning of the embryo results in a single amniotic sac for both twins. Many monoamniotic twins do not survive.

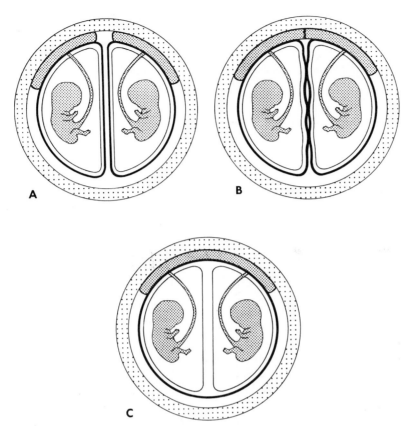

Figure 16-1 Diagrammatic representation of some common arrangements of placenta and fetal membrane in twins. Uterine wall lightly stippled; chorion, heavy line; amnion, light line.

A. Separate placentas and membranes, common in DZ twins and relatively rare in MZ twins.

B. Separate but secondarily fused placentas and chorions. The two halves of the placenta have separate circulations. Common in both MZ and DZ twins.

C. A single placenta with a common circulation and a single chorion but separate amniotic sacs. This arrangement occurs only in MZ twins and is diagnostic of MZ twinning.

4. Rarely, still later twinning of the embryo occurs after the formation of the umbilical cord, resulting in conjoined ("Siamese") twins. Conjoined twins are monoamniotic and usually have a common, branched umbilical cord.

Though superficially similar, the fused dichorionic placenta and fetal membranes can be clearly distinguished from the monochorionic type if the membrane between the two amniotic cavities is examined. In a dichorionic placenta, the membrane is opaque, being composed of two layers of chorion and two layers of amnion, as shown in Figure 16-1B. The two chorions can be readily separated. In contrast, in a monochorionic placenta, the two amniotic cavities are separated only by a semitransparent layer composed of two thin amnions, as demonstrated in Figure 16-1C. (According to an old tale, if twins are of different sexes but share a placenta, a heavy curtain is drawn between them to protect their modesty.)

In summary, the only kind of placenta that is diagnostic of the type of

TABLE 16-2 TYPES OF PLACENTAS AND FETAL MEMBRANES IN TWINS

	Monochorionic		Dichorionic	
	Monoamniotic	Diamniotic	Single Placenta (by secondary fusion)	Two Placentas
MZ	Rare	65%	25%	10%
DZ	—	—	40%	60%

twinning is the monochorionic type, and it is diagnostic of MZ twinning. Separate placentas and membranes, or secondarily fused placentas without common circulation, can be found with either type of twin set. It is helpful to examine and record the nature of the placenta and membranes at like-sexed twin births, since the information about twin zygosity then available may not be easy to obtain in later life, except by extensive tests.

GENETIC MARKERS

Monozygotic twins are always alike in sex and in blood groups and other genetic markers. Dizygotic twins, like ordinary sibs, may be alike or different in these characteristics. To determine whether a twin pair is MZ or DZ, the twins are compared with respect to as many traits as possible. A single difference in any genetic marker proves twins to be DZ. However, it is impossible to prove monozygosity, since any two children of the same parents resemble one another in many genetic traits; even so, it may be possible to show that the probability of monozygosity is very high.

The probability that twins are MZ can best be worked out if the genotypes of the parents and sibs are known. The following example shows how to use the method in an oversimplified case. The data are taken from Race and Sanger (1975), where a fuller discussion may be found.

Blood Group Genotypes	
Father	Mother
A_2O	OO
$MsNS$	$MSMS$
P_1P_2	P_1P_2
Twin A	Twin B
A_2O	A_2O
$MSMs$	$MSMs$
P_2P_2	P_2P_2

	DZ	MZ
Relative frequency in the population	0.70	0.30
Chance that twins will be alike in sex	0.50	1.00
Chance that twins born to these parents will be alike in ABO groups	0.50	1.00
in MNSs groups	0.50	1.00
in P groups	0.375	1.00
Product of chances	0.033	0.30

$$\text{Probability that twins are MZ} = \frac{0.30}{0.033 + 0.30} = 0.90$$

This probability (90 percent) is not high enough to make it safe to assume that the twins are MZ, for indeed there is one chance in 10 that they are not. But with the numerous additional markers that can be used it may be possible to increase the probability close to the 100 percent level, or on the other hand to show that the twins are DZ. Useful tables of probabilities for situations in which the parental genotypes are unknown are given by Race and Sanger (1975) and Corney and Robson (1975).

DERMATOGLYPHICS

Dermatoglyphics can often be helpful in the determination of zygosity, even though their inheritance is multifactorial rather than Mendelian. There are useful tables for comparison of a twin pair in terms of the total fingerprint ridge count or the maximal atd palmar angle (Maynard-Smith et al., 1961; Corney and Robson, 1975). (Dermatoglyphic terms are defined in Chapter 17.) The probabilities given in these tables can be combined with the probabilities from the other sources just quoted.

LIMITATIONS OF TWIN STUDIES

The chief drawback of the twin method is that, though it tells something about the strength of the genetic predisposition to develop a disorder, it gives no insight into the genes concerned, their mode of action or their pattern of transmission. Perhaps this problem is becoming less acute as methods of assessing multifactorial inheritance are becoming better known. Many of the traits in which twin comparisons are used are multifactorial (e.g., congenital malformations). If a trait is determined by a single autosomal dominant gene, it should be half as common in a sib or in a DZ co-twin of the propositus as in an MZ co-twin; if it is an autosomal recessive, it should affect one-quarter as many sibs or DZ co-twins as MZ co-twins. If the proportions vary by much from these ratios, multiple factors (genetic or environmental) are probably involved. In most of the common malformations, even the MZ concordance rate is well below 50 percent, indicating that environmental factors operative before birth are important in causing them.

Although the twin method assumes that postnatal environmental differences are constant for both types of twins, this assumption is unwarranted in many cases. MZ twins, because they are more alike, may seek the same environment and develop along much the same paths. DZ twins, who may even be of different sexes, probably have more different environments than do MZ twins. In many studies a comparison can be made more fairly between MZ twins and like-sexed DZ twins than between all MZ pairs and all DZ pairs. It is noteworthy that DZ twins become less and less alike as they grow older, whereas MZ twins remain remarkably similar throughout life, aging in much the same way and being subject to the same geriatric disorders.

Still another limitation of the twin method is related to bias of ascertainment. Concordant MZ pairs are much more likely to be reported than any other combination. If a twin series is compiled from the literature, it is likely to include a preponderance of this type. Discordant MZ pairs, who may be very informative, are unlikely to be reported.

EXAMPLES OF TWIN STUDIES

A few examples will indicate ways in which twins provide information about medical genetics.

Chromosomal Aberrations.   If one member of an MZ twin pair has Down syndrome, the other twin is always affected; if one DZ twin is affected, the other is nearly always normal. This is one of the observations that led to the hypothesis that a chromosomal defect might be the cause of mongolism and is

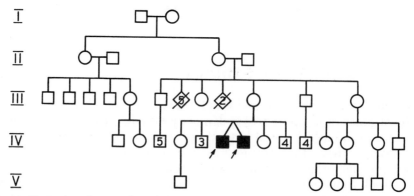

Figure 16–2   A pedigree of Duchenne muscular dystrophy in which the disease appeared only in a pair of MZ twins. The large number of normal males related through females to the propositi (seven brothers, eleven other relatives) makes it unlikely that the mutant gene had been present previously in the kindred. Its occurrence in both members of an MZ twin pair is evidence of its genetic basis, even though it is sporadic. Redrawn from Stephens and Tyler, Am. J. Hum. Genet. 3:111–125, 1951, Kindred 31.

an example of the way in which a twin study can point to an unusual causative mechanism of disease.

Mutation in Sporadic Cases of Disease. Many cases of muscular dystrophy indistinguishable from the X-linked recessive Duchenne type arise as sporadic cases presumably caused by new mutations. The observation of concordance for muscular dystrophy in an MZ twin pair in an otherwise normal family helps to verify the genetic causation of sporadic cases of the disease (Fig. 16–2).

Congenital Malformations. The concordance rates for congenital malformations in MZ and DZ twins allow a comparison of the strength of genetic predisposition. Some typical figures follow:

	Concordance*	
	MZ Pairs	**DZ Pairs**
Congenital dislocation of the hip	0.40	0.03
Cleft lip	0.40	0.05
Club foot (talipes equinovarus)	0.35	0.03
Congenital heart defects	0.05	0.05

*After the data of Carter, 1964.

These few examples show that for some congenital malformations, the concordance rate is rather high in MZ pairs as compared with DZ pairs. For congenital heart defects the risk is low, about the same in both types of twins, so that in this type of malformation genetic factors appear to be of little overall significance, though in a few cases they may be important. Other evidence (see Chapter 12) suggests that most congenital heart defects are multifactorial.

Cerebral Palsy. Cerebral palsy is an example of a disorder in which twin concordance rates can be misleading. Cerebral palsy is usually not genetically determined, but is produced by difficulties at the time of birth, especially prematurity or anoxia. These factors are more prominent in twin births than in single births, and therefore a high frequency of twins, whether MZ or DZ, is found in children with cerebral palsy.

UNUSUAL TYPES OF TWINS

Monozygotic Twins with Different Karyotypes (Heterokaryotic Twins). Rarely, twins are found to have different karyotypes, though evidence from genetic markers, placenta and physical appearance affirms their MZ origin. Among 12 pairs reported, four comprised a normal child and a 21-trisomic Down syndrome member; six comprised a normal child and a 45,X Turner syndrome; the others were a Turner with a Turner mosaic twin, and an X/XXX "normal" child with an X/XXX Turner twin. In almost all these cases, both twins were mosaic, each with a minor population of cells like the co-twin's major population.

Twins such as these are very rare and of great theoretical interest. Presumably they originate in postzygotic nondisjunction (as do chromosomal mosaics) followed or accompanied by twinning.

Conjoined Twins. Conjoined twins are believed to be MZ twins produced by an incomplete split of the original embryo, occurring relatively late in development (i.e., after the eighth day). They are more common than is generally believed, with a frequency of approximately one in 400 MZ twin births, or one in 120,000 births.

Chimeras. In Greek mythology, a chimera was a fire-breathing monster with the head of a lion, the body of a goat and the tail of a serpent. More prosaically, the term is used in human genetics to denote those rare individuals who are composed of a mixture of cells from two separate zygotes. There are two types of naturally occurring chimeras, both exceedingly rare: blood group chimeras and dispermic or "whole-body" chimeras.

Blood group chimeras are produced when DZ twins exchange hematopoietic stem cells in utero. The grafted cells are not recognized as foreign, and are retained. Each twin then has two populations of blood cells, one with his own genetic markers, the other with those of the twin.

Dispermic chimeras are thought to develop from fusion of fraternal twin zygotes in very early development. Separate fertilization of the egg and a polar body is one of several other suggested mechanisms. If a whole-body chimera has developed from fused XX and XY zygotes, true hermaphroditism usually results. Indeed, the very first case to be described was found after a deliberate search for a true hermaphrodite with different-colored eyes (Gartler, Waxman and Giblett, 1962).

GENERAL REFERENCES

Benirschke, K. 1972. Origin and clinical significance of twinning. Clin. Obstet. Gynecol. 15:220–235.

Bulmer, M. G. 1970. *The Biology of Twinning in Man.* Clarendon Press, Oxford.

Maynard-Smith, S., Penrose, L. S., and Smith, C. A. B. 1961. *Mathematical Tables for Research Workers in Human Genetics.* Churchill Livingstone, London.

MacGillivray, I., Nylander, P. P. S., and Corney, G. 1975. *Human Multiple Reproduction.* W. B. Saunders Company, London.

PROBLEMS

1. A pair of like-sexed dichorionic twins each have the following blood group geno-types: AB, MS/MS, CDe/cde, P_1P_2, Xg^a. The father's genotype is AO, MS/Ns, cde/cde, P_1P_1, Xg, and the mother's genotype is BO, MS/MS, CDe/cDE, P_1P_2, Xg^aXg.
   a) What is the probability that the twins are MZ?
   b) If twin A had haptoglobin type Hp 2-1 and twin B had haptoglobin type 1-1, what conclusion could be made as to the zygosity of the twins?

2. In a certain population, 40 percent of all twin pairs are unlike-sexed. What is the proportion of MZ twins in this population?

3. In another population, 55 percent of the twin pairs are like-sexed. What is the pro-portion of monozygotic twins?

17

DERMATOGLYPHICS IN MEDICAL GENETICS

Dermatoglyphics are the patterns of the ridged skin of the palms, fingers, soles and toes. Though these patterns show great diversity in detail and in the combinations found in any one person, they can be systematically classified into a reasonable number of different types. Dermatoglyphics are important in medical genetics chiefly because of the characteristic combinations of pattern types found in Down syndrome and, to a lesser extent, in other chromosomal disorders. Moreover, the dermatoglyphic traits themselves are genetically determined, at least in part; for example, the fingerprint ridge counts offer a classic example of multifactorial inheritance. Dermatoglyphics are a useful aid in the determination of twin zygosity. It is also important to keep in mind that dermal patterns develop in early embryonic life, before the end of the third month; hence dermatoglyphic abnormalities in any syndrome indicate that the syndrome originated prior to that stage.

The flexion creases — the heart, head and life lines of palmistry — are not, strictly speaking, dermal ridges, but they are formed at the same time, during the third fetal month, and affect the course of the dermal ridges. Indeed, flexion creases may themselves be determined in part by the same forces that affect ridge alignment. A single transverse crease (simian crease) in place of the usual two creases is found in 1 percent of normal whites and in a larger percentage of normal Orientals. Simian creases are more common in abnormal individuals, such as children with congenital malformations, even when the dermal patterns themselves are not obviously disturbed. In Down syndrome and other chromosomal disorders, single flexion creases are much more common than in controls; about half the palms of patients with Down syndrome have a single crease (Fig. 17–1).

The history of dermal pattern studies goes back to antiquity, but the scientific classification of patterns used today was proposed by Galton, who was the first to study dermal patterns in families and racial groups. Cummins, who coined the term dermatoglyphics ("writing on the skin"), made great contributions to the methodology and scientific basis of the field. It was

320

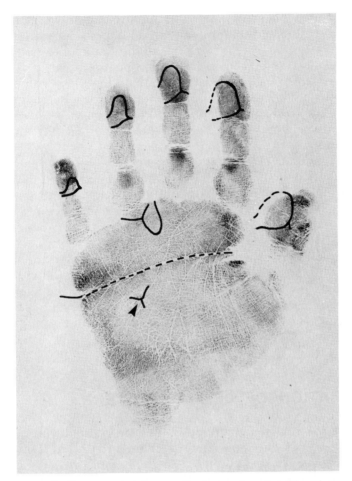

Figure 17–1   Single flexion crease (simian line) on palm of child with Down syndrome. Note also the distal axial triradius *t″* (arrow) and ulnar loops on all digits.

Cummins who first noted that the dermal patterns in Down syndrome show certain distinct trends that differentiate them from those of normal persons. Walker (1958) developed the first dermatoglyphic index for detection of Down syndrome, showing that some 70 percent of Down patients could be distinguished from normals on the basis of the dermal patterns alone. Following the discovery of the chromosomal basis of Down syndrome, interest in dermal patterns intensified, and dermal patterns of patients with other chromosomal disorders were also found to be distinctive (Uchida and Soltan, 1963). Dermal pattern analysis is now recognized as a useful technique for screening patients for chromosome study, and more generally as a research tool for disorders of development.

EMBRYOLOGY

Differentiation of the dermal ridges begins in the third month of prenatal development and is complete in all but minor details by the end of the fourth month. The initial phase of ridge alignment is the most critical, since the patterns then established undergo no further alterations throughout life.

The alignment of the ridges conforms to the shape of the growing hand or foot. At 10 weeks the fetal hand bears conspicuous volar pads, relatively as large as a cherry on an adult fingertip. At about the thirteenth week the pads regress, and meanwhile the dermal ridges differentiate in the thickening skin. Foot patterns develop slightly later, but the sequence of events is the same. Growth disturbances at this stage can produce abnormal dermatoglyphics, but later ones cannot.

Ridge alignments give the impression that a system of parallel lines has been drawn as economically as possible over an irregular terrain. The topology of ridges on nonflat surfaces has intrigued many scientists, including the mathematician Littlewood, who pointed out that when combing a spherical dog, at least four loops or two whorls would have to be produced!

Little is known about the details of abnormal pattern development, but Turner syndrome may give a clue. A fetus with Turner syndrome has generalized edema, with large edematous sacs in the neck region that account for the neck webbing seen postnatally. If in early prenatal life the fingertips are also swollen with excessive fluid, the ridges will have to cover a greater area; this may account for the high fingerprint ridge counts and large patterns seen in Turner syndrome.

CLASSIFICATION OF DERMAL PATTERNS

Prints of the digits, palms and soles can be made by one of several standard methods. In our experience, a technique used in many hospitals for identification of newborns produces excellent results. In very young infants, who have soft skin and fine ridges that are usually obscured by dry, scaling epithelium, clear prints may be extremely difficult to obtain. It is much easier to take a print of a child at least a month old than of a newborn.

The dermatoglyphics of importance in medical genetics are fingerprints (rolled prints of the ridged skin of the distal phalanx of each finger), palm prints and prints of the hallucal area of the sole. Rules for the formulation of dermal patterns in these areas are to be found in Cummins and Midlo (1943, 1961), Penrose (1968) and Holt (1968).

Pattern combinations and frequencies are more significant than pattern types alone as indicators of abnormal development. In Down syndrome, for example, there is no single characteristic of the dermal patterns that does not also occur in controls; but the combination of a number of patterns, most of which are more common in Down syndrome than in normal persons, permits definite recognition of the majority of affected children.

FINGERPRINTS

Fingerprints are classified, according to Galton's system, as **whorls, loops** or **arches.** Examples of each type are shown in Figure 17–2. The classification

Figure 17–2   Three basic fingerprint types: A, arch; B, loop; C, whorl. Ridge counts are made by counting along the indicated lines, excluding the triradius and the pattern core. The ridge count is 16 for the loop, 14–10 for the whorl, and 0 for the arch, which has no triradius.

is made on the basis of the number of **triradii**: two in a whorl (**W**), one in a loop and none in an arch (**A**). (A triradius is a point from which three ridge systems course in three different directions, at angles of 120°; see Fig. 17–1.) Loops are subclassified as radial (**R**) or ulnar (**U**) depending on whether they open to the radial or ulnar side of the finger. In a fingerprint formula, the patterns are conventionally listed in the following order: left 5, 4, 3, 2, 1; right, 1, 2, 3, 4, 5.

The frequency of the different patterns varies greatly from finger to finger; for example, in northern Europeans the frequency of radial loops is about 20 percent on the second digit but less than 0.3 percent on the fifth.

The size of a finger pattern is expressed as the ridge count, i.e., the number of ridges that touch a line drawn from the triradial point to the pattern core. An arch has a count of zero, since it has no triradius. The line of count for a loop and the two lines for a whorl are shown in Figure 17–2. Pattern size is especially important in assessing conditions characterized by a high frequency of arches (e.g., trisomy 18), since small loop patterns with low ridge counts merge into arches. The **total ridge count (TRC)** of the 10 digits is a useful dermatoglyphic parameter. In a normal British series, the mean TRC is 145 (standard deviation 50) for males, and 126 (standard deviation 52) for females (Holt, 1968).

PALM PATTERNS

Figure 17–3 indicates the chief landmarks of the palm. Palm patterns are defined chiefly by five triradii: four **digital triradii**, near the distal border of the palm, and an **axial triradius**, which is commonly over the fourth metacarpal near the base of the palm but sometimes displaced distally, especially in Down syndrome and other chromosomal disorders. Interdigital patterns (loops or whorls) may be formed by the recurving of ridges between the digital triradii. Hypothenar or thenar patterns may be present. Commonly, in a normal palm the ridges course obliquely toward the proximal portion of the ulnar side. Palm prints also show whether a single transverse crease is present.

The position of the axial triradius is perhaps the single most important feature, because it is distally displaced in many abnormal conditions. Its

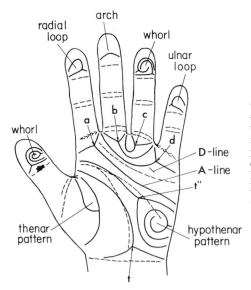

Figure 17–3 The nomenclature of the dermatoglyphics of the palm and fingers. There are four digital triradii (a, b, c and d) and an axial triradius t (t″ indicates its distal location). The main lines A and D are traced from the corresponding distal triradii. Thenar, hypothenar and digital patterns are also shown. From Smith and Berg, *Down's Anomaly.* 2nd ed., Churchill Livingstone, Edinburgh, London and New York, 1976, p. 81, by permission.

location may be measured either as a fraction of the total length of the palm, or as the **atd angle,** shown in Figure 17–4. An axial triradius in a position of 0.40 or more, or an *atd* angle greater than 57 degrees, is much more common in patients with Down syndrome and several other chromosomal syndromes than in the general population.

Figure 17–4 Measurement of the *atd* angle of the palm. Here the angle measures 60°. If there is more than one axial triradius, the distal one is used. From Holt, *The Genetics of Dermal Ridges.* Charles C Thomas, Publisher, Springfield, Ill., 1968, p. 86, by permission.

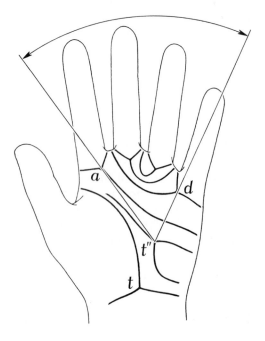

SOLE PATTERNS

Sole (plantar) patterns have been studied less extensively than palm patterns, chiefly because soles are more difficult to print and to classify. Only in the hallucal area have distinctive patterns have described in clinical syndromes. The unusual "arch tibial" (A^t) pattern, which is found in nearly 50 percent of all cases of Down syndrome and very rarely (0.3 percent) in controls, is the single most useful dermal pattern in Down syndrome (Fig. 17–5), and probably the most useful one in clinical dermatoglyphics.

CLINICAL APPLICATIONS

CHROMOSOME DISORDERS

In nearly all the chromosome disorders the dermatoglyphic patterns are unusual. This is not surprising because abnormal karyotypes lead to multiple morphological abnormalities, and the dermatoglyphic patterns are much influenced by the shape of the underlying structures. Table 17–1 lists some of the chief variants seen in the classic chromosome syndromes. The differences described are chiefly in the TRC (which may be either higher or lower than normal in different conditions), in the flexion creases of the palm (often reduced to a single crease) and in the position of the axial triradius (usually distally displaced). It must be emphasized that dermal patterns are highly variable, and none of these dermatoglyphic features is in itself abnormal, but their different frequency in patients and controls and the combination of several different unusual features in a single patient is distinctive, and in the case of Down syndrome may even be diagnostic.

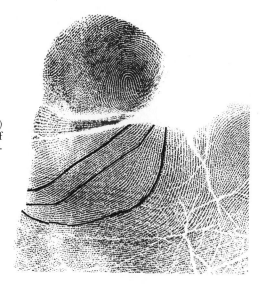

Figure 17–5 The hallucal arch tibial (A^t) pattern, which occurs in about 50 percent of individuals with Down syndrome but is otherwise very rare.

TABLE 17-1 CHARACTERISTIC DERMATOGLYPHICS IN CLASSIC CHROMOSOME SYNDROMES

	Fingers	Palms	Soles
5p–	Many arches Low TRC	t″ Single flexion crease	Open field
Trisomy 13	Many arches Low TRC	t″ or t‴ Single flexion crease Thenar pattern	Large pattern, arch fibular or loop tibial
Trisomy 18	6–10 arches (also on toes) Very low TRC	Single flexion crease	—
Down syndrome	Many ulnar loops (usually 10) Radial loop on 4th and/or 5th digits	t″ in 85% Single flexion crease in 50%	Arch tibial (50%) or small loop distal (35%)
45,X	Large loops or whorls High TRC	t slightly more distal	Very large loop or whorl
XXY	Many arches Low TRC	t slightly more proximal	—
XYY	Normal	Normal	Normal
Other syndromes with extra X and Y chromosomes	Many arches Low TRC The more sex chromosomes, the lower the TRC		

Abbreviations: t, axial triradius; t″, distal displacement of axial triradius; t‴, more extreme displacement; TRC, total ridge count.

Down Syndrome

Figure 17–6 illustrates a set of typical patterns of a Down syndrome patient, with a control. No single dermatoglyphic feature of Down syndrome patients is unique (except possibly the arch tibial hallucal pattern, which is very rare in controls), but the frequencies are often quite different. For example, on the second finger there is a radial loop in 20 percent of controls but only 2 percent of Down patients, whereas on the fifth finger there is a radial loop in only 3 per 1000 controls but in 4 percent of Down patients.

Typically, the palm in Down syndrome has the axial triradius displaced distally to about the center of the palm. The dermal ridges are transversely oriented, rather than obliquely as in most people. A single flexion crease is present in about half the patients.

The highly characteristic pattern in the hallucal area of the sole has already been referred to. Over 80 percent of the feet of Down patients have either an arch tibial (Fig. 17–5) or a very small "loop distal", which is a transitional pattern type closely similar to an arch tibial.

There are a number of methods for using dermal patterns in a diagnosis of Down syndrome, only one of which is described here. References to others can be found in the General References to this chapter. Figure 17–7 illustrates the Walker index (Walker, 1958), based on the relative frequency of certain finger, palm and sole patterns in a large series of Down patients and controls. The ratios are converted to logs and summed. The histogram in Figure 17–7

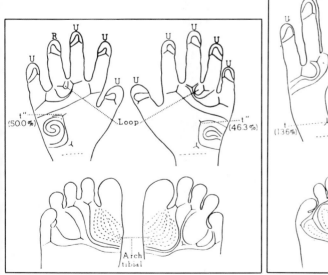
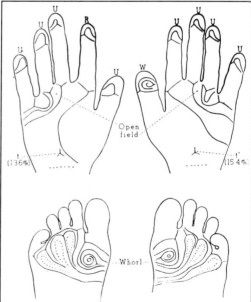

Figure 17–6 Typical dermal patterns in a Down syndrome patient (left) and a control (right). Compare the position of the axial triradius, the hypothenar pattern in the patient that is absent in the control, the fingerprints (the radial loop on the *fourth* digit is a useful diagnostic sign in Down syndrome) and the sole patterns. For further discussion, see text.

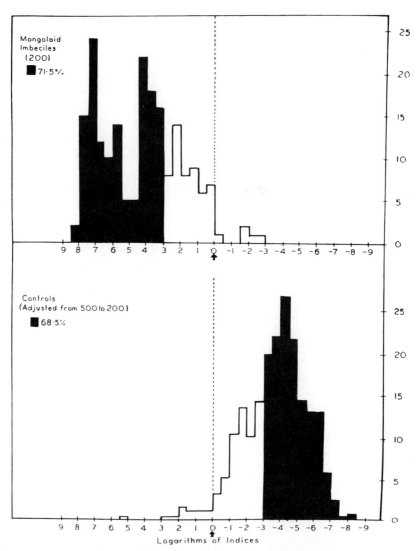

Figure 17–7 Histograms of the Walker index, based on dermal patterns in 200 Down syndrome patients and 500 controls. An index higher than +3 occurs in 70 percent of Down patients and in almost no controls. The method of calculation of the index is given in the original reference (Walker, Pediatr. Clin. North Am. 5:531–543, 1958).

shows that approximately 70 percent of each group can be unequivocally classified by this method.

Dermal pattern analysis cannot replace karyotyping in the diagnosis of Down syndrome, but it has its place as a screening technique or as an objective morphological criterion to be used in conjunction with other morphological criteria when an immediate diagnosis is required. Since dermal patterns show ethnic variation, patients and controls should be matched for ethnic background; the Walker index is appropriate for whites of North European origin.

OTHER DISORDERS

Unusual dermatoglyphics have been described in a wide variety of disorders, including several in which there is no reason to expect dysmorphism. Many of these reports remain unconfirmed or have been challenged; the differences described may in some cases reflect small sample size or inappropriate controls.

Gross distortion of the patterns can occur in association with any limb malformation. In arthrogryposis multiplex congenita, joint contractures are present at birth. In some but not all patients, the dermatoglyphics are very unusual, with longitudinal orientation of the ridges and other peculiarities. Such findings suggest that the condition was already present, or developing, at the time of ridge differentiation. As a general rule, abnormal dermatoglyphics point to early prenatal origin of a defect.

Curious digital patterns are seen in digits that have no fingernails, for example, in anonychia, nail-patella syndrome and brachydactyly Type B. Figure 17–8 illustrates the ridge arrangement on a nailless finger.

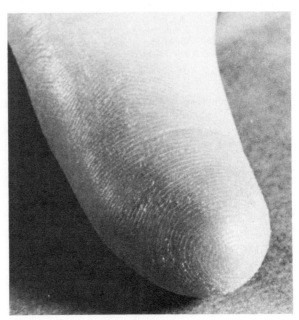

Figure 17–8　The ridge arrangement on a nailless finger of an individual with Type B brachydactyly. Reprinted from Battle et al., Mackinder's hereditary brachydactyly. Ann. Hum. Genet. 36:415–424, 1973, by permission.

"Fluctuating Dermatoglyphic Asymmetry"

As a rule, dermatoglyphics show reasonable bilateral symmetry. "Fluctuating asymmetry" is defined as random deviation from bilateral symmetry. In a study of cleft lip with or without cleft palate (CLP), Adams and Niswander (1967) reported that when the patients were divided into two groups, one with and one without a family history of the disorder, dermatoglyphic asymmetry was present in the group with a family history but not in the sporadic group. Later Woolf and Gianas (1977) reported that the first-degree relatives of CLP patients also show more dermatoglyphic asymmetry than controls.

If these studies are confirmed, two conclusions seem inescapable: (1) Familial and sporadic cases of CLP are different entities, with different etiology. (2) Whatever the genetic mechanism is that causes familial CLP, it must also be responsible for fluctuating dermatoglyphic asymmetry.

Thus it seems that the usefulness of dermatoglyphics in assessing disorders of development and their origin may be entering a new phase. If so, the many biases and pitfalls inherent in dermatoglyphic studies will have to be taken into account. Perhaps the chief problem is the control series; it is extremely difficult to obtain a control series of suitably large size, matched in ethnic background, sex and age with the patient group. Age-matching is essential because the *atd* angle changes with age, though otherwise the dermatoglyphics are permanent records.

GENERAL REFERENCES

Cummins, H., and Midlo, C. 1961. *Fingerprints, Palms and Soles; An Introduction to Dermatoglyphics.* Dover Publications, Inc., New York.

Holt, S. B. 1968. *The Genetics of Dermal Ridges.* Charles C Thomas, Publisher, Springfield, Illinois.

Penrose, L. S. 1963. Fingerprints, palms and chromosomes. Nature 197:933–938.

Penrose, L. S. 1969. Dermatoglyphics. Sci. Am. 221:72–84.

Schauman, B., and Alter, M. 1976. *Dermatoglyphics in Medical Disorders.* Springer-Verlag, New York.

Walker, N. F. 1958. The use of dermal configurations in the diagnosis of mongolism. Pediatr. Clin. North Am. 5:531–543.

Woolf, C. M., and Gianas, A. D. 1977. A study of fluctuating dermatoglyphic asymmetry in the sibs and parents of cleft lip propositi. Am. J. Hum. Genet. 29:503–507.

18

OVERVIEW

In the foregoing pages we have outlined some of the areas in which genetics plays a role in medicine and some of the mechanisms by which genetic disorders are produced. In the present chapter, we plan not to recapitulate, but to expand on some topics mentioned previously and to draw together some of the earlier topics, stressing their implications for medical research and practice.

BEHAVIOR GENETICS

Very little is known about the genetics of human behavior, either normal or abnormal. The critical problems that have frustrated research efforts are that few behavioral traits are well defined and few segregate in families in a clear Mendelian way. There have been three main approaches: attempts to work out the relative contributions of genetics and environment to biometrical traits, such as intelligence; descriptive studies, beginning with a specific gene mutation or chromosomal error and delineating the disturbances of behavior determined by it; and analyses of the genetic background of pathological traits such as schizophrenia, which itself may be the most poorly characterized and genetically heterogeneous trait ever subjected to genetic investigation.

Both mental retardation and mental illness, as well as their normal counterparts, fall within the behavior geneticist's field of study. In each class more is known about the inheritance of the abnormalities than of the normal. It is well known that mental retardation may result from the action of a single mutant gene or chromosome abnormality; but the psychological components of normal intelligence and their genetic basis, in particular the specific genes that underlie that basis and the way their gene products differ in individuals of different IQ, are still obscure. Turning to mental illness, again we have more information about the genetics of abnormal behavior than about how normal behavior is encoded in the gene loci. This is not to say that very much is known about the genetics of abnormal behavior. It is true that a number of single-gene and chromosomal disorders lead to phenotypic changes in behavior; examples include the well-known pathological behavior of some XYY males, the bizarre tendency to self-mutilation that is such a distressing aspect

331

of Lesch-Nyhan syndrome and the personality changes evident in Huntington chorea even before the characteristic choreic movements begin. In none of these examples, however, is the pathogenesis of the behavioral disorder known in any detail.

Failure to learn to speak or read at an appropriate age, when development is otherwise normal, are puzzling behavioral disorders that are beginning to become clear in the light of research into their origin and thus to be amenable to genetic investigation. **Developmental aphasia** appears to be caused by the inability of the aphasic child to process auditory stimuli when they are presented at a normal rate. This ability does eventually develop, but at a later age than average. In our own experience, admittedly limited, autosomal dominant transmission seems likely. But if so, where is the gene, what is its product, and what is the pathogenesis of the trait? **Specific reading disability** (dyslexia) in otherwise normal children is, in the majority of cases, autosomal dominant with some degree of sex limitation. (Males are far more frequently affected.) Again, one wonders at the nature of the gene and its product, the more so because reading is a learned skill that man did not have to perform until very recently, in terms of evolution.

Schizophrenia is extremely common, with incidence in the 1 percent range, and may form a continuous spectrum with the even more common trait known as "schizoid personality". In family studies, 2 to 5 percent of the parents and 6 to 10 per cent of the sibs are affected. If a propositus and parent are both affected, the risk for sibs is higher. Concordance is much greater for MZ than for DZ twin pairs. Finally, children of schizophrenic parents raised by their own parents and by adoptive parents show the same incidence of schizophrenia, a finding that seems to establish the role of heredity in schizophrenia beyond question. However, questions of genetic heterogeneity, the role of environmental stress and the nature of the biochemical abnormalities associated with schizophrenia must be resolved before the genetic aspects can be thoroughly understood. At present it does not seem reasonable to suspect a single gene-determined basic defect.

The **affective disorders** are somewhat similar to schizophrenia in terms of population incidence and frequency within families. Two separate types are recognized: **bipolar** disease, in which the patient alternates between depression and mania, and **unipolar** disease, in which there is depression but not mania. On the whole, these types run true within families. It has been suggested that bipolar disease is an X-linked dominant and unipolar disease an autosomal dominant, but so simple a hypothesis seems implausible on the basis of current evidence. Biochemical evidence of a single-gene basis for either the bipolar or the unipolar type is still lacking.

Finally, a reminder that reference has been made in a number of earlier chapters to topics that fall within the behavioral genetics area. For example, variation in response to drugs has been mentioned; this may play a part in such behavioral traits as **alcoholism** and drug addiction. One of the loci for alcohol dehydrogenase (ADH_2) is polymorphic; the "usual" phenotype is found in 94 percent of Europeans, but the "atypical" phenotype (in which the activity of the enzyme is much higher) has a frequency of about 90 percent in Japanese. It is not yet known how large a part this metabolic difference plays in determining the well-known difference in alcohol tolerance between Ori-

entals and whites or in the risk of fetal alcohol syndrome in offspring of alcoholic mothers.

GENETICS AND CANCER

The name "cancer" comes from the distinctive appearance of a solid
tissue malignant tumor, a central mass with crab-like claws extending into the
surrounding normal tissue (*cancer*, crab). Both Hippocrates and Galen noted
the resemblance and used the term. In cancer, one or more cells escape from
whatever factors control differentiation and proliferation. The undifferentiated or partly differentiated cells proliferate uncontrolled and in solid tissue
may form a tumor at the expense of the normal surrounding tissue. If the
process occurs in blood-forming elements rather than in solid tissue, the
consequence is leukemia.

The causes of the malignant transformation that leads to cancer are incompletely understood. Ionizing radiation and certain chemicals are carcinogenic (cancer-causing), and many carcinogenic agents are also mutagenic. In
experimental animals some types of cancer are caused by viruses, and much
current cancer research is directed toward the possibility of a viral origin. The
radiation, chemical or virus may be thought of as a triggering agent that
produces somatic mutation that allows the cell to escape from the normal
control of cell division.

INHERITANCE OF SUSPECTIBILITY TO SOME FORMS OF CANCER

It is well established that some forms of cancer are heritable. Perhaps the
best known example is **retinoblastoma**, a tumor of the retina that appears in
early childhood and may be either bilateral or unilateral. It is quite common,
with a population incidence of about one in 25,000, and accounts for about 5
percent of childhood blindness. Originally its genetic pattern was puzzling
because it appeared to be sometimes inherited, sometimes sporadic, all the
bilateral and some of the unilateral cases being inherited as autosomal dominants, whereas most of the unilateral cases were sporadic. The explanation
provided by Knudson (1971), and now generally accepted, is that development of the malignancy must require two successive mutations. In the hereditary cases, the first mutation is in the germ line and consequently is present in
all somatic cells; the second mutation is somatic, occurring in a retinal cell.
Since the second change is a random change, it might occasionally fail, in
which case the tumor would appear in only one eye. In the sporadic cases, two
somatic mutations in the same somatic cell line are required to initiate the
tumor. It is very unlikely that this would happen in both eyes, so the tumor is
unilateral and, since the germ cells are not affected, it is not heritable. The
difference in onset age (about 14 months in hereditary but 30 months in
sporadic cases) supports this concept. Further support is given by the observation that children with retinoblastoma have a high risk of other forms of tumor,
as though the initial event can be followed by a somatic mutation in more than

one tissue. The occurrence of sporadic retinoblastoma in some infants with a deletion of a specific part of chromosome 13q could perhaps be explained as a hemizygous expression of a retinoblastoma mutation, but this is speculation.

Genetic counseling in retinoblastoma must take account of the origin. If the tumor is unilateral and there is no family history of retinoblastoma, the recurrence risk for sibs is negligible, and the risk for offspring of the affected child should also be negligible, though there is a slight chance that he or she is a new mutant. If the tumor is bilateral or there is a family history or both, the risk of the same disorder in subsequent sibs or offspring is 50 percent. Since these tumors occasionally regress spontaneously, the parents should be given ophthalmological examinations as part of the genetic counseling workup.

Another example of an autosomal dominant form of tumor is familial polyposis of the colon. Here too there is increased susceptibility to primary tumors of other tissues.

The **chromosome breakage syndromes,** Fanconi anemia (FA), ataxia telangiectasia (AT) and Bloom syndrome (BL), are so characterized because, in addition to their typical clinical features that are congenital or expressed in early childhood, they share a propensity to chromosome breakage that can be seen in cultured cells. In all three, there is an increased risk of neoplasia, particularly of leukemia. Curiously, for Fanconi anemia and perhaps for the other disorders as well, the heterozygotes are also at a considerably increased risk of cancer.

Twin studies show the concordance rate for total incidence of cancer to be similar in monozygotic (MZ) and dizygotic (DZ) twin pairs; but if cancer of a particular organ is studied, MZ twins show a higher concordance rate than do DZ twins. The difference in concordance rate is less for MZ versus like-sexed DZ twins than it is for MZ versus all DZ pairs.

Family studies also indicate that the role of heredity is relatively minor, but where it is a factor the site is usually the same in members of a family group. The incidence of cancer of a specific organ is raised in relatives of propositi, especially in MZ twins, although the increase is relatively slight.

CHROMOSOME ABNORMALITIES IN CANCER

Specific chromosome abnormalities are seen in connection with some forms of cancer, and less specific cytogenetic changes are seen in others. The best known example is the **Philadelphia chromosome,** a deleted chromosome 22 with translocation of the deleted part to chromosome 9, seen in marrow cells of patients with chronic myelogenous leukemia (Rowley, 1973). As this disease progresses and reaches its final acute stage, additional karyotypic changes appear: the chromosome number may alter; there is often a second Philadelphia chromosome; there may be an extra chromosome 8 and an isochromosome for 17q, and there are other, less specific additions or losses of chromosomes. In acute myelogenous leukemia there are also nonrandom karyotypic changes. Other examples, less consistent than the Philadelphia chromosome, are 22q− in meningioma and 20q−, trisomy 1 or other variants in polycythemia vera and some other blood disorders, these changes being present before chemotherapy is instituted.

A different kind of association of a specific chromosome with cancer comes from the demonstration that an antigen characteristic of human cells transformed to malignancy by SV40 (simian virus 40) is specifically associated, usually, with chromosome 7. This observation was made by somatic cell hybridization techniques (see Chapter 10).

In summary, though very few forms of cancer show straightforward Mendelian transmission, cancer has an intimate association with the human genome and its mutations.

SCREENING FOR GENETIC DISEASE

Genetic screening programs, unknown until a few years ago, are now an important part of public health programs. Population screening originally had the objective of identifying newborns with genetic disorders that could be treated if diagnosed promptly. The prototype disorder is phenylketonuria, but a number of other metabolic disorders can also be diagnosed in newborns if urine is screened for abnormal metabolites, or blood for amino acids. More recently, heterozygote screening programs have been introduced. As a consequence, perhaps, of overenthusiasm and haste in putting some programs into operation prematurely, not all the programs have been successful, and some have generated public concern about the legal and ethical aspects of screening, which has been called "an arguably unique intrusion into the domain of personal freedom" (Reilly, 1975), but also "a major philosophical advance in medicine" (Levy, 1973).

NEWBORN SCREENING PROGRAMS

The principles of an adequate newborn screening program are obvious. The condition should be clearly defined and treatable and should have a reasonable frequency in the population; the screening test itself should be one that can be done rapidly and inexpensively on a large scale; the test should yield few false positives and, ideally, no false negatives; follow-up for definitive diagnosis and initiation of treatment should be well organized and prompt.

If the condition does not meet these requirements, problems may arise. The question of the validity of the test results is very important; false positives cause great concern to the parents, and false negatives vitiate the whole reason for the project.

Not everyone agrees that conditions that are not treatable should not be screened, basing their augment on the need for early identification so that genetic counseling can be provided and births of additional affected children in the family can be prevented. This argument is made in particular for Duchenne muscular dystrophy (Zellweger and Antonik, 1975). In this severe and burdensome condition, identification can be made in newborns, whereas the disease may not be evident clinically until another affected boy has been born, but the benefit of early diagnosis for genetic counseling must be balanced against the consideration of whether it is humane to inform the parents of the child's problem so far in advance of its manifestation.

HETEROZYGOTE SCREENING

There are at least three conditions that are common enough in particular geographic areas or ethnic subgroups of the population to be appropriate candidates for genetic screening of heterozygotes. These are Tay-Sachs disease, sickle cell anemia and thalassemia.

Tay-Sachs disease is virtually restricted to Ashkenazi Jews, though it is also common in a few small inbred communities. The practicality and acceptability of Tay-Sachs screening have been amply demonstrated. Carriers can be identified by a screening test, their condition can be verified by a more accurate test, and they can be advised of their risk and of the possibility of monitoring their pregnancies for the homozygous affected fetuses. Termination of these pregnancies is considered to be the logical outcome of the program.

In **sickle cell anemia**, in contrast, there is still no generally available prenatal diagnostic test, although an experimental method has been developed. Heterozygotes for sickle cell anemia are at least twice as common among American blacks (8 to 14 percent) as are Tay-Sachs heterozygotes among Jewish Americans (3 to 7 percent). Until prenatal diagnosis is widely available and widely accepted, the value of routine testing for heterozygotes is questionable, since the anxiety caused by heterozygote identification has no compensatory benefit.

Thalassemia may be a common problem in parts of the United States and Canada where there is a large population of Mediterranean descent. The incidence is particularly high in Greece and Cyprus. At present the prenatal test has been developed, but it has not yet become routinely available for screening of high-risk pregnancies.

OTHER POSSIBLE SCREENING PROGRAMS

It is theoretically possible to screen adult populations to identify members who are at risk for one of several kinds of disorders that could be forestalled by diet, by avoidance of smoking, or by other health-promoting measures that are matters of life-style rather than of medicine. For example, population screening for **hyperlipoproteinemia** would allow identification of people at high risk of coronary artery disease, and population screening for **alpha-1 antitrypsin deficiency** would identify not only homozygotes susceptible to severe lung and liver disease, but also heterozygotes whose lung function might be impaired by smoking. Public health officials who struggle to promote health by programs aimed at the whole population might have better success if they were able to concentrate on the identification and persuasion of high-risk groups. But this is for the future.

THE NEGATIVE IMPLICATIONS OF SCREENING PROGRAMS

The objective of genetic screening is to improve the public health, and the insidious aspects that have since been pointed out by lawyers and philosophers come as something of a surprise to those who had instituted the programs for what they saw as altruistic reasons. Invasion of privacy is seen as a major problem, and opposition has already led to abandonment of a program aimed at establishing a data base on the characteristics of XYY males. Concern is also expressed about stigmatization of individuals as abnormal on the basis of the tests; about whether informed consent is obtained, or whether compulsion, overt or implied, is exercised; about "the right not to know" about one's inborn handicaps; about whether the data will be assembled in data banks that are not totally confidential; and so on. Obviously, anything that has to do with human reproduction is a sensitive subject that will inevitably arouse public concern and controversy.

GENETIC COUNSELING

Genetic counseling has been defined as "a communication process which deals with the human problems associated with the occurrence, or the risk of occurrence, of a genetic disorder in a family." Usually it begins with parents who have had an affected child and want to know the risk that any subsequent child might have the same problem. Less often, the "counselee" is a member of a family in which some hereditary disease such as Huntington chorea is known or someone who plans to marry a cousin and wants to know the genetic risk of consanguineous marriage. Other kinds of problems, such as evaluation of a child for adoption, arise from time to time. More and more frequently, genetic counseling is provided as a prelude to prenatal diagnosis, or is in response to a request for a hoped for but unavailable prenatal diagnostic test.

Different genetic counselors have different philosophies. Some believe that their function is complete when a recurrence risk has been estimated. Others lay great stress on the supportive role. The writers' personal view is between these extremes. There is a responsibility to help the patient to understand the genetic aspects, to evaluate the various options available for dealing with the problem, and to take the appropriate action. We would not go as far as some others in offering psychological support, sharing the view that "people are usually resilient and strong, even in the face of serious adverse news, for the most part they prefer having it to being ignorant, and they move constructively to deal with it" (Childs et al., 1976).

Much genetic counseling is done by clinicians as an integral part of their overall management of patients with genetic disorders. For more complex cases, or if the clinician does not have the time or the genetic knowledge required, referral to a genetics center may be desirable. Most medical schools and major teaching hospitals now have medical genetics centers in which genetic counseling is available.

RECURRENCE RISK

When the diagnosis and the family history are known, it is often possible to classify the disorder as single-gene, chromosomal, multifactorial or nongenetic and to determine the recurrence risk either on a Mendelian basis or as an empiric risk. Examples have been given in previous chapters.

It may be difficult or impossible to estimate a recurrence risk if the specific diagnosis is in doubt, if the trait is rare and its pattern of inheritance is unknown, or if phenocopies (nongenetic traits mimicking genetic ones) exist. Reduced penetrance or variable expressivity may blur a pedigree pattern, even to the point of making it difficult to know whether a child's disorder is a new mutation or has been inherited; this is an important problem, because if the child is a new mutant there is virtually no recurrence risk for his sibs.

Diabetes mellitus is a condition that has earned the sobriquet of "geneticist's nightmare" because of its intractability to genetic definition. Empiric risk tables are available for diabetes, giving the risk in terms of sex, family history, and onset age of affected relatives; examples are given by Simpson (1968), Darlow et al. (1973) and others. However, now that the part played by the HLA complex in determining susceptibility to juvenile diabetes is becoming clear, it seems that it may be necessary to make HLA typing a part of the genetic counseling process in families of juvenile diabetics.

BURDEN

"Burden" refers to the total cost of a condition to affected persons, their families and society, in both economic and psychological terms. Whereas for risk the question is, "What is the chance that it will happen?", for burden the question is, "If it happens, how bad will it be?" For conditions in the same risk category, the burden may be quite different, both in severity and in duration. Fig. 18–1 illustrates the concepts of risk and burden, both of which must be evaluated by parents in reaching a decision about whether to have another child, to terminate a pregnancy, to have prenatal diagnosis, to have sterilization, or take some other option.

HETEROZYGOTE DETECTION

Tests for carrier detection can be useful in genetic counseling in a number of ways. For some conditions, there are reliable biochemical tests to use either for relatives of affected persons or sometimes, as in Tay-Sachs disease, for certain high-risk populations. These tests are particularly useful

BURDEN (arbitrary units)

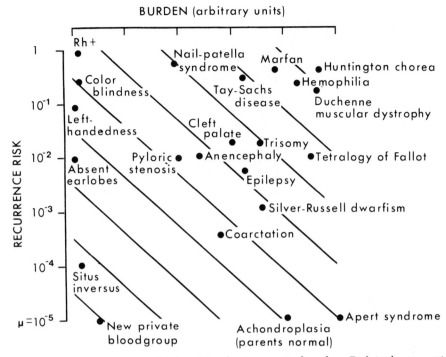

Figure 18–1 The concepts of risk and burden in genetic disorders. Risk is shown vertically on a log scale from μ (mutation rate) to 1. Burden is shown in arbitrary units on the horizontal axis. Parental decisions may depend in part on the parents' view of the product of risk and burden. The diagonal lines enclose conditions in which the product of risk and burden is roughly equal. From E. A. Murphy, 1973. Probabilities in genetic counseling. Birth Defects: Original Article Series 9:19–33; reprinted in E. A. Murphy and G. A. Chase, 1975. *Principles of Genetic Counseling.* Year Book Medical Publishers, Inc., Chicago.

in X-linked conditions; in autosomal recessives, the disorder is usually restricted to a single sibship anyway, unless the family belongs to a relatively inbred community.

For autosomal dominants, tests for heterozygote detection are of two kinds: tests to identify those who have the gene either before the typical onset age or when the expression of the trait is mild.

As noted previously, autosomal dominants with reduced penetrance or variable expressivity present difficulties in counseling because it may be very hard to distinguish between sporadic cases (new mutants) and inherited cases (subjects who have inherited the gene from a minimally affected parent). Crouzon's craniofacial dysostosis is such a condition. In typical cases this disorder is readily diagnosed by the characteristic facies (Fig. 18–2). The defect is burdensome because it is disfiguring, and extensive plastic surgery is needed to correct the hypertelorism and exophthalmos of the affected child in order to make his appearance more or less normal. Some cases (perhaps one-fourth of the total) are new mutants, and in these of course the only member of the family capable of passing on the defect is the affected person. Other cases are inherited but may be expressed so mildly that it is hard to determine whether either parent (or any other fami-

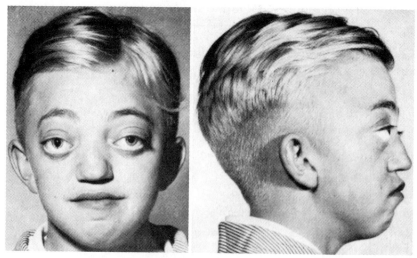

Figure 18-2 Crouzon's craniofacial dysostosis, an autosomal dominant defect discussed in the text. Note exophthalmos and hypertelorism. From Gorlin, Pindborg and Cohen, *Syndromes of the Head and Neck.* 2nd ed., McGraw-Hill Book Company, New York, 1976, p. 221, by permission.

ly member) has the gene, yet in these instances the risk for subsequent children is 1/2.

Carriers of chromosomal aberrations form a special group of high-risk individuals who can often be identified by karyotyping. Prenatal diagnosis can help such people to have normal children.

BAYESIAN METHODS IN GENETIC COUNSELING

Bayes' theorem (first published in 1763) gives a method of assessment of the relative probability of each of two alternative possibilities. This theorem can be usefully applied to certain problems in genetic counseling (Murphy and Chase, 1975). To illustrate, we will consider some pedigrees of Duchenne muscular dystrophy.

In Figure 18-3, II-3 is the daughter of a definite carrier of the Duchenne gene. The *prior* probability that she is a carrier is 1/2, and the prior probability that she is a noncarrier is also 1/2.

This woman has three normal sons. Now if she is a carrier, the *conditional* probability that all three sons would be normal is 1/8; and if she is normal, the *conditional* probability that all three boys would be normal is 1 (or very close to 1, since she might have a new mutant child).

We can now consider the *joint* probability, which is the product of the prior and conditional probabilities; the joint probability that she is a carrier is $1/2 \times 1/8 = 1/16$, and the joint probability that she is not a carrier is $1/2 \times 1 = 1/2$.

The posterior probability that she is a carrier is therefore:

$$\frac{1/16}{1/16 + 1/2} = 1/9$$

and the posterior probability that she is not a carrier is:

$$\frac{1/2}{1/16 + 1/2} = 8/9$$

In summary:

PROBABILITY THAT II–3 OF FIGURE 18–3A IS A CARRIER

	II–3 a Carrier	II–3 Not a Carrier
Prior probability	1/2	1/2
Conditional probability	1/8	1
Joint probability	1/16	1/2
Posterior probability	$\frac{1/16}{1/16 + 1/2} = 1/9$	$\frac{1/2}{1/16 + 1/2} = 8/9$

Now the probability that the woman is a carrier (1/9) can be applied to genetic counseling. The risk that her next child will be an affected male is $1/9 \times 1/4 = 1/36$. This is appreciably below the prior probability of 1/8 estimated when the genetic evidence provided by her children is not taken into consideration. Figure 18–4 is a further example, in which the information from several generations of normal males is used.

Next, consider the mother of a sporadic case who may or may not be a new mutant (Fig. 18–3B). Here the prior probabilities are quite different. The *prior* probability that any woman is a carrier is 4μ (μ = mutation rate). (This probability is calculated as follows: the chance that she received a new mutation from one of her parents is μ for each parent, and the chance that she received the gene from her mother who is a carrier is 2μ.) The prior probability that she is not a carrier is $1 - 4\mu$ which is very close to 1. The *conditional* probability of an affected son is 1/2 if she is a carrier, but only μ

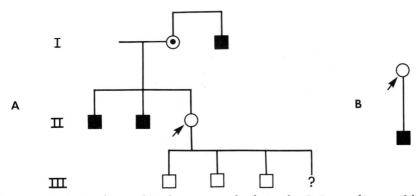

Figure 18–3 A, A pedigree of Duchenne muscular dystrophy. B, A sporadic case of the same disorder. These figures are used in the text to illustrate the application of Bayesian methods to determine the probability that the mother in each figure (shown by arrow) is a carrier.

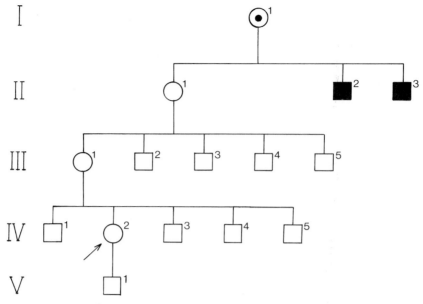

Figure 18-4 A pedigree of an X-linked recessive lethal, in which the prior probability that IV-2 is a carrier is 1/8, but Bayesian analysis gives a probability of 1/2115. From E. A. Murphy, 1973. Probabilities in genetic counseling. Birth Defects: Original Article Series 9:19–33; reprinted in E. A. Murphy and G. A. Chase, 1975. *Principles of Gentic Counseling.* Year Book Medical Publishers, Inc., Chicago.

if she is not a carrier. The *joint* probabilities are $4\,\mu \times 1/2 \times 2\mu$ and $1 \times \mu$, respectively. The posterior probability that she is a carrier is therefore $2\mu/2\mu + \mu = 2/3$, and the corresponding probability that she is not a carrier is 1/3.

Haldane (1935) showed that the proportion of all patients with an X-linked recessive disease who are new mutants is given by the rule

$$m = \frac{(1 - f)\mu}{2\mu + \nu}$$

where m is the proportion of all cases that are new mutants, f is fitness, and μ and ν are the mutation rates in females and males respectively. For Duchenne muscular dystrophy, $f = 0$, and the evidence is that $\mu = \nu$; thus 1/3 of all cases should be new mutants. Note that this calculation gives a result equivalent to the one reached by Bayesian analysis.

Additional information may be provided by other family members, as shown in the examples above, or by carrier testing. Creatine phosphokinase (CPK) activity in serum is elevated in two-thirds of Duchenne carriers, but not in all. If the woman in question has normal CPK activity, this information can be included and the probability that she is a carrier then becomes 2/5. In summary:

	Carrier	Not a Carrier
Prior probability	4μ	$1 - 4\mu \cong 1$
Conditional probability		
Genetic evidence	$1/2$	μ
CPK test normal	$1/3$	1
Joint probability	$2/3\ \mu$	μ
Posterior probability	$\dfrac{2/3\ \mu}{2/3\ \mu + \mu} = 2/5$	$\dfrac{\mu}{2/3\ \mu + \mu} = 3/5$

Bayesian probability also underlies two other examples we have used: determining the risk of Huntington chorea at various onset ages (Chapter 4) and determining the relative probability of monozygosity in twins (Chapter 16).

TREATMENT OF GENETIC DISORDERS

Although there is a persistent impression that labeling a disease as genetic is labeling it as incurable, actually many genetic disorders can be treated with a reasonable degree of success. For genetic disorders of metabolism, possible forms of therapy include:

1. Restriction of a substance that the patient cannot metabolize (phenylalanine in PKU, galactose in galactosemia).
2. Product replacement (hormones in hereditary hormone deficiencies, antihemophilic globulin for hemophilia).
3. Vitamin supplementation, to enhance enzyme activity by increasing ingestion of its coenzyme (vitamin D in vitamin D-dependent rickets).
4. Enzyme replacement (still in the experimental stage).
5. Surgical intervention, including transplantation of organs or marrow.

In some genetic disorders, patients may be at risk only under certain environmental conditions; e.g., patients with G6PD deficiency develop hemolysis only when exposed to primaquine or certain other drugs. Here, "treatment" is merely a matter of keeping the patient away from the precipitating agent.

A model program for diagnosis, counseling and treatment of genetic diseases has been in operation in Quebec province since 1971 (Clow et al. 1973, Laberge 1976). In addition to the traditional type of genetic counseling supported by a network of university-based genetics centers, this program has pioneered in the delivery of "continuous" counseling and management to patients with genetic metabolic disorders treatable by special diets. More such programs are needed to bridge the gap between theory and practice in medical genetics.

PRENATAL DIAGNOSIS

The development of prenatal diagnosis has added a new dimension to genetic counseling; now, instead of counseling in terms of probabilities, for some conditions the physician can establish definitely whether or not a particular fetus has a particular disease and can do so early enough to allow selective abortion of defective fetuses.

Prenatal diagnosis owes much to the recent change in the public attitude toward abortion, which was formerly proscribed but is now widely accepted. However, there are many areas where abortion is proscribed or restricted by law, and some parents are opposed to it on legal or ethical grounds. On the whole, parents who seek prenatal diagnosis have usually decided in advance that they will accept termination of the pregnancy if the test shows the fetus to be abnormal.

AMNIOCENTESIS

In amniocentesis, the amnion is tapped transabdominally and amniotic fluid is withdrawn. The technique was originally developed by Liley (1962) for treatment of fetuses suffering from the consequences of Rh incompatibility. Cells present in the amniotic fluid sample can be cultured and used for karyotyping or, after longer culture, for biochemical assays. (Microassays are being developed, and may reduce the time required for cell culture.) The amniotic fluid itself can also be assayed, especially for alpha fetoprotein, a protein formed in the fetal liver as early as the sixth week of gestation. The use of alpha fetoprotein assay in the prenatal detection of neural tube malformations has already been referred to.

The following conditions are generally considered to be indications for amniocentesis:

Late maternal age (40 in some centers, as low as 35 in others), because of the risk of a trisomic child.

Previous trisomic child. The recurrence risk, regardless of maternal age, is about 1 percent.

Chromosome translocation or other structural chromosome abnormality in either parent, which can lead to an unbalanced karyotype in the child.

Women who are carriers of X-linked recessive disorders and are willing to terminate any pregnancy in which the fetus is male.

Certain rare biochemical defects, when the family history indicates that the fetus is at risk of a specific defect.

Neural tube defects.

The risks of amniocentesis seem very small, especially when the procedure is carried out by experienced obstetricians guided by ultrasonography. However, like any surgical procedure, it carries a small risk of infection, and there may be a slight increase in the probability of spontaneous abortion. These risks must be weighed against the benefits of the procedure, especially if consideration is given to testing women outside the categories listed above.

ULTRASONOGRAPHY

Recently, technological developments in ultrasonography have made this technique a major tool for prenatal diagnosis. Among the fetal abnormalities that can be visualized are anencephaly, hydrocephaly, fetal ascites and renal agenesis. Figure 18–5 illustrates how much detail of the fetal structure can be seen under favorable circumstances.

FETOSCOPY

Direct visualization of the fetus by fetoscopy, using fiber optics, is now practicable. It is possible to scan an entire fetus, looking for morphological abnormalities such as cleft lip or palate, limb malformations, or ear deformities that might be clues to the presence of a genetic syndrome. A problem with the present instrumentation is that only a very small part of a fetus can be seen at a time (Fig. 18–6). The procedure takes much more time than amniocentesis, and has a correspondingly higher risk of infection or spontaneous abortion, but as experience grows these problems become less serious.

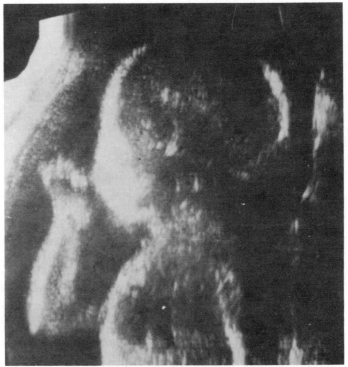

Figure 18–5 An ultrasound scan of a 30-week-old fetus, sucking his thumb. Note that the radius and ulna are clearly visible. Photograph courtesy of Stuart Campbell.

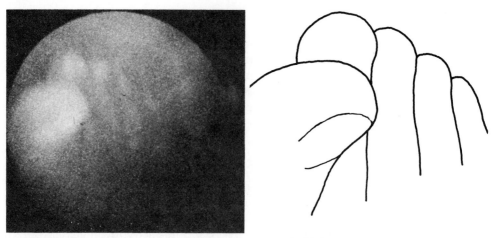

Figure 18–6 The distal part of a foot of a 16-week-old fetus, as visualized by fetoscopy. Photograph courtesy of R. Benzie.

Fetoscopy can also be used to allow sampling of fetal blood from the placenta or cord, and this opens up a new range of diagnostic possibilities. Mention has been made of the use of fetal blood to diagnose sickle cell anemia and thalassemia major.

CONCLUDING COMMENTS

The remarkable successes of molecular genetics have opened the prospect of spectacular developments in the manipulation of the human genome. The possibility of using **recombinant DNA techniques** to cure human disease has already been mentioned. Predetermination of sex by separation of X and Y sperm should soon be possible and might be accepted by many people who would not use abortion to have children of a chosen sex. **In vitro fertilization** of human ova is feasible, and 1978 saw the birth of the first "test tube baby" as a result of the pioneering work of Steptoe and Edwards. The prospect of **cloning,** that is, inducing a single somatic cell to develop into an exact replica of the individual who provided the cell, has caused great public interest; it has already been done in frogs, by taking out the nucleus of an egg cell and slipping the nucleus of an intestinal cell into its place. There is no reason to think it cannot eventually be done in man, though first there are a number of technical problems to be solved; meanwhile we can be thinking about the possible consequences, and the ethics of using these techniques for humans. These and other methods of manipulation of the human genome are often spoken of as **genetic engineering**.

There is much public concern about these approaches and the dangers inherent in them. It is hard to know how serious the imagined dangers really are; some of the possibilities are so remote as to be in the realm of science fiction. In any case, it is hard to see that the human genetic endowment would be impaired by such measures as long as the genetic diversity in the species is maintained.

Even while the broader ethical and social issues of the applications of genetics to man are being debated, there is much that can be done to bridge the gap between what is known about medical genetics and how it is applied to genetic disease. Screening programs, genetic counseling and prenatal diagnosis with selective abortion will be important to the practice of genetic medicine, but by no means the whole of it. Because genes interact with environment, a large part of the task is to find ways of manipulating the environment in order to allow harmful genes to be viable. Will this lead to "genetic pollution"? This is a hard question and one that will have to be resolved, but it is not a new question; medical ethics has always been on the side of the welfare of the individual if it conflicts with the welfare of society as a whole. The application of genetic knowledge to improve human health is the ultimate goal of genetics in medicine.

GENERAL REFERENCES

Childs, B., and Simopoulous, A. P., eds. 1975. *Genetic Screening: Programs, Principles, and Research.* National Academy of Sciences, Washington, D.C.

Milunsky, A. 1977. *Know Your Genes.* Houghton Mifflin Company, Boston.

Murphy, E. A., and Chase, G. A. 1975. *Principles of Genetic Counseling.* Year Book Medical Publishers, Inc., Chicago.

Nyhan, W. L., and Sakati, N. O. 1976. *Genetic and Malformation Syndromes in Clinical Medicine,* Year Book Medical Publishers, Inc., Chicago.

Stevenson, A. C., and Davison, B. C. C. 1976. *Genetic Counselling.* 2nd ed. William Heinemann Medical Books Limited, London.

GLOSSARY

Acentric A chromosome fragment lacking a centromere.

Allele Alleles are alternative forms of a gene occupying the same locus on homologous chromosomes. Alleles segregate at meiosis, and a child normally receives only one of each pair of alleles from each parent. See also *multiple alleles, isoalleles.*

Allograft A graft in which donor and host are members of the same species but not genetically identical.

Amino acids The building blocks of protein, for which DNA carries the genetic code. Abbreviations for the amino acids are listed in Table 3–1.

Amniocentesis Needle puncture of the uterus and amniotic cavity through the abdominal wall to allow amniotic fluid to be withdrawn by syringe. The term is often inappropriately applied to prenatal diagnosis by culture and analysis of amniotic fluid cells.

Amorph A gene that has no detectable product; an apparently inactive gene.

Anaphase The phase of mitosis or meiosis at which the chromosomes leave the equatorial plate and pass to the poles of the cell.

Aneuploid A chromosome number that is not an exact multiple of the haploid number; or an individual with an aneuploid chromosome number.

Antibody An immunoglobulin molecule formed by immune-competent cells in response to an antigenic stimulus and reacting specifically with this antigen.

Anticipation The term used to describe the apparent tendency of certain diseases to appear at earlier onset ages and with increasing severity in successive generations.

Antigen Any macromolecule that can elicit antibody formation by immune-competent cells and react specifically with the antibody so produced.

Ascertainment The method of selection of families for inclusion in a genetic study.

Association The presence together of two or more phenotypic characteristics in members of a kindred more often than expected by chance.

Assortative mating Selection of a mate with preference for a particular genotype, i.e., nonrandom mating. Preference for a mate of the same genotype is *positive* assortative mating; preference for a spouse of a different genotype is *negative* assortative mating.

Assortment The random distribution of different combinations of the chromosomes to the gametes. At anaphase of the first meiotic division, one member of each pair passes to each pole, and the gametes thus contain one chromosome of each type but *this chromosome may be of either paternal or*

349

maternal origin. Thus, nonallelic genes assort independently to the gametes. Exception: linked genes.

Autograft A graft of the host's own tissue.

Autoimmunity The ability to form antibodies against one or more of one's own antigens.

Autosome Any chromosome other than the sex chromosomes. Man has 22 pairs of autosomes.

B cells Small lymphocytes that respond to antigenic stimulation by the production of humoral antibodies.

Backcross Mating of a heterozygote to a recessive homozygote ($Aa \times aa$), in which the progeny (1/2 Aa, 1/2 aa) reveal the genotype of the heterozygous parent. The double backcross mating ($AaBb \times aabb$) is the most useful mating for linkage analysis in family studies.

Banding A technique of staining chromosomes in a characteristic pattern of cross bands. See G bands, Q bands and other types described in the text.

Barr body The sex chromatin as seen in female somatic cells. Named for Murray Barr, who with his student E. G. Bertram first described sexual dimorphism in somatic cells.

Base pairing In nucleic acids, adenine must always pair with thymine (or, in RNA, with uracil) and guanine with cytosine. The specificity of base pairing is fundamental to DNA replication and to its transcription into RNA.

Bivalent A pair of homologous chromosomes as seen at metaphase of the first meiotic division.

Blood group A genetically determined antigen on a red cell. The antigens formed by a set of allelic genes make up a blood group system.

Burden In clinical genetics the burden is the total impact of a genetic disorder on the patient, his family and society as a whole.

Carrier An individual who is heterozygous for a normal gene and an abnormal gene that is not expressed phenotypically, though it may be detectable by appropriate laboratory tests.

Centric fusion Fusion of the long arms of two acrocentric chromosomes at the centromere; Robertsonian translocation.

Centriole One of the pair of organelles that form the points of focus of the spindle during cell division. The centrioles lie together outside the nuclear membrane at prophase and migrate to opposite poles of the cell during cell division.

Centromere The heterochromatic region within a chromosome by which the chromatids are held together and to which the kinetochore is attached. Also called the *primary constriction.*

Chiasma Literally, a chiasma is a cross. The term refers to the crossing of chromatid strands of homologous chromosomes, seen at diplotene of the first meiotic division. Chiasmata either result in or are evidence of interchanges of chromosomal material between members of a chromosome pair. Such an exchange is called a *crossover.*

Chimera An individual composed of cells derived from different zygotes. In human genetics the term is used with reference to blood group chimerism—a phenomenon in which dizygotic twins exchange hematopoietic stem cells in utero and continue to form blood cells of both types—or to dispermic chimerism—a phenomenon in which two separate zygotes are fused into one individual.

Chromatid   A chromosome at prophase and metaphase can be seen to consist of two parallel strands held together by the centromere. Each strand is a chromatid. A single-stranded chromosome replicates during the DNA synthesis stage of the cell cycle and is then composed of two chromatids until the next mitotic division, at which each chromatid becomes a chromosome of a daughter cell.

Chromatin   The nucleoprotein fibers of which chromosomes are composed.

Chromomere   A densely coiled region of chromatin on a chromosome. Chromomeres give the extended chromosome a beaded appearance.

Chromosomal aberration   An abnormality of chromosome number or structure.

Clone   A cell line derived by mitosis from a single ancestral diploid cell.

Codominant   If both alleles of a pair are expressed in the heterozygote, the traits determined by them are codominant.

Codon   A triplet of three bases in a DNA or RNA molecule specifying a single amino acid.

Coefficient of inbreeding *(F)*   The probability that an individual has received both alleles of a pair from an identical ancestral source; or the proportion of loci at which he is homozygous.

Coefficient of relationship   The probability that two persons have inherited a certain gene from a common ancestor; or the proportion of all their genes that have been inherited from common ancestors.

Colinearity   Term used to describe the parallel relationship between the base sequence of the DNA of a gene (or the RNA transcribed from it) and the amino acid sequence of the polypeptide determined by that gene.

Complementation   In genetics, complementation is the ability of two different genetic defects to correct for one another, thus demonstrating that the defects are not in the same gene. This can happen when two persons, each homozygous for a recessive defect (e.g., deafness), produce children without the defect. More commonly, the term is now applied to the complementation test, in which two mutant cell lines grown together (co-cultivated) mutually correct one another's biochemical defect or in which a somatic cell hybrid made from two genetically defective cells lacks the defects of the parent lines.

Compound   A genotype in which the two alleles are different mutations from the wild type; or an individual who carries two different mutant alleles at a locus.

Concordant   A term often used in twin studies to describe a twin pair in which both members exhibit a certain trait.

Congenital defect   A defect present at birth; it may be genetically determined or may result from some environmental insult during prenatal development.

Consanguinity   Relationship by descent from a common ancestor.

Coupling   Two genes at different loci on the same chromosome are linked in *coupling* or in *cis* configuration. Antonym: *repulsion.*

Crossover or crossing over   Exchange of genetic material between members of a chromosome pair. The chiasmata seen at diplotene are the physical evidence of crossing over.

Cytogenetics   The study of the relationship of the microscopic appearance of chromosomes and their behavior during cell division to the genotype and phenotype of the individual.

Deletion   A form of chromosomal aberration in which a portion of a chromosome is lost.

Dermatoglyphics   The patterns of the ridged skin of the palms, fingers, soles and toes.

Dicentric   A structurally abnormal chromosome with two centromeres.

Dictyotene   The stage of the first meiotic division in which the human oocyte remains from late fetal life until ovulation.

Diploid   The number of chromosomes in most somatic cells, which is double the number found in the gametes. In man, the diploid chromosome number is 46.

Discordant   A term used in twin studies to describe a twin pair in which one member shows a certain trait and the other does not.

Dispermy   Fertilization of a duplicated egg nucleus, or egg and polar body, by two sperm. This event can produce a rare type of chimera in which two separate zygotes fuse to form a single individual.

Dizygotic twins   Twins produced by two separate ova, separately fertilized. Fraternal twins.

DNA (Deoxyribonucleic acid)   The nucleic acid of the chromosomes, which carries the genetic code.

Dominant   A trait is dominant if it is expressed in the heterozygote when the allele of the gene that determines this trait is not expressed. Modifications of this definition are discussed in the text.

Duplication   Presence of part of a chromosome in duplicate. Duplication may involve whole genes, series of genes or only part of a gene

Dysmorphism   Morphological developmental abnormality, as seen in many syndromes of genetic or environmental etiology.

Empiric risk   Estimate that a trait will occur or recur in a family based on past experience rather than on knowledge of the causative mechanism.

Euchromatin   Chromatin is differentiated into euchromatin and heterochromatin by its staining properties. See *heterochromatin*.

Eukaryote   An organism in which the cells have a nucleus with a nuclear membrane and other advanced characteristics. For further discussion, see text.

Expressivity   The extent to which a genetic defect is expressed. If there is variable expressivity, the trait may vary in expression from mild to severe but is never completely unexpressed in individuals who have the corresponding genotype.

F   See *coefficient of inbreeding*.

F_1 ("F one")   The first-generation progeny of a mating.

Fetoscopy   A technique for direct visualization of the fetus used for prenatal diagnosis.

Fingerprint   1. The pattern of the ridged skin on the distal phalanx of a finger. 2. A method of combining electrophoresis and chromatography to separate the components of a protein such as hemoglobin.

Fitness   The ability to transmit one's genes to the next generation and have them survive in that generation to be passed on to the next.

Forme fruste   An expression of a genetic trait so mild as to be of no clinical significance.

G bands   The dark and light cross-bands formed on chromosomes after treatment with trypsin and Giemsa stain.

Gamete   A reproductive cell (ovum or sperm) with a haploid chromosome number.

Gene   A segment of a DNA molecule coded for the synthesis of a single polypeptide. For further discussion, see text.

Gene(s) in common   Those genes inherited by two individuals from a common ancestral source.

Gene flow   Gradual diffusion of genes from one population to another.

Gene map   A representation of the human karyotype showing the chromosomal locations of the genes that have been mapped. See Figure 12–5.

Gene pool   All the genes present at a given locus in the population.

Genetic code   The base triplets that specify the 20 different amino acids. See Table 3–1.

Genetic counseling   Provision of information bearing upon the problems related to the occurrence, or risk of occurrence, of a genetic disorder in a family. The process is concerned with the risk and burden of the disorder and the options available for dealing with it.

Genetic drift   Random fluctuation of gene frequencies in small populations.

Genetic lethal   A genetically determined trait in which affected individuals do not reproduce.

Genetic marker   A trait can be used as a genetic marker in studies of cell lines, individuals, families and populations if it is genetically determined, can be accurately classified, has a simple unequivocal pattern of inheritance and has heritable variations common enough to allow it to be classified as a genetic polymorphism.

Genetic screening   Testing on a population basis to identify individuals at risk of having a specific genetic disorder or of having a child with a specific genetic disorder. Screening tests normally are applied only when some method of treatment or intervention is available.

Genetic trait   A trait determined genetically; to be distinguished from *congenital.*

Genome   The full set of genes.

Genotype   The genetic constitution (genome), or more specifically the alleles present at one locus.

Germ line   The cell line that produces gametes.

Haploid   The chromosome number of a normal gamete, with only one member of each chromosome pair. In humans, $n = 23$.

Haplotype   A group of alleles from closely linked loci, usually inherited as a unit. Example: inheritance of alleles at the closely linked loci that make up the HLA complex.

Hardy-Weinberg law   The law that relates gene frequency to genotype frequency. See text.

Hemizygous   A term which applies to the genes on the X chromosome in a male. Since males have only one X, they are hemizygous (not homozygous or heterozygous) with respect to X-linked genes.

Heritability   A statistical measure of the degree to which a trait is genetically determined. See text.

Heterochromatin   Chromatin that remains compacted and stains deeply in interphase. See *euchromatin.*

Heterogeneity If a certain phenotype (or very similar phenotypes) can be produced by different genetic mechanisms, the phenotype is genetically heterogeneous.

Heteroploid Any chromosome number other than the normal.

Heterozygote An individual who has two different alleles, one of which is the normal allele, at a given locus on a pair of homologous chromosomes. Adjective: *heterozygous.*

Histocompatibility A host will accept a particular graft only if it is histocompatible, i.e., if it contains no antigens that the host lacks.

Histones Proteins associated with DNA in the chromosomes, rich in basic amino acids (lysine or arginine) and virtually invariant throughout eukaryote evolution.

Homologous chromosomes A "matched pair" of chromosomes, one from each parent, having the same gene loci in the same order.

Homozygote An individual possessing a pair of identical alleles at a given locus on a pair of homologous chromosomes. Adjective: *homozygous.*

Immune reaction The specific reaction between antigen and antibody.

Immunological homeostasis The characteristic condition of a normal adult, who has certain antigens and the ability to react to antigens by producing antibodies, but who does not produce antibodies to his own antigens.

Immunological tolerance The inability to respond to a specific antigen resulting from previous exposure to that antigen, especially during embryonic life.

Inborn error A genetically determined biochemical disorder in which a specific enzyme defect produces a metabolic block that may have pathological consequences.

Inbreeding The mating of closely related individuals. The progeny of close relatives are said to be *inbred.* In laboratory mice, brother-sister inbreeding over many generation has been used to produce *inbred lines.*

Insertion A structural chromosomal aberration in which part of an arm of one chromosome is inserted into the arm of a nonhomologous chromosome.

Interphase The part of the cell cycle between two successive cell divisions.

Inversion A chromosomal aberration in which a segment of a chromosome is reversed end-to-end. See *paracentric inversion* and *pericentric inversion* in text.

Isoalleles Allelic genes that are "normal" and can be distinguished from one another only by their differing phenotypic expression when in combination with a dominant mutant allele.

Isochromosome An abnormal chromosome with duplication of one arm, forming two arms of equal length and bearing the same loci in reverse sequence, with deletion of the other arm of the normal chromosome.

Isograft A tissue graft between two individuals who have identical genotypes.

Isolate A subpopulation in which matings take place exclusively with other members of the same subpopulation.

Isozymes Multiple molecular forms of an enzyme, which may be in the same or different tissues within the same individual.

Karyotype The chromosome set. The term is often used for a photomicrograph of the chromosomes of an individual, arranged in the standard classification; or as a verb to indicate the process of preparing such a photo-micrograph.

Kindred An extended family.

Kinetochore A structure beside the centromere to which the spindle fibers are attached.

Lethal equivalent A gene carried in the heterozygous state which, if homozygous, would be lethal; or a combination of two genes in the heterozygous state each of which, if homozygous, would cause the death of 50 percent of homozygotes; or any equivalent combination.

Linkage Linked genes have their loci within measurable distance of one another on the same chromosome. See *synteny*.

Locus The position of a gene on a chromosome. Different forms of the gene (alleles) are always found at the same position on the chromosome. A *complex locus* is a locus within which mutation and recombination can occur at more than one site.

Lyon hypothesis (Lyonization; X-inactivation) Random and fixed inactivation of one X chromosome in the somatic cells of female mammals during early embryonic life, which leads to dosage compensation, heterozygote variability and female mosaicism.

Manifesting heterozygote A female heterozygous for an X-linked disorder, in whom, because of X-inactivation, the trait is expressed clinically with approximately the same degree of severity as in hemizygous affected males.

Meiosis The special type of cell division occurring in the germ cells by which gametes containing the haploid chromosome number are produced from diploid cells. Two meiotic divisions occur, the *first* and *second* (meiosis I and meiosis II). Reduction in number takes place during meiosis I. To be distinguished from *mitosis*.

Meiotic drive Term used for the situation in which, in heterozygotes, one allele is significantly more likely than the other to be transmitted to the progeny, i.e., prezygotic selection is operating in favor of one allele at the expense of the other. For example, see *T* alleles in text.

Metaphase The stage of mitosis or meiosis in which the chromosomes have reached their maximum condensation and are lined up on the equatorial plane of the cell, attached to the spindle fibers. This is the stage at which chromosomes are most easily studied.

MHC Abbreviation for the major histocompatibility complex, H-2 in the mouse or HLA in humans.

Mitosis Some cell division resulting in the formation of two cells, each with the same chromosome complement as the parent cell. To be distinguished from *meiosis*.

Mitotic cycle The cycle of a cell between two successive mitoses in which four periods are distinguished: G_1, S (DNA synthesis), G_2 and mitosis.

Monosomy A condition in which one chromosome of one pair is missing, as in 45,X Turner syndrome. Partial monosomy may occur.

Monozygotic twins Twins derived from a single fertilized ovum. Identical twins.

Mosaic An individual or tissue with at least two cell lines differing in genotype or karyotype, derived from a single zygote.

Multifactorial Determined by multiple factors, genetic and possibly also nongenetic, each with only a minor effect. See also *polygenic*.

Multiple alleles   Though only two alleles at a locus can be present in any one normal individual, there may be more than two different alleles (multiple alleles) at that locus in the population as a whole.

Mutant   A gene in which a mutation has occurred; or an individual carrying such a gene.

Mutation   A permanent heritable change in the genetic material. Usually defined as a change in a single gene (point mutation), although the term is sometimes used more broadly for a structural chromosomal change.

Mutation rate   The rate at which mutations occur at a given locus, expressed for individuals as mutations per gamete per locus per generation.

Nondisjunction   The failure of two members of a chromosome pair to disjoin during anaphase of cell division, so that both pass to the same daughter cell.

Nucleoside   A purine or pyrimidine base attached to a 5-carbon sugar (deoxyribose or ribose).

Nucleotide   A nucleoside (see preceding definition) attached to a phosphate group. A nucleic acid molecule is a polymer of many nucleotide units.

Operon   A postulated unit of gene action, consisting of an operator and the closely linked structural gene(s) whose action it controls.

p   1. The short arm of a chromosome (from the French *petit*). 2. In population genetics, used to indicate the frequency of the more common allele of a pair.

Pedigree   In medical genetics, a diagrammatic representation of a family history, indicating the affected individuals and their relationship to the propositus.

Penetrance   When the frequency of expression of a genotype is less than 100 percent, the trait is said to exhibit reduced penetrance. In an individual who has a genotype that characteristically produces an abnormal phenotype but who is phenotypically normal, the trait is said to be nonpenetrant.

Pharmacogenetics   The area of biochemical genetics concerned with drug responses and their genetically controlled variations.

Phenocopy   A copy of a phenotype that is usually determined by one specific genotype, produced instead by the interaction of some environmental factor with a different genotype.

Phenodeviant   A member of a population differing significantly in phenotype from the population as a whole.

Phenotype   The entire physical, biochemical and physiological nature of an individual, as determined by his genotype and the environment in which he develops; or, in a more limited sense, the expression of some particular gene or genes, as classified in some specific way.

Philadelphia chromosome   The structurally abnormal chromosome 22 typically occurring in a proportion of the bone marrow cells in most patients with chronic myelogenous leukemia. The abnormality is a deletion of the distal portion of the long arm (22q−), the deleted portion being translocated to the long arm of chromosome 9 (9q+).

Pleiotropy   If a single gene or gene pair produces multiple effects, it is said to exhibit pleiotropy (or to have pleiotropic effects).

Polygenic   Determined by many genes at different loci, with small additive effects. Also termed *quantitative.* To be distinguished from *multifactorial,* in which environmental as well as genetic factors may be involved.

Polymorphism The occurrence together in a population of two or more genetically determined alternative phenotypes, each with appreciable frequency. Arbitrarily, if the rarer allele has a frequency of at least 0.01, so that the heterozygote frequency is at least 0.02, the locus is considered to be polymorphic. Chromosome polymorphism (the presence of at least two variants of a given chromosome, each with appreciable frequency) also exists for several chromosomes of the human karyotype.

Polypeptide A chain of amino acids, held together by peptide bonds between the amino group of one and the carboxyl group of an adjoining one. A protein molecule may be composed of a single polypeptide chain or of two or more identical or different polypeptides.

Polyploid Any multiple of the basic haploid chromosome number, other than the diploid number; thus, $3n$, $4n$ and so forth.

Prenatal diagnosis Determination of the sex, karyotype and certain phenotypic features of the fetus, usually prior to the 20th week of gestation. The procedure is used when there is reason to suspect that the fetus might have a detectable abnormality.

Proband See *propositus.*

Prokaryote A simple unicellular organism, such as a bacterium, lacking a nuclear membrane and simpler than eukaryotic cells in other ways.

Prophase The first stage of cell division, during which the chromosomes become visible as discrete structures and subsequently thicken and shorten. Prophase of the first *meiotic* division is further characterized by pairing (synapsis) of homologous chromosomes.

Propositus The family member who first draws attention to a pedigree of a given trait. Also called *index case* or *proband.*

q 1. The long arm of a chromosome. 2. Often used to indicate the frequency of the rarer allele of a pair.

Q bands The pattern of bright and dim cross-bands seen on chromosomes under fluorescent light after quinacrine staining.

Quasicontinuous variation The type of variation shown by a multifactorial trait that has a threshold effect and thus appears to have a discontinuous distribution.

Quasidominant The pattern of inheritance produced by the mating of a recessive homozygote with a heterozygote, so that recessively affected members appear in two successive generations and the frequency of affected persons in the second generation is 1/2.

Random mating Selection of a mate without regard to the genotype of the mate. In a randomly mating population, the frequencies of the various matings are determined solely by the frequencies of the genes concerned.

Recessive Strictly, a trait that is expressed only in homozygotes. Less precisely, a *gene* that is expressed only when homozygous. For further discussion, see text.

Recombinant An individual who has a new combination of genes not found together in either parent. Usually applied to linked genes.

Recombinant DNA Artificially synthesized DNA in which a gene or part of a gene from one organism is inserted into the genome of another.

Recombination   The formation of new combinations of linked genes by cross-ing over between their loci.

Recurrence risk   The probability that a genetic disorder present in one or more members of a family will recur in another member of the same or a subsequent generation.

Reduction division   The first meiotic division, so called because at this stage the chromosome number per cell is reduced from diploid to haploid.

Regulator genes   Genes that control the rate of production of the product of other genes by synthesis of a substance that inhibits the action of an operator.

Repulsion   See *coupling*. (Also known as *trans* configuration.)

Ribosomes   Cytoplasmic organelles composed of ribosomal RNA and pro-tein, on which polypeptide synthesis from messenger RNA occurs.

Ring chromosome   A structurally abnormal chromosome in which the end of each arm has been deleted and the broken arms have reunited in ring formation.

RNA (Ribonucleic acid)   A nucleic acid formed upon a DNA template and taking part in the synthesis of polypeptides. Four forms are recognized: messenger RNA (mRNA), which is the template upon which polypeptides are synthesized; transfer RNA, which in cooperation with the ribosomes brings activated amino acids into position along the mRNA template; ribosomal RNA, a component of the ribosomes that functions as a non-specific site of polypeptide synthesis and heterogeneous RNA (HnRNA).

Robertsonian translocation.   See *centric fusion*.

Segregation   In genetics, the separation of allelic genes at meiosis. Since allelic genes occupy the same locus on homologous chromosomes, they pass to different gametes, i.e., they segregate.

Selection   1. In population genetics, the operation of forces that determines the relative fitness of a genotype in the population. 2. The manner in which kindreds are chosen for study, i.e., ascertainment.

Sex chromatin   A chromatin mass in the nucleus of interphase cells of fe-males of most mammalian species, including humans. It represents a single X chromosome that is inactive in the metabolism of the cell. Normal fe-males have sex chromatin and thus are *chromatin positive*; normal males lack it, and thus are *chromatin negative*. Synonym: *Barr body*.

Sex chromosomes   Chromosomes responsible for sex determination. (In humans: XX in female, XY in male.)

Sex-influenced   A trait which is not X-linked in its pattern of inheritance but which is expressed differently (either in degree or in frequency) in males and females.

Sex-limited   A trait which is expressed in only one sex though the gene determining it is not X-linked.

Sex linkage   X-linkage.

Sib, sibling   Brother or sister.

Silent allele   An allele that has no detectable product.

Sister chromatid exchange   Exchange of segments of DNA between sister chromatids. Occurs with particularly high frequency in patients with Bloom syndrome.

Somatic cell genetics   The study of genetic phenomena in cultured somatic cells.

Spindle Intracellular microtubules involved in the organization of the chromosomes on the metaphase plate and their segregation at anaphase.

Sporadic A trait is said to be sporadic when it occurs in a single individual within a kindred and has no known genetic basis.

Synteny Presence together on the same chromosome of two or more gene loci, whether or not they are close enough together for linkage to be demonstrated. Adjective: *syntenic*.

T cells Small lymphocytes committed by the influence of the thymus gland to be responsible for cell-mediated response to antigens.

Telophase The stage of cell division that begins when the daughter chromosomes reach the poles of the dividing cell and lasts until the two daughter cells take on the appearance of interphase cells.

Teratogen An agent that produces or raises the incidence of congenital malformations.

Transformation A form of recombination of genetic material in bacteria in which a bacterium incorporates DNA extracted from other bacteria into its own genetic material.

Translocation The transfer of a segment of one chromosome to a nonhomologous chromosome. If two nonhomologous chromosomes exchange pieces, the translocation is *reciprocal*. See also *centric fusion*.

Triploid A cell with three times the normal haploid chromosome number; or an individual made up of such cells.

Triradius In dermatoglyphics, a point from which the dermal ridges course in three directions at angles of approximately 120 degrees.

Trisomy The state of having three of a given chromosome instead of the usual pair, as in trisomy 21 (Down syndrome).

Ultrasonography A technique in which high frequency sound waves are used to outline structures of the body, very useful in prenatal diagnosis.

Wild type Term used especially in experimental genetics to indicate the normal allele (often symbolized as +) or the normal phenotype.

X-inactivation See *Lyon hypothesis*.

X-linkage Genes on the X chromosome, or traits determined by such genes, are X-linked.

Xenograft Graft from a donor of one species to a host of a different species.

Zygosity Twins may be either monozygotic or dizygotic. To determine which type a certain twin pair represents is to determine the zygosity of the pair.

REFERENCES

Adams, M. S., and Niswander, J. D. 1967. Developmental "noise" and a congenital malformation. *Genet. Res.* 10:313–317.

Aird, I., Bentall, H. H., and Roberts, J. A. F. 1953. A relationship between cancer of the stomach and the ABO blood groups. *Brit. Med. J.* 1:799–801.

Albright, F., Butler, A. M., and Bloomberg, E. 1937. Rickets resistant to vitamin D therapy. *Am. J. Dis. Child.* 54:529–547.

Alexander, J. W., and Good, R. A. 1977. *Fundamentals of Clinical Immunology.* W. B. Saunders Company, Philadelphia.

Alper, C. A., and Rosen, F. S. 1976. Genetics of the complement system. *Adv. Hum. Genet.* 7:141–188.

Ames, B. N., Durston, W. E., Yamasaki, E., and Lee, F. D. 1973. Carcinogens are mutagens: a simple test system combining liver homogenates for activation and bacteria for detection. *Proc. Natl. Acad. Sci. USA* 70:2281–2285.

Ampola, M. G., Mahoney, M. J., Nakamutz, E., and Tanaka, K. 1975. Prenatal therapy of a patient with vitamin B12 responsive methylmalonicacidemia. *N. Engl. J. Med.* 293:313–317.

Avery, O. T., MacLeod, C. M., and McCarty, M. 1944. Studies on chemical nature of substance inducing transformation of pneumococcal types: induction of transformation by desoxyribonucleic acid fraction isolated from pneumococcus type III. *J. Exp. Med.* 79:137–158.

Bach, G., Eisenberg, F., Cantz, M., and Neufeld, E. F. 1973. The defect in the Hunter syndrome: deficiency of sulfoiduronate sulfatase. *Proc. Natl. Acad. Sci. USA* 70:2134–2138.

Bach, G., Friedman, R., Weissmann, B., and Neufeld, E. F. 1972. The defect in the Hurler and Scheie syndromes: deficiency of α-L-iduronidase. *Proc. Natl. Acad. Sci. USA* 69:2048–2051.

Bahr, G. F. 1977. Chromosomes and chromatin structure. In: *Molecular Structure of Human Chromosomes.* (Yunis, J. J., ed.) Academic Press, Inc., New York.

Bain, B., and Lowenstein, L. 1964. Genetic studies on the mixed leucocyte reaction. *Science* 145:1315–1316.

Baker, T. G. 1963. A quantitative and cytologic study of germ cells in human ovaries. *Proc. R. Soc. London [Biol.],* 158:417–433.

Barr, M. L. 1960. Sexual dimorphism in interphase nuclei. *Am. J. Hum. Genet.* 12:118–127.

Barr, M. L., and Bertram, E. G. 1949. A morphological distinction between neurones of the male and female, and the behaviour of the nucleolar satellite during accelerated nucleoprotein synthesis. *Nature* 163:676–677.

Barski, G., Sorieul, S., and Cornefer, T. F. 1960. Production dans des cultures *in vitro* de deux souches cellulaires en association, de cellules de caractère "hybride". *C. R. Hebd. Séances Acad. Sci.* 251:1825–1827.

Bateson, W., and Punnett, R. C. 1905–1908. Experimental studies in the physiology of heredity. Reports 2, 3, 4 to the Evolution Committee of the Royal Society. Reprinted in: *Classic Papers in Genetics.* (Peters, J. A., ed.) Prentice-Hall, Englewood Cliffs, New Jersey.

Battle, H. I., Walker, N. F., and Thompson, M. W. 1973. Mackinder's hereditary brachydactyly. *Ann. Hum. Gent.* 36:415–424.

Beadle, G. W., and Tatum, E. L. 1941. Genetic control of biochemical reactions in *Neurospora. Proc. Natl. Acad. Sci. USA* 27:499–506.

Benirschke, K. 1972. Origin and clinical significance of twinning. *Clin. Obstet. Gynecol.* 15:220–235.

Bennett, D. 1977. Developmental antigens and differentiation. *Fifth International Conference on Birth Defects.*(Littlefield, J. W., and de Grouchy, J., eds.) Excerpta Medica, Amsterdam, p. 6.

Bodmer, W. F., and Cavalli-Sforza, L. L. 1976. *Genetics, Evolution, and Man.* W. H. Freeman & Company, San Francisco.

Bonaiti-Pellié, C., and Smith, C. 1974. Risk tables for genetic counselling in some common congenital malformations. *J. Med. Genet.* 11:374–377.

Bondy, P. K., and Rosenberg, L. E., eds. 1979 *Duncan's Diseases of Metabolism.* 8th ed. W. B. Saunders Company. Philadelphia.

Boué, J., Boué, A., and Lazar, P. 1975. Retrospective and prospective epidemiological studies of 1500 karyotyped spontaneous human abortions. *Teratology* 12:11–26.

Boveri, T. 1904. *Ergenbnisse Über die Konstitution der Chromatischen des Zellerns.* Fischer, Jena.

Brimhall, B., Duerst, M., Hollan, S. R., Stenzel, P., Szelenyi, J., and Jones, R. T. 1974. Structural characterization of hemoglobins J-Buda [α61 (E10) lys-asn] and G-Pest [α74 (EF3) asp$\rightarrow$asn]. *Biochim. Biophys. Acta* 336:344–360.

Brown, M. S., and Goldstein, J. L. 1974. Familial hypercholesterolemia: defective binding of lipoproteins to cultured fibroblasts associated with impaired regulation of 3-hydroxy-3-methylglutaryl coenzyme A reductase activity. *Proc. Natl. Acad. Sci. USA* 71:788–792.

Bulmer, M. G. 1970. *The Biology of Twinning in Man.* Clarendon Press, Oxford.

Carr, D. H. 1971. Chromosomes and abortion. *Adv. Hum. Genet.* 2:201–258.

Carr, D. H., and Gedeon, M. 1977. Population genetics of human abortuses. In: *Population Cytogenetics: Studies in Humans.* (Hook, E. B., and Porter, I. H., eds.) Academic Press, Inc., New York.

Carter, C. O. 1964. The genetics of common malformations. In: *Congenital Malformations: Papers and Discussions Presented at the Second International Conference on Congenital Malformations.* (Fishbein, M., ed.) International Medical Congress. New York, pp. 306–313.

Carter, C. O. 1965. The inheritance of common congenital malformations. *Prog. Med. Genet.* 4:59–84.

Carter, C. O. 1967. Risk to offspring of incest. *Lancet* 1:436.

Carter, C. O. 1969. Genetics of common disorders. *Brit. Med. Bull.* 25:52–57.

Carter, C. O. 1976. Genetics of common single malformations. *Brit. Med. Bull.* 32:21–26.

Carter, C. O. 1977. *Human Heredity,* 2nd ed. Penguin Books, New York.

Caspersson, T., Zech, L., Johansson, C., and Modest, E. J. 1970. Identification of human chromosomes by DNA-binding fluorescent agents. *Chromosoma* 30:215–227.

Cavalli-Sforza, L. L., and Bodmer, W. F. 1971. *The Genetics of Human Populations.* W. H. Freeman & Company, San Francisco.

Childs, B., Finucci, J. M., Preston, M. S., and Pulver, A. E. 1976. Human behaviour genetics. *Adv. Hum. Genet.* 7:57–97.

Chu, E. H. Y., and Powell, S. S. 1976. Selective systems in somatic cell genetics. *Adv. Hum. Genet.* 7:189–258.

Claiborne, R., and McKusick, V. A., eds. 1973. *Medical Genetics.* Hospital Practice Publishing Co., New York.

Cleaver, J. E. 1968. Defective repair replication of DNA in xeroderma pigmentosum. *Nature* 218:652–656.

Clermont, Y. 1972. Kinetics of spermatogenesis in mammals: seminiferous epithelium cycle and spermatogonial renewal. *Physiol. Rev.* 52:198–236.

Clow, C. L., Fraser, F. C., Laberge, C., and Scriver, C. R. 1973. On the application of knowledge to the patient with genetic disease. *Prog. Med. Genet.* 9:159–214.

Cohen, M. M., and Rattazzi, M. C. 1971 Cytological and biochemical correlation of late X-chromosome replication and gene inactivation in the mule. *Proc. Natl. Acad. Sci. USA* 68:544–548.

Comings, D. E. 1972. The structure and function of chromatin. *Adv. Hum. Genet.* 3:237–431.

Coombs, R. R. A., Mourant, A. E., and Race, R. R. 1946. In-vivo isosensitization of red cells in babies with haemolytic disease. *Lancet* 1:264–266.

Cori, G. T., and Cori, C. F. 1952. Glucose-6-phosphatase of liver in glycogen storage disease. *J. Biol. Chem.* 199:661–667.

Corney, G., and Robson, E. B. 1975. Types of twinning and determination of zygosity. In: *Human Multiple Reproduction.* (MacGillivray, I., Nylander, P. P. S., and Corney, G., eds.) W. B. Saunders Company, Philadelphia.

Creagan, R. P., and Ruddle, F. H. 1977. New approaches to human gene mapping by somatic cell genetics. In: *Molecular Structure of Human Chromosomes* (Yunis, J. J., ed.) Academic Press, Inc., New York.

Croce, C. M., and Koprowski, H. 1978. The genetics of human cancer. *Sci. Am.* 238:117–125.

Crow, J. F. 1965. Problems of ascertainment in the analysis of family data. In: *Genetics and the Epidemiology of Chronic Diseases* (Neel, J. V., Shaw, M. W., and Schull, W. J., eds.) U.S. Department of Health, Education and Welfare, Public Health Service Publication 1163.

Cummins, H., and Midlo, C. 1943. *Fingerprints, Palms and Soles: An Introduction to Dermatoglyphics.* Blakiston, Philadelphia. Reprinted in paperback by Dover, New York, 1961.

Curtis, E., Fraser, F. C., and Warburton, D. 1961. Congenital cleft lip and palate. Risk figures for counseling. *Am. J. Dis. Child.* 102:853–857.

Darlington, G. J., Bernhard, H. P., and Ruddle, F. H. 1974. Human serum albumen phenotype activation in mouse hepatoma-human leukocyte cell hybrids. *Science* 185:859–862.

Darlow, J. M., Smith, C., and Duncan, L. P. 1973. A statistical and genetical study of diabetes. III. Empiric risks to relatives. *Ann. Hum. Genet.* 37:157–174.

Davidson, R. G., Nitowsky, H. M., and Childs, B. 1963. Demonstration of two populations of cells in the human female heterozygous for glucose-6-phosphate dehydrogenase variants. *Proc. Natl. Acad. Sci. USA* 50:481–485.

Denver Conference, 1960. A proposed standard system of nomenclature of human mitotic chromosomes. *Am. J. Hum. Genet.* 12:384–388.

Deol, M. S., and Whitten, N. K. 1972. X-chromosome inactivation: does it occur at the same time in all cells of the embryo? *Nature [New Biol.]* 240:277–279.

Dronamraju, K. R. 1960. Hypertrichosis of the pinna of the human ear. Y-linked pedigrees. *J. Genet.* 57:230–244.

Dubowitz, V. 1978. *Muscle Disorders in Childhood.* W. B. Saunders Company, Philadelphia.

Edwards, J. A., and Gale, R. P. 1972. Camptobrachydactyly, a new autosomal dominant trait with two probable homozygotes. *Am. J. Hum. Genet.* 24:464–474.

Edwards, J. H. 1960. The simulation of Mendelism. *Acta Genet.* (Basel) 10:63–70.

Edwards, J. H. 1969. Familial predisposition in man. *Br. Med. Bull.* 25:58–64.

Edwards, J. H., Harnden, D. G., Cameron, A. H., Crosse, V. M., and Wolff, O. H. 1960. A new trisomic syndrome. *Lancet* 1:787–790.

Edwards, Y. H., and Hopkinson, D. A. 1977. Developmental changes in the electrophoretic patterns of human enzymes and other proteins. In: *Isozymes: Current Topics in Biological and Medical Research.* (Rattazzi, M. C., Scandalios, J. G., and Whitt, G. S., eds.) Alan R. Liss, Inc., New York.

Eicher, E. M. 1970. X-autosome translocations in the mouse: total inactivation versus partial inactivation of the X-chromosome. *Adv. Genet.* 15:175–259.

Emery, A. E. H. 1975. *Elements of Medical Genetics,* 4th ed. Churchill Livingstone, Edinburgh, London, and New York.

Emery, A. E. H. 1976. *Methodology in Medical Genetics: An Introduction to Statistical Methods.* Churchill Livingstone, Edinburgh, London, and New York.

Epel, D. 1977. The program of fertilization. *Sci. Am.* 237:128–138.

Epstein, C. J. 1969. Mammalian oöcytes: X-chromosome activity. *Science* 163:1078–1079.

Epstein, C. J., and Motulsky, A. G. 1965. Evolutionary origins of human proteins. *Prog. Med. Genet.* 4:85–127.

Falconer, D. S. 1965. The inheritance of liability to certain diseases, estimated from the incidence among relatives. *Ann. Hum. Genet.* 29:51–71.

Falconer, D. S. 1967. The inheritance of liability to disease with variable age of onset, with particular references to diabetes mellitus. *Ann. Hum. Genet.* 31:1–20.

Figura, K. von, and Kresse, H. 1972. The Sanfilippo B corrective factor: an N-acetyl-α-D-glucosaminidase. *Biochem. Biophys. Res. Commun.* 48:262–269.

Fisher, R. A. 1944. *Statistical Methods for Research Workers.* Oliver and Boyd, Edinburgh.

Fisher, R. A. 1970. *Statistical Methods for Research Workers,* 14th ed. Oliver and Boyd, Edinburgh.

Fogh-Anderson, P. 1942. Inheritance of hare-lip and cleft-palate. *Op. Dom. Biol. Hered. Hum. Kbh.* 4. Munksgaard, Copenhagen.

Fölling, A. 1934. Über Ausscheidung von Phenylbrenztraubensäure in den Harn als Stoffwechselanomalie in Verbindung mit Imbezillität. *Hoppe-Seyeler's Z. Physiol. Chem.* 227:169–176.

Ford, C. E., Jones, K., Polani, P., de Almeida, J., and Briggs, J. 1959. A sex chromosome anomaly in a case of gonadal dysgenesis (Turner's syndrome). *Lancet* 1:711–713.

Franklin, R. E., and Gosling, R. G. 1953. Molecular configuration in sodium thymonucleate. *Nature* 171:740–741.

Fraser, F. C. 1963. Taking the family history. *Am. J. Med.* 34:585–593.

Fraser, F. C. 1970. The genetics of cleft lip and cleft palate. *Am. J. Hum. Genet.* 22:336–352.

Fraser, F. C. 1976. The multifactorial/threshold concept — uses and misuses. *Teratology* 14:267–280.

Fraser, G. R. 1976. *The Causes of Profound Deafness in Childhood.* Johns Hopkins University Press, Baltimore.

Fraser, G. R., and Mayo, O. 1975. *Textbook of Human Genetics.* Blackwell Scientific Publications, Oxford.

Fredrickson, D. S., Goldstein, J. L., and Brown, M. S. 1978. The familial hyperlipoproteinemias. In: *The Metabolic Basis of Inherited Disease* (Stanbury, J. B., Wyngaarden, J. B., and Fredrickson, D. S., eds.) 4th ed. McGraw-Hill Book Company, New York.

Gardner, L. I., ed. 1975. *Endocrine and Genetic Diseases of Childhood and Adolescence,* 2nd ed. W. B. Saunders Company, Philadelphia.

Gardner, R. L., and Lyon, M. F. 1971. X-chromosome inactivation studied by injection of a single cell into the mouse blastocyst. *Nature* 231:385–386.

Garrod, A. E. 1902. The incidence of alkaptonuria: a study in chemical individuality. *Lancet* 2:1616–1620.

Gartler, S. M., and Andina, R. J. 1976. Mammalian X-chromosome inactivation. *Adv. Hum. Genet.* 7:99–140.

Gartler, S. M., Waxman, S. H., and Giblett, E. R. 1962. An XX/XY human hermaphrodite resulting from double fertilization. *Proc. Nat. Acad. Sci. USA* 48:332–335.

Giblett, E. R. 1969. *Genetic Markers in Human Blood.* Blackwell Scientific Publications, Oxford.

Giblett, E. R., Anderson, J. E., Cohen, F., Pollara, B., and Meuwissen, H. 1972. Adenosine deaminase deficiency in two patients with severely impaired cellular immunity. *Lancet* 2:1067–1069.

Goldstein, J. L., Hazzard, W. R., Schrott, H. G., Bierman, E. L., and Motulsky, A. G. 1973. Hyperlipidemia in coronary heart disease. I. Lipid levels in 500 survivors of myocardial infarction. *J. Clin. Invest.* 52:1533–1543.

Goldstein, J. L., Schrott, H. G., Hazzard, W. R., Bierman, E. L., and Motulsky, A. G. 1973. Hyperlipidemia in coronary heart disease. II. Genetic analysis of lipid levels in 176 families and delineation of a new inherited disorder, combined hyperlipidemia. *J. Clin. Invest.* 52:1544–1568.

Gorlin, R. J., Pindborg, J. J., and Cohen, M. M. 1976. *Syndromes of the Head and Neck,* 2nd ed. McGraw-Hill Book Company, New York.

Greulich, W. W. 1934. Heredity in human twinning. *Am. J. Phys. Anthropol.* 19:391–431.

Griffith, F. 1928. The significance of pneumococcal types. *J. Hyg.* 27:113–159.

Grüneberg, H. 1947. *Animal Genetics and Medicine.* Paul B. Hoeber, New York.

Grüneberg, H. 1952. Genetical studies on the skeleton of the mouse. IV. Quasicontinuous variation. *J. Genet.* 51:95–114.

Guthrie, R., and Susi, A. 1963. A simple phenylalanine method for detecting phenylketonuria in large populations of newborn infants. *Pediatrics* 32:338–343.

Haldane, J. B. S. 1935. The rate of spontaneous mutation of a human gene. *J. Genet.* 31:317–326.

Hamerton, J. L. 1971. *Human Cytogenetics.* Vol. 1, *General Cytogenetics.* Vol. 2, *Clinical Cytogenetics.* Academic Press, Inc., New York.

Harper, P. S., and Dyken, P. R. 1972. Early-onset dystrophia myotonica: evidence supporting a maternal environmental factor. *Lancet* 2:53–55.

Harris, H. 1975. *The Principles of Human Biochemical Genetics,* 2nd ed. North-Holland Publishing Company, Amsterdam and London; American Elsevier Publishing Co., New York.

Harris, H., and Hirschhorn, K., eds. 1970–1977. *Advances in Human Genetics.* Vols. 1–8. 1970. Plenum Press, New York.

Harris, H., and Hopkinson, D. A. 1972. Average heterozygosity per locus in man: an estimate based on enzyme polymorphism. *Ann. Hum. Genet.* 36:9–20.

Harris, H., and Kalmus, H. 1949. Measurement of taste sensitivity to phenylthiourea (P.T.C.). *Ann. Eugen.* 15:24–31.

Harris, Henry, and Watkins, J. F. 1965. Hybrid cells derived from mouse and man: artificial heterokaryons of mammalian cells from different species. *Nature* 205:640–646.

Harrison, G. A., Weiner, J. S., Tanner, J. M., Barnicot, M. A., and Reynolds, V. 1977. *Human Biology,* 2nd ed. Oxford University Press, Oxford.

Hazzard, W. R., Goldstein, J. L., Schrott, H. G., Motulsky, A. G., and Bierman, E. L. 1973. Hyperlipidemia in coronary heart disease. III. Evaluation of lipoprotein phenotypes of 156 genetically defined survivors of myocardial infarction. *J. Clin. Invest.* 52:1569–1577.

Hecht, F., Bryant, J. S., Gruber, D., and Townes, P. L. 1964. The nonrandomness of chromosomal abnormalities. *N. Engl. J. Med.* 271:1081–1086.

Heinonen, O. P., Sloane, D., and Shapiro, S. 1977. *Birth Defects and Drugs in Pregnancy.* Publishing Sciences Group, Inc., Littleton, Massachusetts.

Heistö, H., van der Hart, M., Madsen, G., Moes, M., Noades, J., Pickles, M. M., Race, R. R., Sanger, R., and Swanson, J. 1967. Three examples of a new red cell antibody, anti-Co[a]. *Vox. Sang.* 12:18–24.

Holt, S. B. 1961. Quantitative genetics of finger print patterns. *Br. Med. Bull.* 17:247–250.

Holt, S. B. 1968. *The Genetics of Dermal Ridges.* Charles C Thomas, Publisher, Springfield, Illinois.

Hook, E. B. 1978. Rates of Down's syndrome in live births and at midtrimester amniocentesis. *Lancet* 1:1053–1054.

Hook, E. B., and Chambers, G. M. 1977. Estimated rates of Down syndrome in live births by one year maternal age intervals in a New York State study—implications of the risk figures for genetic counseling and cost-benefit analysis of prenatal diagnosis programs. *Birth Defects: Orig. Art. Ser.* 13(3A):123–141.

Hook, E. B., and Hamerton, J. L. 1977. The frequency of chromosome abnormalities detected in consecutive newborn studies — differences between studies — results by sex and by severity of phenotypic involvement. In: *Population Cytogenetics: Studies in Humans* (Hook, E. B., and Porter, I. H., eds.) Academic Press, New York, San Francisco and London.

Hook, E. B., and Porter, I. H., eds. 1977. *Population Cytogenetics: Studies in Humans.* Academic Press, Inc., New York, San Francisco and London.

Hopkinson, D. A., Edwards, Y. H., and Harris, H. 1976. The distribution of subunit numbers and subunit sizes of enzymes: a study of the products of 100 human gene loci. *Ann. Hum. Genet.* 39:383–411.

Hoppe, P. C., and Whitten, W. K. 1972. Does X chromosome inactivation occur during mitosis of first cleavage? *Nature* 239:520.

Huberman, J. A., and Riggs, A. D. 1968. On the mechanism of DNA replication in mammalian chromosomes. *J. Mol. Biol.* 32:327–341.

Ingram, V. M. 1956. A specific chemical difference between the globins of normal human and sickle cell anaemia haemoglobin. *Nature* 178:792–794.

Ingram, V. M. 1963. *The Hemoglobins in Genetics and Evolution.* Columbia University Press, New York.

Jacob, F., and Monod, J. 1961. Genetic regulatory mechanisms in the synthesis of proteins. *J. Mol. Biol.* 3:318–356.

Jacobs, P. A., Melville, M., Ratcliffe, S., Keay, A. J., and Syme, J. 1974. A cytogenetic survey of 11,680 newborn infants. *Ann. Hum. Genet.* 37:359–376.

Jacobs, P. A., Price, W. H., Court-Brown, W. M., Brittain, R. P., and Whatmore, P. B. 1968. Chromosome studies on men in a maximum security hospital. *Ann. Hum. Genet.* 31:339–358.

Jacobs, P. A., and Strong, J. A. 1959. A case of human intersexuality having a possible XXY sex-determining mechanism. *Nature* 183:302–303.

Jeanes, C. W. L., Schaefer, O., and Eidus, L. 1972. Inactivation of isoniazid by Canadian Eskimos and Indians. *Canad. Med. Assoc. J.* 106:331–335.

Jervis, G. A. 1953. Phenylpyruvic oligophrenia: deficiency of phenylalanine oxidizing system. *Proc. Soc. Exp. Biol. Med.* 82:514–515.

Kalow, W. 1962. *Pharmacogenetics: Heredity and Response to Drugs.* W. B. Saunders Company, Philadelphia.

Kalow, W., and Britt, B. A. 1973. Inheritance of malignant hyperthermia. In: *International Symposium of Malignant Hyperthermia* (Gordon, R. A., Britt, B. A., and Kalow, W., eds.) Charles C Thomas, Publishers, Springfield, Illinois.

Kalow, W., and Staron, N. 1957. On distribution and inheritance of atypical forms of human serum cholinesterase, as indicated by dibucaine numbers. *Canad. J. Biochem. Physiol.* 35:1305–1320.

Kimura, M., and Ohta, T. 1971. *Theoretical Aspects of Population Genetics.* Princeton University Press, Princeton, New Jersey.

Kirk, R. L. 1968. *The Haptoglobin Groups in Man.* Karger, Basel.

Klinefelter, H. F., Reifenstein, E. C., and Albright, F. 1942. Gynecomastia, aspermatogenesis without aleydigism and increased excretion of follicle-stimulating hormone. *J. Clin. Endocrinol.* 2:615–627.

Knudson, A. G. 1971. Mutation and cancer: statistical study of retinoblastoma. *Proc. Natl. Acad. Sci. USA* 68:820–823.

Kresse, H., and Neufeld, E. F. 1972. The Sanfilippo A corrective factor: purification and mode of action. *J. Biol. Chem.* 247:2164–2170.

Laberge, C. 1969. Hereditary tyrosinemia in a French Canadian isolate. *Am. J. Hum. Genet.* 21:36–45.

Laberge, C. 1976. Population genetics and health care delivery: the Quebec experience. *Adv. Hum. Genet.* 6:323–374.

La Du, B. N., Zannoni, V. G., Laster, L., and Seegmiller, J. E. 1958. The nature of the defect in tyrosine metabolism in alkaptonuria. *J. Biol. Chem.* 230:251–260.

Latt, S. A. 1973. Microfluorimetric detection of deoxyribonucleic acid replication in human metaphase chromosomes. *Proc. Natl. Acad. Sci. USA* 70:3395–3399.

Lawler, S. D. 1954. Family studies showing linkage between elliptocytosis and the Rhesus blood group system. *Caryologia,* Suppl. 6:1199.

Lejeune, J., Gautier, M., and Turpin, R. 1959. Étude des chromosomes somatiques de neuf enfants mongoliens. *C.R. Acad. Sci.* (Paris) 248:1721–1722.

Lejeune, J., LaFourcade, J., Bergen, R., Vialatte, J., Boesweillwald, M., Seringe, P., and Turpin, R. 1963. Trois cas de deletion partielle du bras court du chromosome 5. *C.R. Acad. Sci.* (Paris) 257:3098–3102.

Lesch, M., and Nyhan, W. L. 1964. A familial disorder of uric acid metabolism and central nervous system function. *Am. J. Med.* 36:561–570.

Levine, P., Robinson, E., Celano, M., Briggs, O., and Falkinburg, L. 1955. Gene interaction resulting in suppression of blood group substance B. *Blood* 10:1100–1108.

Levitan, M., and Montagu, A. 1977. *Textbook of Human Genetics*, 2nd ed. Oxford University Press, Inc., New York and London.

Levy, H. L. 1973. Genetic screening. *Adv. Hum. Genet.* 4:1–104.

Li, C. C. 1963. Genetic aspects of consanguinity. *Am. J. Med.* 34:702–714.

Li, C. C., and Mantel, N. 1968. A simple method of estimating the segregation ratio under complete ascertainment. *Am. J. Hum. Genet.* 20:61–81.

Lifschytz, E., and Lindsley, D. L. 1972. The role of X-chromosome inactivation during spermatogenesis. *Proc. Natl. Acad. Sci. USA* 69:182–186.

Littlefield, J. 1964. Selection of hybrids from mating of fibroblasts in vitro and their presumed recombinants. *Science* 145:709–710.

Littlewood, J. E. 1953. A Mathematician's Miscellany. Methuen, London.

Longster, G., and Giles, C. M. 1976. A new antibody specificity, anti-Rg^a, reacting with a red cell and serum antigen. *Vox Sang.* 30:175–180.

Lowry, R. B., Jones, D. C., Renwick, D. H. G., and Trimble, B. K. 1976. Down syndrome in British Columbia, 1952–1973: incidence and mean maternal age. *Teratology* 14:29–34.

Lubs, H., and Cruz, F. de la, eds. 1977. *Genetic Counseling*. Raven Press, New York.

Lyon, M. F. 1961. Gene action in the X-chromosome of the mouse (*Mus musculus* L.) *Nature* 190:372–373.

Lyon, M. F., and Hawkes, S. G. 1970. X-linked gene for testicular feminization in the mouse. *Nature* 225:1217–1219.

Macvie, S. I., Morton, J. A., and Pickles, M. M. 1967. The reactions and inheritance of a new blood group antigen, Sd.^{a} *Vox Sang.* 13:485–492.

Markert, C. L. 1963. Lactate dehydrogenase isozymes: dissociation and recombination of subunits. *Science* 140:1329–1330.

Markert, C. L., and Ursprung, H. 1971. *Developmental Genetics*. Prentice-Hall, Englewood Cliffs, New Jersey.

Markovic, V., Worton, R. G., and Berg, J. M. 1978. Evidence for the inheritance of silver-stained nucleolus organizer regions. *Hum. Genet.* 41:181–187.

Marx, J. L. 1978. Antibodies. I. New information about gene structure. II. Another look at the diversity problem. *Science* 202:298–299 and 412–415.

Matalon, R., and Dorfman, A. 1972. Hurler's syndrome, an α-L-iduronidase deficiency. *Biochem. Biophys. Res. Commun.* 47:959–964.

Maynard-Smith, S., Penrose, L. S., and Smith, C. A. B. 1961. *Mathematical Tables for Research Workers in Human Genetics*. Churchill, London.

Mayr, E. 1963. *Animal Species and Evolution*. Harvard University Press, Cambridge, Massachusetts.

MacGillivray, I., Nylander, P. P. S., and Corney, G. 1975. *Human Multiple Reproduction*. W. B. Saunders Company, London.

McKusick, V. A. 1964. *On the X Chromosome of Man*. Amer. Inst. Bio. Sciences, Washington.

McKusick, V. A. 1969. *Human Genetics*, 2nd ed. Prentice-Hall, Englewood Cliffs, New Jersey. (A Study Guide to this book was published in 1972.)

McKusick, V. A. 1972. *Heritable Disorders of Connective Tissue*, 4th ed. The C. V. Mosby Company, St. Louis.

McKusick, V. A. 1978. *Mendelian Inheritance in Man:* Catalogs of Autosomal Dominant, Autosomal Recessive and X-linked Phenotypes, 5th ed. The Johns Hopkins Press, Baltimore.

McKusick, V. A., Howell, R. R., Hussels, J. E., Neufeld, E. F., and Stevenson, R. E. 1972. Allelism, non-allelism, and genetic compounds among the mucopolysaccharidoses. *Lancet* 1:993–996.

McKusick, V. A., and Ruddle, F. H. 1977. The status of the gene map of the human chromosomes. *Science* 196:390–405.

Mendel, G. 1865. Experiments in plant hybridization. Translation. In: *Classic Papers in Genetics*. (Peters, J. A., ed.) Prentice-Hall, Englewood Cliffs, New Jersey, pp. 1–20.

Meselson, M., and Stahl, F. W. 1958. The replication of DNA in *Escherichia coli*. *Proc. Natl. Acad. Sci. USA* 44:671–682.

Mettler, L. E., and Gregg, T. G. 1969. *Population Genetics and Evolution*. Prentice-Hall, Inc., Englewood Cliffs, New Jersey.

Miller, O. J., and Beatty, B. R. 1969. Portrait of a gene. *J. Cell Physiol.* 74, Suppl. 1:225–232.

Milunsky, A. 1977. *Know Your Genes*. Houghton Mifflin Company, Boston.

Mintz, B. 1967. Gene control of mammalian pigmentary differentiation. I. Clonal origin of melanocytes. *Proc. Natl. Acad. Sci. USA* 58:344–351.

Mohr, J. 1951. A search for linkage between the Lutheran blood group and other hereditary characters. *Acta Path. Microbiol. Scand.* 28:207–210.

Moody, P. A. 1975. *Genetics of Man*, 2nd ed. W. W. Norton & Co., Inc., New York.

Moore, K. L. 1977. *The Developing Human: Clinically Oriented Embryology*, 2nd ed. W. B. Saunders Company, Philadelphia.

Moore, K. L., and Barr, M. L. 1955. Smears from the oral mucosa in the detection of chromosomal sex. *Lancet* 2:57–58.

Morgan, W. T. J., and Watkins, W. M. 1969. Genetic and biochemical aspects of human blood group A-, B-, H-, Le^a- and Le^b-specificity. *Brit. Med. Bull.* 25:30–34.

Morton, N. E. 1959. Genetic tests under incomplete ascertainment. *Am. J. Hum. Genet.* 11:1–16.

Morton, N. E. 1972. The future of human population genetics. *Prog. Med. Genet.* 8:103–124.

Morton, N. E., Crow, J. F., and Muller, H. J. 1956. An estimate of mutational damage in man from consanguineous matings. *Proc. Natl. Acad. Sci. USA* 421:855–863.

Mourant, A. E. 1954. *The Distribution of the Human Blood Groups*. Blackwell Scientific Publications, Oxford.

Muller, H. J. 1927. Artificial transmutation of the gene. *Science* 66:84–87.

Murphy, E. A. 1973. Probabilities in genetic counseling. *Birth Defects: Original Article Series* 9:19–33. The National Foundation, New York.

Murphy, E. A., and Chase, G. A. 1975. *Principles of Genetic Counseling*. Year Book Medical Publishers, Inc., Chicago.

Nance, W. E. 1964. Genetic tests with a sex-linked marker: glucose-6-phosphate dehydrogenase. *Symp. Quant. Biol.* 29:415–425.

Nance, W. E. 1968. *Genetic Studies of Human Serum and Erythrocyte Polymorphisms. Glucose-6-Phosphate Dehydrogenase, Haptoglobin, Hemoglobin, Transferrin Lactate Dehydrogenase and Catalase*. Ph.D. Thesis, University of Wisconsin.

Nance, W. E., and Uchida, I. 1964. Turner's syndrome, twinning and an unusual variant of glucose-6-phosphate dehydrogenase. *Am. J. Hum. Genet.* 16:380–392.

Neel, J. V. 1949. The inheritance of sickle cell anemia. *Science* 110:64–66.

Neel, J. V., Kato, H., and Schull, W. J. 1974. Mortality in the children of atomic bomb survivors and controls. *Genetics* 76:311–326.

Neel, J. V., and Rusk, M. L. 1963. Polydactyly of the second metatarsal with associated defects of the feet: a new, simply inherited skeletal deformity. *Am. J. Hum. Genet.* 15:288–291.

Neel, J. V., and Schull, W. J. 1954. *Human Heredity*. University of Chicago Press, Chicago.

Neel, J. V., and Schull, W. J. 1962. The effect of inbreeding on mortality and morbidity in two Japanese cities. *Proc. Natl. Acad. Sci. USA* 48:573–582.

Neufeld, E. F., and Fratantoni, J. C. 1970. Inborn errors of mucopolysaccharide metabolism. *Science* 169:141–146.

Newcombe, H. B. 1964. Epidemiologic studies: discussion. In: *Congenital Malformations: Papers and Discussions Presented at the Second International Conference on Congenital Malformations*. (Fishbein, M., ed.) International Medical Congress, New York, pp. 306–313.

Nichols, W. W. 1963. Relationships of viruses, chromosomes, and carcinogenesis. *Hereditas* 50:53–80.

Nora, J. J. 1968. Multifactorial inheritance hypothesis for the etiology of congenital heart diseases. The genetic-environmental interaction. *Circulation* 38:604–617.

Nora, J. J., and Fraser, F. C. 1974. *Medical Genetics: Principles and Practice*. Lea & Febiger, Philadelphia.

Novitski, E. 1977. *Human Genetics*. Macmillan Publishing Company, Inc., New York.

Nyhan, W. L., Bakay, B., Connor, J. D., Marks, J. F., and Keele, D. K. 1970. Hemizygous expression of glucose-6-phosphate dehydrogenase in erythrocytes of heterozygotes for the Lesch-Nyhan syndrome. *Proc. Natl. Acad. Sci. USA* 65:214–218.

Nyhan, W. L., and Sakati, N. O. 1976. *Genetic and Malformation Syndromes in Clinical Medicine*. Year Book Medical Publishers, Inc., Chicago.

O'Brien, J. S. 1972. Sanfilippo syndrome: profound deficiency of α-acetylglucosaminidase activity in organs and skin transplants from Type B patients. *Proc. Natl. Acad. Sci. USA* 69:1720–1722.

Ohno, S. 1967. *Sex Chromosomes and Sex-Linked Genes*. Springer-Verlag New York, Inc., New York.

Ohno, S., Klinger, H. P., and Atkin, N. B. 1962. Human oogenesis. *Cytogenetics* 1:42–51.

Oriol, B., Cartron, J., Bedrossian, J., Duboust, A., Bariety, J., Gluckman, J. C., and Gagnadoux, M. F. 1978. The Lewis system: new histocompatibility antigens in renal transplantation. *Lancet* 1:574–575.

Painter, T. S. 1921. The Y-chromosome in mammals. *Science* 53:503–504.

Painter, T. S. 1923. Studies in mammalian spermatogenesis. *J. Exp. Zool.* 37:291–334.

Paris Conference (1971): Standardization in Human Cytogenetics. 1972. *Birth Defects: Original Article Series* 8:7. The National Foundation, New York.

Paris Conference (1971): Standardization in Human Cytogenetics, Supplement. 1975. *Birth Defects: Original Article Series* 9:9. The National Foundation, New York.

Partington, M. W. An English family with Waardenburg's syndrome. *Arch. Dis. Child.* 34:154–157, 1959.

Patau, K., Smith, D. W., Therman, E., Inhorn, S. L., and Wagner, H. P. 1960. Multiple congenital anomaly caused by an extra autosome. *Lancet* 1:790–793.

Pauling, L., Itano, H. A., Singer, S. J., and Wells, I. C. 1949. Sickle cell anemia, a molecular disease. *Science* 110:543–548.

Penrose, L. S. 1948. The problem of anticipation in pedigrees of dystrophia myotonica. *Ann. Eugen.* 14:125–132.

Penrose, L. S. 1963. Fingerprints, palms and chromosomes. *Nature* 197:933–938.

Penrose, L. S. 1968. *Memorandum on Dermatoglyphic Nomenclature. Birth Defects Original Article Series* 4(3). The National Foundation, New York.

Penrose, L. S. 1969. Dermatoglyphics. *Sci. Am.* 221:72–84.

Penrose, L. S. 1972. *The Biology of Mental Defect*, 4th ed., revised by Berg, J., and Lang-Brown, H. Sidgwick and Jackson, London.

Perutz, M. F. 1965. Structure and function of haemoglobin. I. A tentative atomic model of horse oxyhaemoglobin. *J. Mol. Biol.* 13:646–668.

Perutz, M. F. 1976. Structure and mechanism of haemoglobin. *Brit. Med. Bull.* 32:195–208.

Race, R. R., and Sanger, R. 1975. *Blood Groups in Man.* 6th ed. Blackwell Scientific Publications, Oxford.

Reed, T. E. 1969. Caucasian genes in American Negroes. *Science* 165:762–768.

Reed, T. E., and Chandler, J. H. 1958. Huntington's chorea in Michigan. *Am. J. Hum. Genet.* 10:201–225.

Reilly, P. 1975. Genetic screening legislation. *Adv. Hum. Genet.* 5:310–376.

Renton, P. H., Howell, P., Ikin, E. W., Giles, C. M., and Goldsmith, K. L. G. 1967. Anti-Sd,[a] a new blood group antibody. *Vox Sang.* 13:485–492.

Renwick, J. H. 1956. Nail-patella syndrome: evidence for modification by alleles at the main locus. *Ann. Hum. Genet.* 21:159–169.

Renwick, J. H. 1969. Progress in mapping human chromosomes. *Br. Med. Bull.* 25:65–73.

Renwick, J. H. 1971. The mapping of human chromosomes. *Ann. Rev. Genet.* 5:81–120.

Renwick, J. H., and Lawler, S. D. 1955. Genetical linkage between the ABO and nail-patella loci. *Ann. Hum. Genet.* 19:312–331.

Richards, B. W. 1975. Observations on the familial appearance of diseases associated with metabolic disorders of the mother. *Ann. Hum. Genet.* 39:189–191.

Roberts, J. A. F., and Pembrey, M. E. 1978. *An Introduction to Medical Genetics*, 7th ed. Oxford University Press, Oxford.

Rosenberg, L. E. 1967. Genetic heterogeneity in cystinuria. In: *Amino Acid Metabolism and Genetic Variation.* (Nyhan, W. L., ed.) McGraw-Hill Book Company, New York.

Rosenberg, L. E. 1979. Inborn errors of metabolism. In: Bondy, P. K., and Rosenberg, L. E. *Duncan's Diseases of Metabolism.* 8th ed. W. B. Saunders Company, Philadelphia.

Rosenberg, L. E. 1976. Vitamin-responsive inherited metabolic disorders. *Adv. Hum. Genet.* 6:1–74.

Rowley, J. D. 1973. A new consistent chromosomal abnormality in chronic myelogenous leukemia identified by quinacrine fluorescence and Giemsa staining. *Nature* 243:290–293.

Rudd, N. L., Gardner, H. A., and Worton, R. G. 1977. Mosaicism in amniotic fluid cell cultures. *Birth Defects: Original Article Series* 13(3D): 249–258. The National Foundation, New York.

Russell, W. L., Russell, L. B., and Cupp, M. B. 1959. Dependence of mutation frequency on radiation dose rate in female mice. *Proc. Natl. Acad. Sci. USA* 45:18–23.

Salmon, C., Salmon, D., Liberge, G., André, R., Tippett, P., and Sanger, R. 1961. Une nouvel antigène de groupe sanguin erythrocytaire présent chez 80% des sujets de race blanche. *Nouv. Rev. Fr. Hèmat.* 1:649–661.

Sanger, R., and Race, R. R. 1958. The Lutheran-secretor linkage in man: support for Mohr's findings. *Heredity* 12:513–520.

Sank, D. 1963. Genetic aspects of early total deafness. In: *Family and Mental Health Problems in a Deaf Population.* (Ranier, J. D., et al., eds.) New York State Psychiatric Institute, New York.

Schaumann, B., and Alter, M. 1976. *Dermatoglyphics in Medical Disorders.* Springer-Verlag, New York, Heidelberg and Berlin.

Scheinfeld, A., and Schachter, J. 1961. Bio-social effects on twinning incidence: I. Intergroup and generation differences in the United States in twinning incidence and MZ:DZ ratios. In: *Proc. II Int. Congr. Hum. Genet.* 1:300–302 (Gedda, L., ed.) Instituto G. Mendel, Rome.

Schull, W. J., and Neel, J. V. 1958. Radiation and the sex ratio in man. *Science* 128:343–348.

Seegmiller, J. E., Rosenbloom, F. M., and Kelley, W. N. 1967. Enzyme defect associated with a sex-linked human neurological disorder and excessive purine synthesis. *Science* 115:1682–1684.

Shepard, T. H. 1976. *Catalog of Teratogenic Agents,* 2nd ed. The Johns Hopkins University Press, Baltimore and London.

Siminovitch, L. 1976. On the nature of hereditable variation in cultured somatic cells. *Cell* 7:1–11.

Simpson, J. L. 1976. *Disorders of Sexual Differentiation: Etiology and Clinical Delineation.* Academic Press, Inc., New York.

Simpson, N. E. 1968. Diabetes in the families of diabetics. *Canad. Med. Assoc. J.* 98:427–432.

Smith, D. W. 1976. *Recognizable Patterns of Human Malformation,* 2nd ed. W. B. Saunders Company, Philadelphia.

Smith, D. W., and Wilson, A. A. 1973. *The Child with Down's Syndrome (Mongolism).* W. B. Saunders Company, Philadelphia.

Smith, E. W., and Torbert, J. V. 1958. Study of two abnormal hemoglobins with evidence for a new genetic locus for hemoglobin formation. *Bull. Johns Hopkins Hosp.* 102:38–45.

Smith, G. F., and Berg, J. M. 1976. *Down's Anomaly,* 2nd ed. Churchill Livingstone, Edinburgh.

Smithies, O. 1955. Zone electrophoresis in starch gels: group variations in the serum proteins of normal human adults. *Biochem. J.* 61:629–641.

Smithies, O., Connell, G. E., and Dixon, G. H. 1962. Chromosomal rearrangements and the evolution of haptoglobin genes. *Nature* 196:232–236.

Snell, G. D., Dausset, J., and Nathenson, S. 1976. *Histocompatibility.* Academic Press, Inc., New York.

Spencer, N., Hopkinson, D. A., and Harris, H. 1964. Quantitative differences and gene dosage in the human red cell acid phosphatase polymorphism. *Nature* 201:299–300.

Spielman, R. S., Mennuti, M. T., Zackai, E. H., and Mellman, W. J. 1978. Aneuploidy and the older gravida: which risk to quote. *Lancet* 1:1306–1307.

Srb, A. M., Owen, R. D., and Edgar, R. S. 1965. *General Genetics,* 2nd ed. W. H. Freeman & Co., Publishers, San Francisco.

Stanbury, J. B., Wyngaarden, J. B., and Fredrickson, D. S., eds. 1978. *The Metabolic Basis of Inherited Disease.* 4th ed. McGraw-Hill Book Company, New York.

Stanners, C. P., and Till, J. E. 1960. DNA synthesis in individual L-strain mouse cells. *Biochim. Biophys. Acta* 37:406–419.

Steinberg, A. G., ed. 1974. *Progress in Medical Genetics,* Vol. 10. Grune & Stratton, Inc., New York.

Steinberg, A. G., and Bearn, A. G., eds. 1961–1973. *Progress in Medical Genetics.* Vols. 1–9. Grune & Stratton, Inc., New York.

Steinberg, A. G., et al., eds. 1976 and 1977. *Progress in Medical Genetics, New Series.* Vols. 1 and 2. W. B. Saunders Company, Philadelphia.

Stephens, F. E., and Tyler, F. H. 1951. Studies in disorders of muscle. V. The inheritance of childhood progressive muscular dystrophy in 33 kindreds. *Am. J. Hum. Genet.* 3:111–125.

Stern, C. 1957. The problem of complete Y-linkage in man. *Am. J. Hum. Genet.* 9:147–166.

Stern, C. 1973. *Principles of Human Genetics,* 3rd ed. W. H. Freeman & Co., Publishers, San Francisco.

Stevenson, A. C., and Cheeseman, E. A. 1956. Hereditary deaf mutism, with particular reference to Northern Ireland. *Ann. Hum. Genet.* 20:177–231.

Stevenson, A. C., and Davison, B. C. C. 1976. *Genetic Counselling,* 2nd ed. William Heinemann Medical Books, London.

Sutton, H. E. 1961. *Genes, Enzymes, and Inherited Diseases.* Holt, Rinehart, & Winston, New York.

Sutton, H. E. 1975. *An Introduction to Human Genetics,* 2nd ed. Holt, Rinehart & Winston, New York.

Sutton, W. S. 1903. The chromosomes in heredity. *Biol. Bull.* 4:231–248.

Tachdjian, M. O. 1972. Pediatric Orthopedics, Vol. I. W. B. Saunders Company, Philadelphia.

Thompson, M. W. 1965. Genetic consequences of heteropyknosis of an X-chromosome. *Canad. J. Genet. Cytol.* 7:202–213.

Ting, A., and Morris, P. J. 1978. Matching for B-cell antigens of the HLA-DR series in cadaver renal transplantation. *Lancet* 1:575–577.

Trimble, B. K., and Baird, P. A. 1978. Maternal age and Down syndrome: age-specific incidence rates by single year intervals. *Am. J. Med. Gen.* 2:1–5.

Turner, H. H. 1938. A syndrome of infantilism, congenital webbed neck and cubitus valgus. *Endocrinology* 23:566–574.

Uchida, I. A. 1977. Maternal radiation and trisomy 21. In: *Population Genetics: Studies in Humans* (Hook, E. B., and Porter, I. A., eds.) Academic Press, Inc., New York.

Uchida, I. A., and Lin, C. C. 1972. Identification of triploid genome by fluorescence microscopy. *Science* 176:304–305.

Uchida, I. A., and Soltan, H. C. 1963. Evaluation of dermatoglyphics in medical genetics. *Pediatr. Clin. N. Amer.* 10:409–422.

Vogel, F. 1963. Mutations in man. In: *Genetics Today*, Vol. 3. Proceedings of the 11th International Congress of Genetics. (Geerts, S. J., ed.) Pergamon Press, New York.

Vogel, F. 1970. ABO blood groups and disease. *Am. J. Hum. Genet.* 22:464–475.

Vogel, F., and Rathenberg, R. 1975. Spontaneous mutation in man. *Adv. Hum. Genet.* 5:223–318.

Waardenburg, P. J. 1932. *Das menschliche Auge und seine Erbanlagen.* Nijoff, The Hague.

Walker, N. F. 1958. The use of dermal configurations in the diagnosis of mongolism. *Pediat. Clin. N. Am.* 5:531–543.

Wallace, B. 1968. *Topics in Population Genetics.* W. W. Norton & Company, Inc., New York.

Warkany, J. 1971. *Congenital Malformations: Notes and Comments.* Year Book Medical Publishers, Inc., Chicago.

Watson, J. D. 1976. *Molecular Biology of the Gene*, 3rd ed. W. A. Benjamin Inc., New York.

Watson, J. D., and Crick, F. H. C. 1953. Molecular structure of nucleic acids — a structure for deoxyribose nucleic acid. *Nature* 171:737–738.

Weatherall, D. J., ed. 1976. *Haemoglobin: Structure, Function and Synthesis. Br. Med. Bull.* 32(3).

Weiss, M. C., and Green, H. 1967. Human-mouse hybrid cell lines containing partial complements of human chromosomes and functioning human genes. *Proc. Natl. Acad. Sci. USA* 58:1104–1111.

White, C., and Wyshak, G. 1964. Inheritance in human dizygotic twinning. *N. Engl. J. Med.* 271:1003–1005.

Wiesmann, U., and Neufeld, E. F. 1970. Scheie and Hurler syndromes: apparent identity of the biochemical defect. *Science* 169:72–74.

Wood, W. G. 1976. Haemoglobin synthesis during human fetal development. *Br. Med. Bull.* 32:282–287.

Woolf, C. M., and Gianas, A. D. 1977. A study of fluctuating dermatoglyphic asymmetry in the sibs and parents of cleft lip propositi. *Am. J. Hum. Genet.* 29:503–509.

Wright, S. 1934. An analysis of variability in numbers of digits in an inbred strain of guinea pigs. *Genetics* 19:506–536.

Wyrobek, A. J., and Bruce, W. R. 1978. The induction of sperm shape abnormalities in mice and humans. In: *Chemical Mutagens: Principles and Methods for their Detection* Vol. 5, pp. 257–285. (Hollaender, A., and de Serres, F. J., eds.) Plenum Press, New York.

Yunis, J. J., ed. 1977. *Molecular Structure of Human Chromosomes.* Academic Press, Inc., New York, San Francisco and London.

Yunis, J. J., ed. 1977. *New Chromosomal Syndromes.* Academic Press, Inc., New York, San Francisco and London.

Yunis, J. J., and Chandler, M. E. 1977. The chromosomes of man — clinical and biologic significance: a review. *Am. J. Pathol.* 88:466–495.

Zellweger, H., and Antonik, A. 1975. Newborn screening for Duchenne muscular dystrophy. *Pediatrics* 55:30–34.

ANSWERS TO PROBLEMS

CHAPTER TWO

1. a) *A* and *a*   b) 1) At 1st meiotic division   2) At 2nd meiotic division

2. a) 2   b) 32   c) 2^n

3. Two of the 2^{23} possible combinations; one of each of the two parental combinations.

4. *Segregation* of alleles at anaphase of the first or second meiotic division; *random recombination* of nonalleles by the random assortment of non-homologous chromosomes at anaphase of the first meiotic division.

CHAPTER THREE

1.   I, point mutation, UAU to UGU changes tyr to cys.

   II, nonsense mutation, deletion of the first nucleotide of the third codon.

   III, nonsense mutation, insertion of G after the first codon.

2. A. 5, 9, 3, 2, 6, 8, 7, 10, 4, 1

CHAPTER FOUR

1. This question gives the student practice in composing a pedigree chart and some insight into the difficulty of obtaining complete and accurate family history information.

2. 1/64

3. b) Homozygous for X-linked gene (their mother must be a carrier)
   c) 100 percent   d) 50 percent

4. b) Autosomal dominant, autosomal recessive and X-linked recessive are all possible. X-linked dominant inheritance is ruled out.
   c) X-linked recessive
   d) 0
   e) Becker

5. b) Cannot estimate because genetic pattern is not clear.
   c) Now appears to be autosomal dominant, so risk for G's child is 1/4.
   d) G now appears to have the gene, so risk for her child is 1/2.

6. a) 1/24   b) 1/4

7. Father, *aa* XcbY   Mother, *Aa* XXcb   Child, *aa* XcbXcb

8. a) *Dd Ee*
   b) 1/12

9. a) 0   b) 1/4

10. About 30 percent.

11. He has about a 40 percent chance of developing the disease, therefore there is about a 20 percent chance that any child would have the gene.

CHAPTER FIVE

1. These two mutations affect different globin chains. The expected off-spring are:
   1/4 normal
   1/4 Hb M Saskatoon heterozygotes with methemoglobinemia
   1/4 Hb M Boston heterozygotes with methemoglobinemia
   1/4 double heterozygotes with four hemoglobin types: normal, both het-erozygous types, and a type with an abnormality in each chain. The clini-cal consequences are unknown — probably more severe methemoglobi-nemia.

2. a) 1/2
   b) B
   c) 2.5 percent of the daughters (5 percent chance that daughter would receive A— from her father, 50 percent chance that she would receive B— from her mother).

3. Both sickle cell trait.

4. $2/3 \times 2/3 \times 1/4 = 1/9$

5. 1/4

6. a) Patient and uncle hemizygous for Hunter allele, mother heterozygous, father's genotype irrelevant.
   b) Patient a Hurler-Scheie compound, mother Hurler heterozygote, father Scheie heterozygote, maternal uncle Hurler homozygote.

7. Theoretically, no difference, since Tay-Sachs patients never survive to reproductive age.

CHAPTER SIX

1. a) 46 chromosomes including the translocation, with the Down pheno-type, or rarely 45 chromosomes with monosomy 21. The latter would probably not survive until birth.
   b) If she can have a normal child, her translocation is not 21q21q, so it must be 21q22q.

2. a) No, the fetus should have a normal phenotype.
   b) Yes, though it is possible that a relative of the husband is the father.
   c) The risk of a translocation Down child is about 2 percent; at the mother's age there is also an appreciable risk of having a trisomic child. Amniocentesis would be advisable for fetal chromosome analysis.

3. 4–12 percent.

4. About 8 percent.

CHAPTER SEVEN

1. Apparently from the mother, since the father has given her his X with his color-blindness gene.

2. The error must have been in the paternal gamete, since the daughter received neither the paternal X nor the paternal Y.

3. a) XcbXcb Y
   b) Father XY, mother XXcb; probably nondisjunction at meiosis II in mother.

4. a) Theoretically, X and XX.
   b) 1/4 each XX, XY, XXX, XXY.

5. a) Paternal, meiosis I.
   b) Mother, nondisjunction at either meiosis I or meiosis II.

6. No, because the nondisjunctional event in XXY can happen in either parent and at either meiotic division, whereas XYY requires nondisjunction in the male at the second meiotic division only.

CHAPTER EIGHT

1. a) C3H, C57BL/6 and C3B6F1
   b) C3B6F1

2. MZ twin; DZ twin, sib; parent; uncle; cousin.

CHAPTER NINE

1. a) Rr or rr. (Here R represents any Rh-positive gene combination and r any Rh-negative combination.)
   b) Approximately 40 percent, i.e., the frequency of the r allele in fathers.

2. a) A, B or AB, but not O.
   b) All those with non-O fathers, i.e., about 54 percent for a typical white North American population.

3. a) No. b) No. c) No. d) His N blood type excludes him.

4. O child, O × O parents
   A child, A × O parents
   B child, AB × O parents
   AB child, A × AB parents

5. a) Wife *Rr*, her father *RR* or *Rr* her mother *rr*; husband *rr*, his father *Rr* his mother *rr*.
   b) Nil.

CHAPTER TEN

1. a) Three
   b) A, C, F; B, E; D
   c) Mutations at a minimum of three loci can produce the phenotype.

2. Tabulate observations:

	1	2	3	4	7	8	9	10	13	19	20	21	X
I				+	+	+		+			+		+
II	+		+		+							+	+
III							+		+	+		+	+
IV	+			+				+		+			+
V		+	+						+		+		+

Conclusions:
Enzyme A has its locus on chromosome 2.
Enzyme B has its locus on the X chromosome.
Enzyme C has its locus on chromosome 19.
Enzyme D cannot be assigned to any of these 13 chromosomes.

CHAPTER ELEVEN

1. b) III–1 and III–4.
   c) With two recombinants in five (40 percent), the loci are not closely linked.
   d. The new mutation could be on either X, so the recombinant sons could be either III–1 and III–4 or the other three. In either case, the linkage cannot be close.

2. 45 percent group A, nail-patella syndrome; 5 percent group A, normal; 5 percent group B, nail-patella syndrome; 45 percent group B, normal.

3. a) Galactokinase (see Fig. 11–5).
   b) The break in the long arm of chromosome 17 is between the TK and GK loci.

4. 150 percent; 50 percent.

5. a) I–1, *Mymy Sese* (or *sese*); I–2, *mymy Sese;* II–1 (proband) *Mymy Sese;* II–2, *mymy sese*; third generation, five are *Mymy Sese*, one is *Mymy sese* (the recombinant), four are *mymy sese.*
   b) Yes, the two genes *My* and *Se* stay together nine out of 10 times, whereas 50 percent is the expectation if they are unlinked.
   c) One in 10, 10 percent.
   d) The marriage would allow a high probability (90 percent) of identifying a fetus with myotonic dystrophy by secretor status; however, there is a 10 percent probability of error because of recombination.

CHAPTER TWELVE

1. a) Autosomal dominant with reduced penetrance (because risk drops by half with each step of more remote relationship).
   b) Any other evidence for autosomal dominance and against multifactorial inheritance, such as twin data (concordance about 20 percent in MZ twins, 10 percent in DZ twins) or equal risk for relatives of propositi of either sex.

2. c) 12 to 15 percent; a) about 5 percent; b) a little less than 5 percent; e) of the order of 1 percent; the fetus has an affected second-degree relative and an affected third-degree relative; d) well below 1 percent; fetus has a third-degree affected relative only. Probably all except d) would qualify.

3. Data are in Table 12–3.

4. Rules are in text. If the malformation is autosomal recessive, the risk for sibs is much higher than the risk for parents, consanguinity of parents does not affect risk for sibs and the concordance rate is four times as high in MZ as in DZ twins.

5. For X-linkage, other affected members are in the maternal line; sex ratio is $q:q^2$ for X-linked recessive, $q:2pq+q^2$ for X-linked dominants and so forth.

CHAPTER FOURTEEN

1. a) $(1/2)^6 = 1/64$ b) $20(1/2)^3(1/2)^3 = 5/16$

2. a) $1/2$ b) $(1/2)^4 = 1/16$ c) $1 - (1/2)^4 = 15/16$

3. a) $1/4$ b) $(3/4)^5 = 243/1024$ c) $10 (3/4)^3(1/4)^2 = 270/1024$

4. a) $(1/2)^3 = 1/8$ b) $(1/2)^3 = 1/8$ c) $3 (1/2) (1/2)^2 = 3/8$

5. a) 57.1 percent   b) 62.5 percent

CHAPTER FIFTEEN

1. a) A 0.9, a 0.1   b) Same   c) $(0.18)^2 = 0.03$

2. a) In Toronto, the frequency of carrier × carrier matings is expected to be about 0.005 and of affected offspring about 1 in 800. In Washington the corresponding figures are 0.001 and 1 in 3300.
   b) The most likely explanation for the difference is genetic drift.

3. 36 percent A, 64 percent O.

4. 1/8

5. 3/8

6. a) 1/20,000   b) 1/100; 2/100   c) 1/10,000; 1/5,000

7. 25 per cent M, 50 percent MN, 25 percent N

CHAPTER SIXTEEN

1. a) 98.5 percent.   b) They must be dizygotic.

2. 20 percent.

3. 10 percent.

NAME INDEX

SUBJECT INDEX

Page numbers in *italics* indicate illustrations; (t) refers to tables.